Trace fossils in evolutionary palaeoecology

Proceedings of Session 18 (Trace Fossils) of the First International Palaeontological Congress
(IPC 2002), held at Macquarie University, Sydney, Australia,
6–10 July 2002

Edited by Barry D. Webby, M. Gabriela Mángano & Luis A. Buatois

Co-sponsored by the International Palaeontological Association
and the
Australasian Association of Palaeontologists

Contents

Preface

The First International Palaeontological Congress (IPC), under the aegis of the International Palaeontological Association (IPA) and the Association of Australasian Palaeontologists (AAP), was hosted by the Macquarie University Centre for Ecostratigraphy and Palaeobiology (MUCEP) and the Australian Museum, Sydney, from 6 to 10 July 2002. One of the 24 symposia held at Macquarie University during the congress was devoted to the topic "Trace Fossils", and attracted a range of excellent oral presentations and poster displays. Ten of the papers presented at the symposium have been edited, and are published here as *Fossils and Strata* No. 51. The papers cover a wide range of mainly palaeoecological and palaeobiological topics within ichnology. Most are regional studies across differing Phanerozoic ecosystems from continental, marginal-, shallow- to deep-marine settings, and involving many parts of the globe, principally South America (Argentina, Brazil), North America (Alberta, Canada; Georgia, eastern USA) Poland, Japan and New Zealand. The series of articles, however, presents a more extended coverage of a number of closely related fields, for instance: (1) an analysis of trace fossils from cores as a most important component of a sedimentological study to interpret the nature of estuarine depositional environments in subsurface deposits; (2) a specific focus on plant–animal interactions (rather than the more commonly documented animal–sediment relationships); (3) a global survey of the geological history of microborings; and (4) reviews of the trace fossil evidence used to define major evolutionary events, such as the invasion of animals onto the land, and the colonisation of infaunal ecospace. Consequently, the volume illustrates a wide range of current research on trace fossils, and offers exciting new data and interpretations.

The volume owes its existence mainly to the contributors to the session. But it is also important to acknowledge the significant support given by the main organisers of IPC 2002, in particular, John Talent, Ruth Mawson, Glen Brock, and a number of research associates and students at MUCEP, who organised the technical programme and provided accommodation for some participants. We also thank David L. Bruton, Chairman of the Lethaia Foundation, for accepting our submission of the proceedings of the "trace fossil" symposium as a stand-alone issue of *Fossils and Strata*. We especially welcome this decision by the Lethaia Foundation because the trace fossil proceedings represent an important part of the technical presentations of this first IPA-sponsored international congress, and IPA, through the Lethaia Foundation, has maintained the journal *Lethaia* since 1968 and the monograph series *Fossils and Strata* since the early 1970s. Through these internationally renowned publications, the web-based *Directories of Palaeontologists of the World* and the *Fossil Collections of the World*, and its support for international palaeontological research groups, IPA continues to maintain a vitally active international role in the co-operation and encouragement of the palaeontological sciences.

Special thanks are also extended to all the expert reviewers who came from all parts of the world and presented most constructively helpful critiques of the content and style of the submitted papers. They include Richard Bromley, Gerhardt Cadée, Rodney Feldman, Jim Gehling, Murray Gingras, Hans Hofman, Conrad Labandeira, Sören Jensen, Duncan McIlroy, Ricardo Melchor, William Miller III, John Pollard, Andy Rindsberg, Richard Twitchett, Alfred Uchman, Andreas Wetzel, Mark Wilson and one other reviewer who wished to retain his anonymity.

Barry D. Webby [bwebby@laurel.ocs.mq.edu.au],
Centre for Ecostratigraphy and Palaeobiology, Department
of Earth and Planetary Sciences, Macquarie University,
North Ryde, NSW 2109, Australia

Luis A. Buatois [luis.buatois@usask.ca] & M. Gabriela
Mángano [gabriela.mangano@usask.ca], Department of
Geological Sciences, University of Saskatchewan, 114
Science Place, Saskatoon, SK S7N 5E2, Canada

Synopses

This issue covers a wide range of topics related to the potential of trace fossils for reconstructing ancient ecosystems and for recognising evolutionary trends through geological time. Ichnology, the study of traces produced by animal and plants, a commonly overlooked discipline, is providing crucial evidence about many of the major debates in evolutionary palaeobiology, such as the changes across the Pre-Cambrian–Cambrian transition and the Palaeozoic invasion onto the land. The papers of this issue document the ichnology of a broad range of environments, from the deep sea, through marginal-marine to continental forest settings. They cover a wide stratigraphic range from the Proterozoic to the Quaternary, and also focus on poorly explored areas of ichnology, such as arthropod–plant interactions and micro-bioerosion. The scope of the various contributions ranges from local to global scales. All contributions, however, frame a particular study within the broader perspective. Methodological approaches adopted by the different authors also vary, and these illustrate that a broadening of the scope of ichnological research is currently taking place. Consequently, we consider that this issue of *Fossils and Strata* represents an eclectic sampling of current work and trends in the discipline.

The first part of the volume includes two regionally based papers that explore aspects of the environmental expansion of Early Phanerozoic ichnofaunas, from the continental margins to the deep sea. Buatois & Mángano discuss ichnological aspects of the terminal Proterozoic to earliest Cambrian Puncoviscana Formation of northwest Argentina. Earliest Cambrian shallow- and deep-marine trace fossil assemblages are characterised using the ichnoguild approach originally proposed by Bromley. They note that tiering was moderately developed in the Puncoviscana ichnofauna. Although biogenic structures were emplaced mainly within the uppermost few millimetres of the sediment, some deeper forms are also recognised. These Early Cambrian deep-marine ichnofaunas are of "Ediacaran aspect", recording the persistence of relict communities after the onset of the dramatic "Agronomic Revolution" of Seilacher.

The overlying, upper Lower to Middle Cambrian, Campanario Formation of northwest Argentina is the basis for a second paper on Early Phanerozoic ichnofaunas. Mángano & Buatois use the ichnofaunal evidence from this formation to reconstruct the palaeoecological relationships. As in the previous paper, an ichnoguild approach is used as a tool to analyse the ecology of the intertidal to shallow subtidal ichnofaunas. The abundant vertical burrows in high-energy deposits are documented, and discussed in terms of the alternative explanations for the presence of trilobite traces in the tidal-flat deposits. The favoured interpretations are that the patterns of activity represent nesting behaviour or migrations in search of food within the highly nutrient-rich intertidal sediments. The Cambrian tidal flats not only provided an abundant food supply but also provided sites that afforded protection from marine predators, at a time when there was an absence of predators on land and in the air.

The second part of the volume comprises two regionally based papers concerned with deep-sea Mesozoic trace fossils. Uchman documents the ichnofaunas from Lower Cretaceous deep-marine turbidites in the Polish Carpathians and evaluates the importance of palaeo-ecological and evolutionary aspects. His approach is based on a careful characterisation of ichnofabrics and tiering structures. He emphasises the importance of anoxic events in controlling the distribution of trace fossils during Cretaceous time. Other factors analysed include preservation potential and frequency of turbiditic events. He compares Lower and Upper Cretaceous turbidite ichnofaunas, noting an increased diversity of complex grazing traces and graphoglyptids in the Late Cretaceous.

An outline of the ichnofaunas from Triassic and Jurassic radiolarian cherts of Japan is presented by Kakuwa. His documentation of the ichnology of the Mesozoic accretionary complex is particularly useful because there are so few studies dealing with trace fossils of ancient (pre-Cretaceous) deep-sea pelagic deposits. Additionally, he detected the presence of anomalous trace fossil suites in Early Triassic rocks, supporting a major anoxic event. His work may also have implications for the study of the environmental perturbations associated with the end-Permian mass extinction.

The third part of three papers includes further regionally based ichnological studies, but they are focused on markedly different, marginal-marine ecosystems. Hubbard *et al.* give a detailed analysis of the ichnofaunas from estuarine deposits of the subsurface Cretaceous Bluesky Formation in Alberta, Canada. The research group led by Pemberton is pre-eminent in the field of "brackish water" ichnology and in identifying ichnofossil assemblages from core material. Their approach is based on the integration of ichnological data within a sedimentological and sequence stratigraphic framework. These authors provide a comprehensive account of the controlling factors in different areas of a wave-dominated estuary. Salinity, sedimentation rate and energy are given pre-eminence, although several other factors are also addressed. Additionally, their paper represents a very good example of the relevance of trace fossil data to subsurface studies linked to oil exploration.

The next study is of geologically younger, marginal-marine ecosystems by Gregory *et al.* and provides a comparative survey of the Quaternary coastal regions of northern New Zealand and Sapelo Island, Georgia (USA), with accompanying analysis of the various sedimentary structures derived from the complex interactions between plant roots and their invertebrate hosts. The moist and sheltered tree roots provide a protected environment that persists long after tree death, becoming a burrow micro-habitat for a wide variety of terrestrial insects (e.g. nymph burrows of cicadas). This recognition of subterranean communities developed in coastal areas is a promising new area of interest, and should be given more attention in the future.

The third paper is a study of plant–insect interactions, presented by Adami-Rodrigues *et al.*, based on a case study in the Permian *Glossopteris* flora of marginal-marine to coal swamp deposits in southern Brazil. The interactions are preserved on fossil seed plants (*Glossopteris*, *Gangamopteris* and *Cordaites*), including continuous and discontinuous external foliage feeding and punctures related to piercing and sucking. Additionally, they discuss potential trace makers among the phytophagous insects. Their paper is one of the first attempts to evaluate interactions in cooler coastal ecosystems of Gondwana. Hence, it is a welcome addition to the recent literature on arthropod–plant interactions, because most of the previous work has focused on tropical to subtropical coal swamps of Euramerica.

The last part of this special issue includes three general surveys of particular trace fossil groups through geological time. The first is by Glaub & Vogel who analyse the stratigraphic record of microborings based on an extensive worldwide database. These authors conclude that the oldest microborings were made by cyanobacteria in the Proterozoic, that there was a diversity increase during the Early Mesozoic, and that a significant change in taxonomic composition occurred by the beginning of the Cenozoic. Also, they note that a comparatively high percentage of the taxa (biotaxa and ichnotaxa) are "living fossils", having survived from Proterozoic or Palaeozoic times. Longevity was favoured by the omnipresence of protected hard substrates through geological time, the high mobility of endolith reproduction cells and the ability of microborers to react to changing light conditions.

The second and third papers are devoted to exploring the roles of trace fossils in identifying significant evolutionary events. First, Braddy presents evidence for the early incursions of arthropods onto the land. He reviews the Palaeozoic record of terrestrial trace fossils and notes that the trace fossil evidence indicates that the earliest invasion occurred during the Late Cambrian. Invasions occurred through the rest of the Early Palaeozoic with significant colonisation of coastal fluvial settings during the Early Ordovician to Late Silurian, and colonisation of all continental habits by the Carboniferous. The possible reasons for arthropod invasion are evaluated, including predatory pressures in aquatic environments, exploitation of empty or under-utilised ecospace and *en-masse* migration for reproduction.

Finally, Carmona *et al.* evaluate the ichnological record of burrowing decapod crustaceans through the Phanerozoic. In order to analyse changes in abundance and ichnodiversity, a database was compiled, summarising occurrences of trace fossils commonly attributed to decapod crustaceans in post-Palaeozoic deposits. They address the problems of ascribing Palaeozoic burrow systems to decapod producers and note that the Mesozoic trace fossil record shows a direct correlation to the trend shown by the body fossil record. By the Neogene the decapod burrow systems had become the dominant elements in shallow-marine ichnofaunas, and exhibited complex endobenthic tiering patterns.

Luis A. Buatois & M. Gabriela Mángano

Terminal Proterozoic–Early Cambrian ecosystems: ichnology of the Puncoviscana Formation, northwest Argentina

LUIS A. BUATOIS & M. GABRIELA MÁNGANO

Buatois, L.A. & Mángano, M.G. **2004 10 25**: Terminal Proterozoic–Early Cambrian ecosystems: ichnology of the Puncoviscana Formation, northwest Argentina. *Fossils and Strata*, No. 51, pp. 1–16. Canada. ISSN 0300-9491.

The Puncoviscana Formation consists of a thick, folded and slightly metamorphosed succession of sandstones, mudstones and conglomerates of terminal Proterozoic to earliest Cambrian age, representing the metasedimentary basement of northwest Argentina. The Puncoviscana ichnofauna is recorded in lowermost Cambrian strata and includes a wide variety of ichnotaxa, such as *Archaeonassa fossulata*, *Circulichnis montanus*, *Cochlichnus anguineus*, *Didymaulichnus lyelli*, *Diplichnites* isp., *Helminthopsis abeli*, *Helminthopsis tenuis*, *Helminthoidichnites tenuis*, *Multina* isp., *Nereites saltensis*, *Oldhamia antiqua*, *Oldhamia curvata*, *Oldhamia flabellata*, *Oldhamia radiata*, *Oldhamia* isp., *Palaeophycus tubularis*, *Saerichnites* isp., *Treptichnus pollardi* and *Volkichnium volki*. Structures considered in previous studies as graphoglyptids (agrichnia), such as *Protopaleodictyon* and *Squamodictyon*, are reinterpreted as wrinkle marks and elephant skin textures. Vertical dwelling structures (domichnia) are absent. Nearly all the ichnofossils in the association are oriented parallel to the bedding plane, displaying restriction to two-dimensional biotopes, and therefore they do not disturb the primary sedimentary fabric. The recognition of trace fossil associations in both shallow- and deep-marine deposits of the Puncoviscana Formation provides valuable information on evolutionary and ecological controls on Early Cambrian marine infaunal communities. Shallow-marine ichnofaunas are dominated by moderate- to large-sized, shallow grazing and feeding traces of deposit feeders. Deep-marine ichnofaunas are dominated by *Oldhamia* and non-specialised grazing trails and reflect lifestyles related to microbial matgrounds. Early Cambrian deep-marine ichnofaunas are of "Ediacaran aspect", recording persistence of relict communities after the onset of the "Agronomic Revolution" that pervaded shallow-marine ecosystems during those times. Early Cambrian tiering was moderately developed in both shallow- and deep-marine communities. Biogenic structures were emplaced within the uppermost millimetres of the sediment (i.e. micro-tiering), with the exception of burrowers that record post-event colonisation of tempestites and turbidites.

Key words: Proterozoic-Cambrian; ichnology; evolution; ecology; microbial mats; Argentina.

Luis A. Buatois [luis.buatois@usask.ca] & M. Gabriela Mángano [gabriela.mangano@usask.ca], Department of Geological Sciences, University of Saskatchewan, 114 Science Place, Saskatoon, Canada S7N 5E2

Introduction

In recent years, several studies have stressed the anactualistic character of the terminal Proterozoic–Early Cambrian ecosystems (Seilacher 1999; Droser & Li 2001). This change in perspective parallels a trend in the reinterpretation of the Ediacaran biota, from ancestors of modern animal phyla (Glaessner 1984) to having independent relationships –the quilted organisms known as Vendozoa or Vendobionta (Seilacher 1989, 1992) or organisms with other affinities (see Waggoner 1998 for a discussion). Additionally, there has been considerable debate regarding the environmental expansion of early biotas (e.g. Narbonne & Aitken 1990; Seilacher & Pflüger 1992; Crimes & Fedonkin 1994; McIlroy & Logan 1999; MacNaughton & Narbonne 1999; MacNaughton *et al.* 2000; Crimes 2001; Orr 2001; Narbonne & Gehling 2003).

The Puncoviscana Formation consists of a thick, folded and metamorphosed succession of sandstones, mudstones and conglomerates of terminal Proterozoic to earliest Cambrian age and represents the metasedimentary basement of northwest Argentina (Fig. 1). The

Fig. 1. Outcrop map of the Puncoviscana Formation and coeval units, showing the location of the most important ichnofossiliferous localities (modified from Aceñolaza *et al.* 1999). 1 = San Antonio de Los Cobres. 2 = Cuesta Muñano. 3 = Abra Blanca. 4 = Quebrada del Toro. 5 = Rio Corralito. 6 = Sierra de Mojotoro. 7 = Palermo Oeste. 8 = Cachi. 9 = La Ovejería.

Puncoviscana ichnofauna was documented originally during the 1970s (Mirré & Aceñolaza 1972; Aceñolaza & Durand 1973; Aceñolaza 1978) and additional descriptions of ichnotaxa were published subsequently (e.g. Aceñolaza & Durand 1982; Durand & Aceñolaza 1990a, 1992). Recent research allows us to re-evaluate its evolutionary significance in the light of new ideas regarding the anactualistic nature of Precambrian–Cambrian ecosystems and the role of microbial communities in the ecological structure (Buatois *et al.* 2000; Mángano *et al.* 2000; Buatois & Mángano 2002, 2003a,b). However, a detailed palaeoecological analysis of the Puncoviscana ichnofauna is still pending. Additionally, the recognition of both shallow- and deep-marine trace fossil associations in the Puncoviscana Formation provides further evidence of the colonisation of the deep sea and allows characterisation of early benthic ichnofaunas in shelf and deep settings. The aim of this paper is to characterise these two ichno-associations in terms of their palaeoecological significance and to discuss the importance of the Puncoviscana ichnofauna in our understanding of the terminal Proterozoic–Early Cambrian ecosystems.

Geological setting

The Puncoviscana Formation represents the metasedimentary basement of northwest Argentina (Turner 1960; Aceñolaza & Toselli 1981; Omarini *et al.* 1999). It consists of a thick metamorphosed and folded succession of sandstones, mudstones and conglomerates. The Puncoviscana Formation was originally regarded as Precambrian (Turner 1960, 1972). However, subsequent ichnological studies demonstrated that this unit undoubtedly ranges into the Cambrian (Mirré & Aceñolaza 1972; Aceñolaza & Durand 1973). Integration of geochronological evidence and ichnological data based on more recent studies suggested that sedimentation in the Puncoviscana basin spanned the terminal Proterozoic and the earliest Cambrian (Ramos 2000; Hongn *et al.* 2001; Buatois & Mángano 2003a; Mángano & Buatois in press). Several authors (e.g. Mon & Hongn 1988, 1991; Hongn 1996; Moya 1998; Becchio *et al.* 1999; Mángano & Buatois in press) have noted that strata with different degrees of metamorphism and tectonic deformation are included under the name "Puncoviscana Formation", suggesting the possibility of further subdivision. In any case, only the youngest strata (lowermost Cambrian) contain the trace fossils that are discussed in this paper.

The interpretation of the Puncoviscana Formation as having been entirely deposited in deep submarine fans (Omarini & Baldis 1984; Ježek 1990; Aceñolaza et al. 1999) has been challenged recently (Mángano et al. 2000; Buatois et al. 2000; Buatois & Mángano 2003a; Van Staden & Zimmermann 2003). A more complex palaeoenvironmental framework, including not only deep-marine, flysch-type deposits, but also shallow-water settings affected by wave action, has been suggested (Mángano et al. 2000; Buatois & Mángano 2002, 2003a). In general, an eastern belt with the presence of wave-influenced, shallow-marine facies, and a western belt recording deposition in deep-marine environments were proposed (Buatois & Mángano 2003a). However, the intense deformation that pervaded outcrops of the Puncoviscana Formation complicates relationships for establishing a sound stratigraphic subdivision and for proposing detailed correlations. From our fieldwork, undisturbed, continuously measured sections are usually limited to a few tens of metres of exposure at the most. Accordingly, although facies analysis has been performed in selected outcrops, correlation of stratal packages over wide areas is limited.

Composition of the Puncoviscana ichnofauna

Durand & Aceñolaza (1990a) provided descriptions of several ichnotaxa, including *Asaphoidichnus trifidus*, *Cochlichnus* isp., *Didymaulichnus* isp., *Dimorphichnus* isp., *Diplichnites* isp., *Glockeria* isp., *Gordia* isp., *Helminthoida* cf. *H. miocenica*, *Helminthopsis* isp., *Helminthopsis* aff. *tenuis*, *Monomorphichnus* isp., *Neonereites uniserialis*, *Nereites saltensis*, *Oldhamia antiqua*, *Oldhamia flabellata*, *Oldhamia radiata*, *Phycodes pedum*, *Planolites* isp., *Protichnites* isp., cf. *Protopaleodictyon* isp., *Scolicia* isp., *Tasmanadia cachii* and *Torrowangea* isp. To this list we have to add other ichnotaxa, such as *Protovirgularia* isp. (described previously by Aceñolaza & Durand 1973), *Multipodichnus holmi* (Durand & Aceñolaza 1992) and cf. *Squamodictyon* isp. (Durand et al. 1994). These original taxonomic assignments have been maintained unchanged in recent reviews (Aceñolaza et al. 1999). However, other authors (e.g. Vidal et al. 1994) have expressed reservations regarding some of the taxonomic assignments. Additional ichnospecies (e.g. *Treptichnus pollardi*, *Circulichnis montanus*, *Helminthopsis abeli*) have been identified in subsequent studies, and a reassessment of the Puncoviscana ichnofauna has been

presented recently (Buatois et al. 2000; Mángano et al. 2000; Buatois & Mángano 2003a,b). Some of these structures originally assigned at the ichnogeneric level are now classified at the ichnospecific level; these include *Cochlichnus anguineus* and *Didymaulichnus lyelli* (Buatois & Mángano 2003a). Although the focus of the present paper is not systematic, some taxonomic aspects are briefly addressed here because of the present confusion with respect to the composition of the Puncoviscana ichnofauna, and because a sound ichnotaxonomy is a prerequisite for any palaeoecological study.

Specimens from the Puncoviscana Formation compared with the graphoglyptid ichnotaxa *Protopaleodictyon* and *Squamodictyon* (Durand & Aceñolaza 1990a; Durand et al. 1994; Aceñolaza et al. 1999) are not trace fossils, but structures formed by microbial activity (Buatois et al. 2000; Buatois & Mángano 2003a). The specimen assigned to the arthropod trackway *Protichnites* isp. (Durand & Aceñolaza 1990a; Aceñolaza et al. 1999) is in all probability an inorganic tool mark (Buatois & Mángano 2003a). The structure originally compared with the body fossil *Praecambridium* by Durand & Aceñolaza (1990b) and subsequently assigned to the trace fossil *Multipodichnus holmi* by Durand & Aceñolaza (1992) lacks any relevant morphological feature and has been regarded as a pseudofossil (Buatois & Mángano 2003a).

Specimens included in *Glockeria* (for *Glockerichnus*) by Durand & Aceñolaza (1990a) and Aceñolaza et al. (1999) lack the diagnostic branching of this ichnotaxon and should be relocated in another radial ichnogenus, most likely *Volkichnium* (Orr 1996; Buatois & Mángano 2003a). Some of the specimens from the coeval Suncho Formation that were originally referred to *Oldhamia antiqua* (Aceñolaza & Durand 1982, 1984, 1986) are now regarded as *Oldhamia curvata* (Hofmann et al. 1994). As noted by Buatois & Mángano (2003a), the specimen illustrated as *Torrowangea* isp. was mistakenly figured from the Puncoviscana Formation by Durand & Aceñolaza (1990a) and Aceñolaza et al. (1999). This specimen does not belong to this unit, but to the Permian La Dorada Formation of western Argentina, and is a bilobate cruzianiform that occurs in continental redbeds. Re-examination of another specimen assigned to *Torrowangea* and actually collected by Durand & Acenolaza (1990a) from the Puncoviscana Formation fails to reveal the diagnostic transverse constrictions of *Torrowangea*; it is better regarded as a grazing trail comparable with *Helminthopsis* or *Helminthoidichnites*. Another specimen recently described as *Torrowangea?* isp. by Aceñolaza & Tortello (2003) does not display the distinct transverse constrictions, so this also should be removed from the ichnogenus.

Considerable confusion persists with respect to meandering traces that have been originally included in

Helminthoida cf. *H. miocenica* by Durand & Aceñolaza (1990a). Because *Helminthoida* is a junior synonym of *Nereites*, regular meandering traces preserved as positive hyporeliefs should be included in *Helminthorhaphe* (Uchman 1995). Accordingly, *Helminthoida* has been removed from updated ichnotaxonomic lists of the Puncoviscana Formation (Buatois *et al.* 2000; Mángano *et al.* 2000; Buatois & Mángano 2003a). However, *Nereites*, *Helminthoida* and *Helminthorhaphe* have been simultaneously retained in other studies (Aceñolaza & Tortello 2000, 2003; Aceñolaza & Aceñolaza 2001). Similarly, the name *Scolicia* has been recurrently used for very simple trails present in the Puncoviscana Formation (Aceñolaza & Durand 1990a; Aceñolaza *et al.* 1999). However, *Scolicia* is a complex endichnial structure, produced by spatangoid echinoids, and characterised by a meniscate backfill, a double ventral cord or drain, and mucus-lined vertical shafts (Bromley & Asgaard 1975; Plaziat & Mahmoudi 1988; Uchman 1995; Bromley 1996). *Scolicia* occurs in Mesozoic and Cenozoic strata, and Palaeozoic recordings should be transferred to other ichnogenera (Smith & Crimes 1983; Mángano *et al.* 2002). Specimens included in *Gordia* isp. (Aceñolaza 1978; Durand & Aceñolaza 1990a; Aceñolaza *et al.* 1999) have been relocated in *Helminthopsis tenuis* and *Helminthoidichnites tenuis* (Hofmann & Patel 1989; Buatois *et al.* 2000; Mángano *et al.* 2000; Buatois & Mángano 2003a,b). The ichnospecies *Phycodes (Treptichnus) pedum* was documented by Durand & Aceñolaza (1990a) and Aceñolaza *et al.* (1999) based on a single specimen. However, re-examination of the specimen does not reveal the classic branching pattern of this ichnospecies; the structure more likely records false branching resulting from overlap of two *Palaeophycus*-like trace fossils (Buatois & Mángano 2003b). Most of the specimens traditionally included in *Planolites* isp. are thinly lined and have the same infill as the host rock. They, additionally, have been reassigned to *Palaeophycus tubularis* (Buatois & Mángano 2003a). The ichnotaxa *Nereites saltensis* is provisionally retained in this paper, pending further studies being undertaken to clarify its taxonomic status (A. Seilacher 2003, pers. comm.). Also, the taxonomy of arthropod trackways is in a state of flux and there is a need for revision of these. Furthermore, recent studies have allowed the documentation of the presence of *Multina* isp., *Saerichnites* isp. and *Archaeonassa fossulata*, and there are additional ichnotaxa currently under study (A. Seilacher 2003, pers. comm.).

Palaeoecology of the Puncoviscana ichnofauna

The existence of two different trace fossil associations (*Nereites* and *Oldhamia*) in the Puncoviscana Formation

has been recognised since the 1970s (e.g. Aceñolaza *et al.* 1976). However, the implications and significance of both associations have not been clear. The recognition that the Puncoviscana Formation encompasses a wide variety of depositional settings (Buatois & Mángano 2003a) allows us to place the ichnological information within a palaeoenvironmental framework. In this section we characterise the taxonomic composition, ethology, tiering, ichnoguild structure, and palaeoenvironmental implications of both associations. All the ichnofossils are oriented parallel to bedding, displaying restriction to two-dimensional biotopes, and therefore they do not disturb the primary sedimentary fabric. For ichnoguild characterisation, we follow the original proposal of Bromley (1996), who stated that an ichnoguild reflects three parameters: (1) bauplan; (2) food source; and (3) use of space. The palaeoenvironmental subdivision of shallow-marine deposits used in this paper is based on the scheme by Pemberton *et al.* (2001). Both the shallow-marine and deep-marine trace fossil associations are differentiated here.

Shallow-marine trace fossil associations

Shallow-marine deposits in the Puncoviscana Formation are common in the eastern belt of the unit. Excellent outcrops are present in the Quebrada del Toro and Cuesta del Obispo regions and, to a lesser extent, in Sierra de Mojotoro, central Salta Province. Similar ichnofaunas have been recorded in outcrops from the Cachi area (Aceñolaza *et al.* 1976). The three main sedimentary facies have been identified here, in ascending order, as occupying successively more onshore environments, from facies 1 to facies 3.

Facies 1 consists of sharp and erosive-based, laterally extensive, rhythmically interbedded, graded, very fine-grained silty sandstone to mudstone forming couplets commonly less than 1 cm thick. Internally, these rhythmites are massive or parallel laminated. These thinly bedded, fine-grained deposits are occasionally interbedded with 1–3 cm thick very fine-grained sandstone layers having combined-flow ripples and ripple cross-lamination. This facies is well developed in the Sierra de Mojotoro area and contains abundant trace fossils that are currently under study (A. Seilacher 2003, pers. comm.). Graded fine-grained deposits probably represent sedimentation from dilute turbidity currents. Their association with sandstone beds having structures indicative of oscillatory flows suggest that they are most likely shallow-marine storm-generated turbidites, rather than deep, base of slope, turbidites (see also Omarini *et al.* 1999). Graded beds were probably emplaced immediately below the storm wave base in a shelf environment (*sensu* Pemberton *et al.* 2001). The succession probably shallows upwards, as indicated by the presence of the sandstone beds with combined-flow ripples. These were most likely deposited immediately above the storm wave base in a lower offshore environment.

Facies 2 consists of erosive-based, laterally extensive to rarely lenticular, thin, very fine-grained sandstone interbedded with parallel-laminated siltstone. Sandstone beds are 2–25 cm thick and mudstone intervals are from a few centimetres up to 70 cm thick. Sandstone bases commonly contain small tool marks and flute marks. Sandstone beds may show well-preserved bedforms, typically micro-hummocks and symmetrical to near-symmetrical ripples that are interpreted as combined-flow ripples. However, other sandstone beds occasionally have planar tops with patchily distributed wrinkle marks. Internally, sandstone displays combined-flow ripple cross-lamination and micro-hummocky cross-stratification. Hummocky cross-stratification is rare. Sandstone–mudstone ratios range from 1:1 to 1:5. This facies is common in the Quebrada del Toro area and contains abundant trace fossils. The presence of micro-hummocky cross-stratification, combined-flow ripple cross-lamination and symmetrical to near-symmetrical ripples in the sandstone beds suggests that these layers are distal tempestites. The interbedded siltstone intervals mostly record sediment fallout. Facies 2 is therefore thought to record alternating background suspension fallout and distal storm deposition above the storm wave base. The most likely environmental scenario is the upper offshore to the offshore transition.

Facies 3 consists of erosive- to load-based, laterally extensive to lenticular, very fine-grained sandstone that is amalgamated and forms packages up to 5 m thick, or separated by thin (1–10 cm thick) siltstone intervals. Sandstone tops display bedforms, commonly hummocks and symmetrical to asymmetrical ripples with sinuous crests. Internal stratification features are hardly discernible, although hummocky cross-stratification is observed in some beds. Flute marks, tool marks and load casts occur on sandstone bases. This facies is also common in the Quebrada del Toro area. In contrast to facies 1 and 2, no trace fossils have been found in these thick-bedded sandstone units. Thick, amalgamated hummocky sandstone beds represent proximal tempestites that record high-energy oscillatory and combined flows during storms. They were formed by repeated storm events, and wave erosion removed mud layers between sandstone beds, with the exception of residual mud partings. Loaded bases suggest rapid influx of sand during storms. These amalgamated hummocky sandstones were deposited in lower to middle shoreface environments. Outcrops of this eastern belt are thought to record deposition in a wave-dominated shallow platform (Buatois & Mángano 2003a). South of Quebrada del Toro, in Rio Corralito, conglomerate facies have been documented by Durand & Spalletti (1986) and Durand

(1990). The abundance of glauconite in the Rio Corralito outcrops supports deposition in shallow-marine environments (Van Staden & Zimmermann 2003).

Shallow-marine deposits of the Puncoviscana Formation contain an abundant and relatively diverse ichnofauna (Fig. 2), including *Archaeonassa fossulata*, *Cochlichnus anguineus*, *Diplichnites* isp., *Helminthopsis tenuis*, *Helminthoidichnites tenuis*, *Multina* isp., *Nereites saltensis*, *Oldhamia* isp., *Palaeophycus tubularis*, *Treptichnus pollardi* and *Volkichnium volki*. Other arthropod trackways, referred to by previous authors as *Asaphoidichnus trifidus* and *Tasmanadia cachii*, have been recorded in these outcrops. *Nereites saltensis* is clearly the dominant ichnotaxon in the association. Tiering structure and ichnoguilds were recognised for each subenvironment, with the exception of the high-energy, lower to middle shoreface that lacks trace fossils.

In the case of the upper offshore to offshore transition deposits, sandstone beds with well-preserved bedforms and those with planar tops with patchily distributed wrinkle marks are analysed separately. A very well-preserved trace fossil assemblage was detected in one of these sandstone tempestites with planar tops and wrinkle marks. This layer occurs immediately below a low-energy drowning surface and the top of the tempestite, is thought to record an omission surface. Low sediment rate during drowning and paucity of bioturbation by sediment bulldozers were probably conducive to the establishment of a biomat. Interestingly, the ichnofauna of this bed resembles that of the deep-marine deposits (see below), particularly with respect to the dominance of *Oldhamia*. The tiering structure of this bed is very simple (Fig. 3). Three ichnoguilds (*Multina*, *Oldhamia* and *Helminthopsis*) have been recognised. The *Multina* ichnoguild is monospecific and consists of semi-permanent, shallow- to middle-tier, deposit-feeder structures produced by vagile vermiform organisms. *Multina* isp. is preserved at the base of a 5 cm thick tempestite and cross-cut inorganic tool marks, indicating that it represents post-depositional burrows. Undoubtedly, this ichnoguild records the activity of organisms that burrowed into the storm bed and moved along the sand–mud interface, and therefore indicates post-storm colonisation. The *Oldhamia* ichnoguild is also monospecific and is represented by *Oldhamia* isp. This form is different from the *Oldhamia* ichnospecies recorded in the deep-sea deposits and represents a new ichnotaxon. It includes semi-permanent, very shallow-tier, undermat-miner structures produced by stationary vermiform organisms. It is considerably less diverse than its deep-marine equivalent discussed below. The *Helminthopsis* ichnoguild is represented by *Helminthopsis tenuis* and *Helminthoidichnites tenuis*. It records the activity of transitory, near-surface to very shallow-tier, mat-grazer

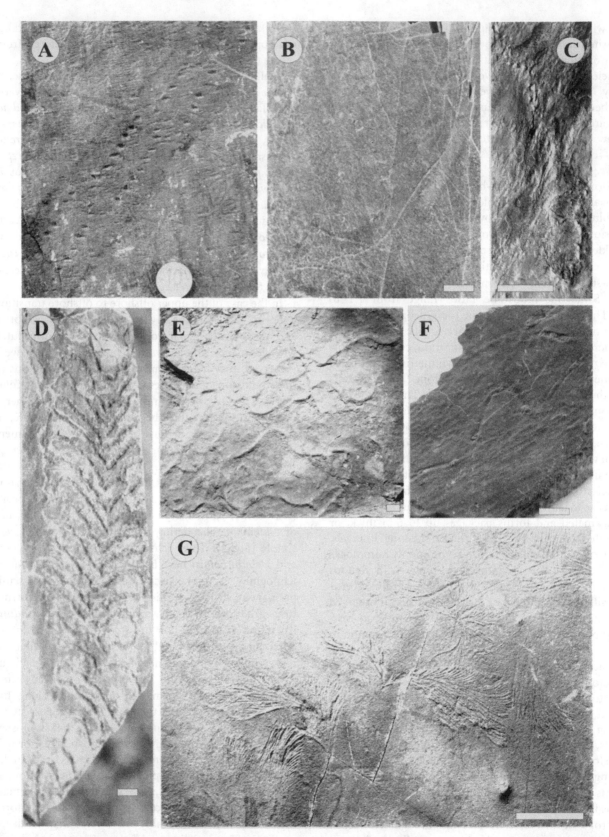

Fig. 2. Selected trace fossils from shallow-marine deposits of the Puncoviscana Formation. A: An arthropod trackway assigned by previous authors to *Tasmanadia cachii*. Top of slab. Cachi. Coin diameter = 1.8 cm. B: *Helminthoidichnites tenuis*. Top of slab. Quebrada del Toro. C: *Saerichnites* isp. Base of slab. Sierra de Mojotoro. D: Guided meandering trace fossil. These grazing traces having angular kinks have been traditionally assigned to *Nereites saltensis*. Top of slab. Cachi. E: *Multina* isp. Base of slab. Quebrada del Toro. F: *Treptichnus pollardi*. Top of slab. Quebrada del Toro. G: *Oldhamia* isp. Top of slab. Quebrada del Toro. All scale bars = 1 cm.

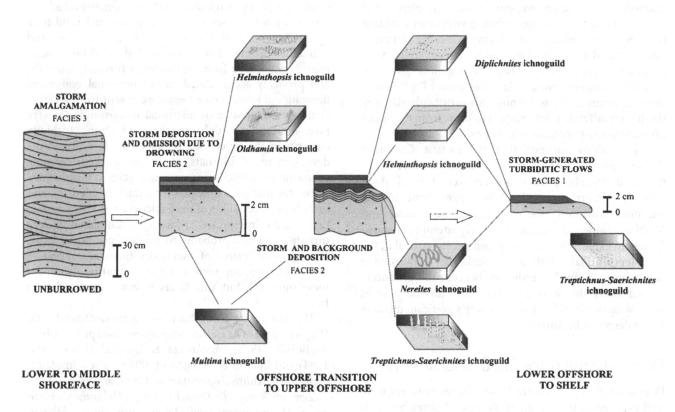

Fig. 3. Tiering structure and ichnoguilds of Early Cambrian shallow-marine ichnofaunas of the Puncoviscana Formation and their environmental distribution.

structures produced by vagile vermiform animals. Both the *Oldhamia* and *Helminthopsis* ichnoguilds are directly associated with structures indicative of microbial mats, such as patchily distributed wrinkle marks and elephant skin textures.

Sandstone beds showing undulatory tops, most commonly micro-hummocks and symmetrical to near-symmetrical ripples, also occur in the upper offshore to offshore transition. In contrast to the flattened top bed, no evidence of significant development of microbial mats has been observed, and *Oldhamia* is not present. The tiering structure and ichnoguilds of these deposits are more complex (Fig. 3). Five ichnoguilds (*Multina, Nereites, Treptichnus–Saerichnites, Helminthopsis* and *Diplichnites*) were recognised. The *Multina* ichnoguild is identical to the one described from the flattened top bed, but it has been observed at the base of several sandstone tempestites, ranging in thickness from 5 to 8 cm, therefore reaching a slightly deeper position into the substrate. The *Treptichnus–Saerichnites* ichnoguild is represented in this facies by the feeding systems *Treptichnus pollardi* and *Saerichnites* isp. This ichnoguild consists of semi-permanent, shallow-tier, deposit-feeder

structures of vagile to semi-vagile vermiform organisms. While most trace fossils from the Puncoviscana Formation reflect horizontal displacement along lithological interfaces (i.e. bedding plane trace fossils), *Treptichnus* and *Saerichnites* are three-dimensional burrow systems with vertical to oblique components, which record a feeding strategy of underground mining. The *Nereites* ichnoguild is represented by *Nereites saltensis* and is made up of transitory, shallow-tier, deposit-feeder structures produced by vagile vermiform animals. The *Helminthopsis* ichnoguild is represented by *Archaeonassa fossulata*, *Helminthopsis tenuis*, *Helminthoidichnites tenuis* and *Cochlichnus anguineus* and is made up of transitory, near-surface to very shallow-tier, deposit-feeder structures produced by vagile vermiform animals. Although the taxonomic composition of this ichnoguild is very similar to that of the deep-marine examples, the *Helminthopsis* ichnoguild of these deposits is remarkably less abundant and no direct association with structures indicative of microbial mats is apparent. The *Diplichnites* ichnoguild includes various types of arthropod trackway and consists of vagile, surface to near-surface structures. It is slightly more diverse than its deep-marine equivalent.

The food source of this ichnoguild is unclear, because trophic types are difficult to infer from trackways.

The rhythmically interbedded, graded, very fine-grained, silty sandstone–mudstone couplets that characterise lower offshore to shelf environments contain trace fossil assemblages displaying a tiering structure similar to that of the shallower storm layers with undulatory tops (Fig. 3). Four ichnoguilds (*Nereites*, *Treptichnus–Saerichnites*, *Helminthopsis* and *Diplichnites*) were recognised. These ichnoguilds do not display any significant difference with respect to those from the upper offshore to offshore transition deposits.

This analysis suggests that the earliest Cambrian shallow-marine tiering was moderately developed. For the most part, biogenic structures were formed along lithologic interfaces, within the upper zone of the sediment. However, the *Treptichnus–Saerichnites* and *Multina* ichnoguilds represent an early attempt of deposit feeders to exploit the infaunal ecospace. In particular, the post-depositional *Multina* isp. indicates a burrowing depth up to 8 cm. Nevertheless, because the producers moved along the sand–mud interface, this increase in burrowing depth did not result in any significant increase in the degree of bioturbation.

Deep-marine trace fossil associations

Deep-marine deposits of the Puncoviscana Formation are well exposed in the region of Sierra de Cobres, which is part of the western belt of the unit. Trace fossils are particularly abundant in outcrops near the town of San Antonio de los Cobres and in Cuesta Muñano, western Salta Province (Aceñolaza 1973; Buatois & Mángano 2003b). In this area, two main sedimentary facies have been identified (Buatois & Mángano 2003b).

Facies 1 consists of thinly interbedded, light grey, current-rippled, fine- to very fine-grained sandstone and light to dark grey, parallel-laminated mudstone. A lower parallel-stratified interval is present in some layers. Individual beds are 1–12 cm thick. Lower boundaries are sharp and erosive. Small- to medium-sized tool marks are present in the thickest beds. This sandstone and mudstone facies represents thin-bedded turbidites that record deposition from low-density turbidity currents coupled with suspension fallout in distal lobe to lobe-fringe settings. The trace fossils occur in these thin-bedded turbidites. Ichnofossils are typically preserved within the finer-grained mudstone divisions and, more rarely, on the top or at the base of the current-rippled sandstone. In contrast to the Early Cambrian Puncoviscana shallow-marine deposits, structures indicative of microbial binding are extremely common in these thin-bedded turbidites.

Facies 2 consists of thick-bedded light grey, fine- to medium-grained sandstone with large load casts and flute marks. Most beds are massive; a subtle parallel stratification is only locally present. Individual beds are generally 10–40 cm thick. They are either amalgamated, forming sandstone packages up to several tens of metres thick, or form discrete layers separated by 1–10 cm thick mudstone layers. Bases are highly irregular due to erosion and sediment loading. The thick-bedded sandstone facies records deposition from high-density turbidity currents, and probably accumulated in the proximal regions of depositional lobes. Trace fossils are absent from facies 2. Structures suggestive of microbial matgrounds are very rare, indicating a high frequency of strong turbidity currents in the inner lobe regions, which inhibited the development of biomats. Beds are stacked, forming coarsening and thickening upwards cycles of thin-bedded turbidites that pass upwards into thick-bedded massive turbidites. Successions of the Puncoviscana Formation in this area are interpreted as having accumulated in deep turbidite systems, probably forming depositional lobes. Structures indicative of oscillatory flows are absent in these outcrops, supporting a deeper marine origin for these outcrops, but this facies is more proximal than facies 1.

The ichnofauna from this region is moderately diverse (Fig. 4). Some of its elements were described previously by Aceñolaza (1973), Aceñolaza & Durand (1973, 1982, 1984) and Buatois & Mángano (2003b). The ichnofauna includes *Circulichnis montanus*, *Cochlichnus anguineus*, *Didymaulichnus lyelli*, *Diplichnites* isp., *Helmithoidichnites tenuis*, *Helminthopsis abeli*, *Helmithopsis tenuis*, *Oldhamia antiqua*, *Oldhamia flabellata*, *Oldhamia radiata*, and *Palaeophycus tubularis* (Buatois & Mángano 2003b). This ichnofauna is dominated by non-specialised grazing trails, such as *Helmithoidichnites tenuis* and *Helmithopsis tenuis*, followed in order of abundance by *Oldhamia* chnospecies, particularly *Oldhamia flabellata* and *Oldhamia radiata*, and less common *Oldhamia antiqua*. The other ichnotaxa are even less abundant.

Four ichnoguilds (*Palaeophycus*, *Oldhamia*, *Helminthopsis* and *Diplichnites*) have been recognised by Buatois & Mángano (2003b) in the thin-bedded turbidites of facies 1 (Fig. 5). The *Palaeophycus* ichnoguild is monospecific and consists of semi-permanent, shallow-tier, suspension-feeder structures produced by vagile vermiform organisms. The trace fossils are preserved at the base of 1–3 cm thick turbidites and cross-cut inorganic sole marks, demonstrating a post-depositional origin. Therefore, they record the activity of organisms that extended into the turbidite bed and moved along the sand–mud interface, indicating post-turbidite colonisation. The *Oldhamia* ichnoguild is represented by the three *Oldhamia* ichnospecies and includes semi-permanent, very shallow-tier, undermat-miner structures produced by stationary vermiform organisms. The feeding trace *Oldhamia* cross-cuts associated grazing trails and is emplaced slightly below

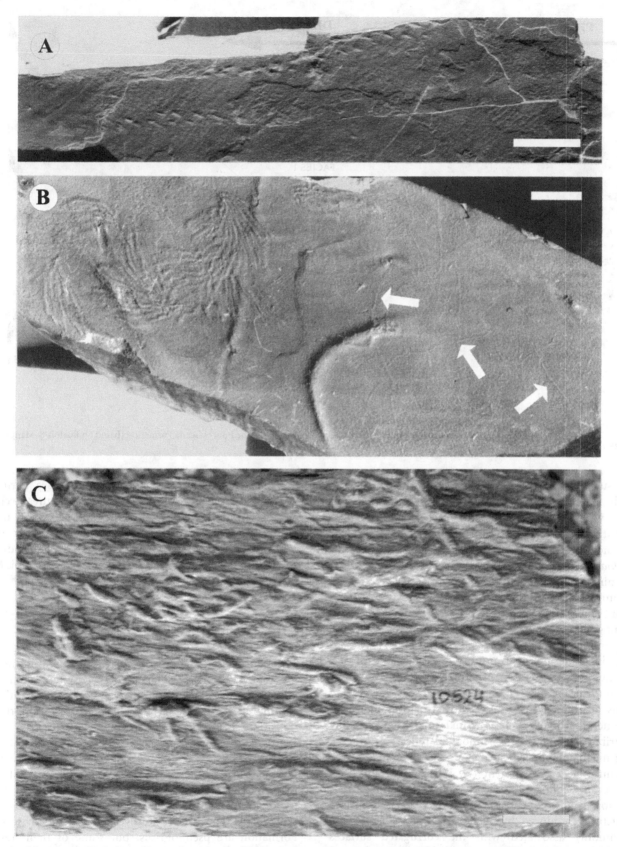

Fig. 4.   Selected trace fossils from deep-marine deposits of the Puncoviscana Formation. A: The arthropod trackway *Diplichnites* isp. Top of slab. B: *Oldhamia flabellata* cutting *Helminthoidichnites tenuis*. *Oldhamia flabellata* is slightly deeper than the grazing trails. Note the presence of tiny, near-surface, non-specialised grazing trails (arrows). C: Parallel-oriented, poorly defined, small tool marks cross-cut by sharp, variably oriented *Palaeophycus tubularis*. Base of a turbidite sandstone. All specimens are from the San Antonio de los Cobres area. All scale bars = 1 cm.

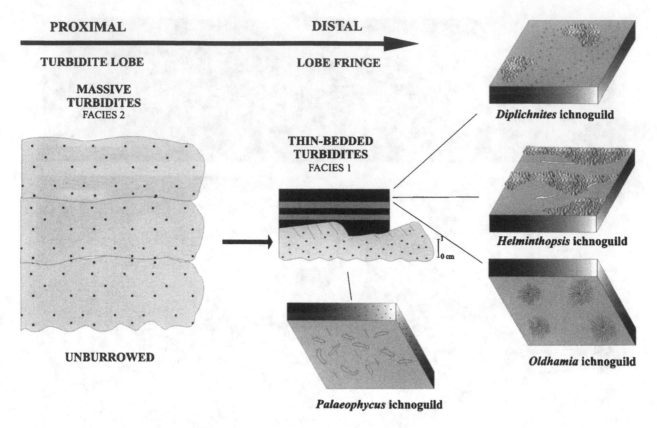

Fig. 5.　Tiering structure and ichnoguilds of Early Cambrian deep-marine ichnofaunas of the Puncoviscana Formation (based on Buatois & Mángano 2003b).

the substrate. Preservation of *Oldhamia* below laminae with palimpsest ripples and in direct association with wrinkled marks support Seilacher's (1997, 1999) hypothesis that this ichnogenus records a feeding strategy of undermat mining. Micro-ripple sets associated with *Oldhamia* are oriented perpendicular to each other, forming palimpsest ripples. These surfaces give evidence of microbial mats that provided the substrate with a thin veneer resistant to erosion. The *Helminthopsis* ichnoguild is represented by *Helminthopsis tenuis*, *Helminthoidichnites tenuis* and *Cochlichnus anguineus* and is made up of transitory, near-surface to very shallow-tier, mat-grazer structures produced by vagile vermiform animals. These non-specialised, tiny grazing trails are directly associated with structures indicative of microbial mats, such as wrinkle marks and patchily distributed ripples. This ichnoguild reflects grazing of organic matter concentrated along microbial mats below a thin veneer of sediment (cf. Ediacaran death mask model of Gehling 1999). The *Diplichnites* ichnoguild is monospecific and consists of surface to near-surface structures produced by vagile arthropods. Tracks are commonly associated with the corrugated surfaces suggestive of microbial mats. Although arthropod trackways are typical in shallow-marine environments, they are also common in Early Palaeozoic deep-marine

deposits (Orr 2001). Early Cambrian deep-sea tiering was moderately developed. Biogenic structures were emplaced within the uppermost millimetres of the sediment (i.e. micro-tiering), with the exception of *Palaeophycus tubularis* that records post-turbidite colonisation below the event sandstone layers (Buatois & Mángano 2003b).

Discussion

The Puncoviscana ichnofauna provides a glimpse into the ecology of Early Phanerozoic ecosystems and is particularly relevant to the understanding of the benthic communities of the terminal Proterozoic–Cambrian transition. Terminal Proterozoic deposits of the Puncoviscana Formation lack trace fossils and contain widespread evidence of biomats. As is common with terminal Proterozoic deposits world-wide, microbial mats are commonly associated with physical structures indicative of oscillatory flows. The supposed medusoid body fossils reported previously from the Puncoviscana Formation are pseudofossils, but some of them may be related to matgrounds. Additionally, structures considered in previous studies as graphoglyptids (agrichnia), such as *Protopaleodictyon* and *Squamodictyon*, have been reinterpreted as wrinkle marks and elephant skin textures (Buatois *et al.* 2000; Mángano *et al.* 2000; Buatois & Mángano 2003a).

Shallow-water ichnofaunal associations

In contrast to the terminal Proterozoic strata, the Early Cambrian Puncoviscana shallow-marine deposits contain a relatively diverse ichnofauna. Behavioural patterns are varied, suggesting a wide range of ethologies in the shallow-marine infaunal communities. Feeding and grazing trace fossils of vermiform animals and crawling traces produced by arthropods are represented. Vertical dwelling structures (domichnia) are absent, but horizontal domiciles of suspension feeders or predators are present. Accordingly, the most representative ethological groups are fodinichnia, pascichnia, repichnia and, to a lesser extent, domichnia. Nearly all the ichnofossils in the association are oriented parallel to the bedding plane, displaying restriction to two-dimensional biotopes; apparently they do not disturb the primary sedimentary fabric. Bedding plane trace fossils mostly reflect shallow to very shallow infaunal grazing by mobile, bilaterian metazoans. Although trace fossils of infaunal organisms are common in these deposits, there is a conspicuous absence of vertical burrows, which are dominant in the overlying upper Lower–Middle Cambrian Mesón Group (Mángano & Buatois 2002, 2004). In contrast to the terminal Proterozoic strata and the diagnosed Early Cambrian Puncoviscana deep-marine deposits, the evidence of microbial matgrounds in the shallow-marine deposits of the Puncoviscana Formation is sparse, suggesting that the infaunal grazers had already disturbed the established biomats of the Early Phanerozoic shallow seas. Lowermost Cambrian ichnofaunas, although diverse, essentially reflect very shallow feeding activities along sand–mud interfaces (cf. McIlroy & Logan 1999).

Although currently referred to the archetypal *Nereites* ichnofacies (Durand & Aceñolaza 1990), the shallow-marine association of the Puncoviscana Formation is best regarded as an example of the *Cruziana* ichnofacies. Spiral, meandering, rosette and network structures that typify post-Cambrian deep-marine deposits were present in shelf environments during the Cambrian, suggesting origination of sophisticated pascichnia and agrichnia in shallow water and subsequent migration to the deep sea in the Ordovician (Crimes & Anderson 1985; Hofmann & Patel 1989; Crimes 1992, 2001; Crimes & Fedonkin 1994; Jensen & Mens 1999; Orr 2001; Mángano & Droser 2004). Terminal Proterozoic and Early Cambrian deep-marine meandering trails do not reflect specialised grazing patterns (e.g. Narbonne & Aitken 1990; Buatois & Mángano 2003b). The presence of these complex patterns in terminal Proterozoic

shallow-marine rocks is debatable because most of the supposedly tightly meandering trails are no longer considered trace fossils (Seilacher *et al.* 2003; Jensen 2003). The only probable exception is a meandering to spiral trail illustrated by Runnegar (1992) and recently discussed by Jensen (2003). The presence of guided meanders in Cambrian shallow-water deposits of the Puncoviscana Formation is consistent with the onshore–offshore model.

Personal data and literature surveys indicate that lowermost Cambrian shallow-marine ichnofaunas are much more diverse and varied than their terminal Proterozoic counterparts. Relatively diverse ichnofaunas dominated by moderate- to large-sized, shallow grazing and feeding traces of deposit feeders are known from the earliest Cambrian of Greenland (Pickerill & Peel 1990; Bryant & Pickerill 1990), Norway (Banks 1970; McIlroy & Logan 1999), Sweden (Jensen & Grant 1998; Moczydłowska *et al.* 2001), Poland (Pacześna 1986, 1996; Orłowski 1989; Orłowski & Żylińska 2002), Spain (Fedonkin *et al.* 1983; García Hidalgo 1993), Mongolia (Goldring & Jensen 1996), Pakistan (Seilacher 1955), China (Crimes & Jiang 1986; Li *et al.* 1997; Zhu 1997), Canada (Young 1972; Crimes & Anderson 1985; Fritz & Crimes 1985; Hofmann & Patel 1989; MacNaughton & Narbonne 1999; Droser *et al.* 2002), USA (Jensen *et al.* 2002), Namibia (Germs 1972; Crimes & Germs 1982; Geyer & Uchman 1995) and Australia (Glaessner 1969; Walter *et al.* 1989).

Deep-water ichnofaunal associations

Analysis of turbidite ichnofaunas provides unequivocal evidence that the deep sea was colonised during the terminal Proterozoic–Early Cambrian transition (Narbonne & Aitken 1990; Seilacher & Pflüger 1992; MacNaughton *et al.* 2000; Crimes 2001; Orr 2001). In fact, our data indicate widespread distribution and a remarkable degree of similarity in taxonomic composition and trophic structure among Early Cambrian deep-marine ichnofaunas. Early Cambrian deep-marine ichnofaunas dominated by *Oldhamia* and non-specialised grazing trails have been recorded in Ireland (Crimes & Crossley 1968; Dhonau & Holland 1974; Crimes 1976; Holland 2001), Belgium (Verniers *et al.* 2001), Alaska (Churkin & Brabb 1965), Canada (Hofmann & Cecile 1981; Lindholm & Casey 1990; Sweet & Narbonne 1993; Hofmann *et al.* 1994), USA (Ruedemann 1942; Neuman 1962) and Antarctica (Seilacher 2003, pers. comm.). More robust, horizontal trace fossils of vagile infaunal organisms, such as *Palaeophycus* or *Planolites*, although subordinate components of Lower Cambrian deep-marine ichnofaunas, are present locally (Hofmann *et al.* 1994). Arthropod trackways were relatively common in deep-marine settings during the Early Palaeozoic (e.g. Crimes *et al.* 1992; Hofmann *et al.* 1994; Orr 2001). As in the shallow-marine association, the ethological groups represented are fodinichnia, pascichnia, repichnia and, to a lesser extent, domichnia.

A remarkably similar ichnofauna to that from the turbidite deposits of the Puncoviscana Formation was documented from the Lower Cambrian to lower Middle Cambrian Grant Land Formation in Arctic Canada by Hofmann *et al.* (1994). These authors recorded a moderately diverse ichnofauna that includes several ichnospecies of *Oldhamia* (*O. antiqua, O. curvata, O. flabellata, O. radiata*), non-specialised grazing trails (*Cochlichnus* isp., *Helminthoidichnites* isp.), locomotion traces (*Didymaulichnus* isp.) and arthropod traces (*Monomorphichnus* isp.). Both the taxonomic composition and the ethology of this Arctic ichnofauna favour comparison with the deep-marine association recorded in the Puncoviscana Formation.

The *Helminthopsis* ichnoguild, particularly widespread in Early Cambrian deep-marine environments, is also the most common component of terminal Proterozoic shallow- to deep-marine environments, where it occurs associated with microbial mats (Gibson 1989; Narbonne & Aitken 1990; Vidal *et al.* 1994; Gehling 1999; MacNaughton *et al.* 2000; Buatois & Mángano 2003b). As noted by Buatois & Mángano (2003b), this ichnoguild became uncommon in Early Phanerozoic shallow-marine settings, experiencing a retreat by the Early Cambrian into deep-marine environments where it occurs together with the *Oldhamia* ichnoguild (e.g. Crimes & Crossley 1968; Crimes 1976; Hofmann & Cecile 1981; Lindholm & Casey 1990; Hofmann *et al.* 1994). Early Cambrian deep-marine ichnofaunas are, in this sense, of "Ediacaran aspect", a notion that is consistent with the idea of archaic relics taking refuge in the deep sea (e.g. Conway Morris 1989). This view is supported by ichnological data from the Cambrian Bray Group of Ireland presented by Crimes (1976). He documented shallow-marine strata with abundant vertical burrows (*Skolithos, Arenicolites*) in the Drumleck and Hippy Hole formations that are replaced upwards by deep-marine deposits having the *Oldhamia* association in the Elsinore Formation. This demonstrates that *Oldhamia*-dominated assemblages in microbial mat ecosystems persisted in the deep sea after the rise of vertical bioturbation in shallow seas of the "Agronomic Revolution" (Seilacher 1999). This fact suggests a gradual closure of the taphonomic window during the Proterozoic–Cambrian transition and is consistent with the recognition of Ediacara-type body fossils in Cambrian strata of different continents (Gehling *et al.* 1998; Jensen *et al.* 1998; Crimes & McIlroy 1999; Hagadorn *et al.* 2000). Integrated ichnological and sedimentological analysis suggests that the deep-sea association of the

Puncoviscana Formation reflects lifestyles related to microbial mats that protected the sediment from erosion. The presence of wrinkled surfaces and palimpsest ripples indicates that stabilisation by microbial binding was a major factor in terminal Proterozoic–Cambrian ecosystems. Some of the ichnofossils (e.g. *Oldhamia*, grazing trails) are directly associated with microbial structures.

Conclusions

The recognition of trace fossil associations in both shallow- and deep-marine deposits of the Puncoviscana Formation provides valuable information on evolutionary and ecological controls on marine infaunal communities near the beginning of the Phanerozoic. Early Cambrian shallow-marine ichnofaunas are much more diverse and varied than their terminal Proterozoic counterparts. These Cambrian shallow-marine ichnofaunas are dominated by moderate- to large-sized, shallow grazing and feeding traces of deposit feeders; arthropod trackways are also present. Turbidite ichnofaunas from the Puncoviscana Formation and coeval units provide unequivocal evidence that the deep sea was colonised by the terminal Proterozoic–Early Cambrian transition. Early Cambrian deep-marine ichnofaunas are dominated by *Oldhamia* and non-specialised grazing trails; horizontal burrows and arthropod trackways are also present. Deep-marine trace fossils of the Puncoviscana Formation reflect lifestyles related to microbial mats that protected the sediment from erosion. Early Cambrian deep-marine ichnofaunas are, in this sense, of "Ediacaran aspect", recording persistence of relict communities after the onset of the "Agronomic Revolution" that pervaded shallow-marine ecosystems during the Early Cambrian. Tiering was moderately developed in both shallow- and deep-marine Early Cambrian communities. Biogenic structures were emplaced within the uppermost millimetres of the sediment (i.e. micro-tiering), with the exception of burrowers that record post-event colonisation below tempestite and turbidite sandstone layers.

Acknowledgements

This project was supported by the Antorchas Foundation and the Research Council of the University of Tucumán (CIUNT). Several colleagues are thanked for valuable discussions and fruitful exchanges on Neoproterozoic–Cambrian ichnofaunas, particularly Richard Bromley, Jim Gehling, Murray Gingras, Sören Jensen, Guy Narbonne, Patrick Orr and Dolf Seilacher. Pamela Aparicio, María Isabel López, Cristina Moya, Eduardo Olivero and Ignacio Sabino assisted during the field work. Florencio Aceñolaza provided access to collections at the Instituto Miguel Lillo. Jim Gehling and Hans Hofmann are thanked for their reviews of the manuscript. Noelia Carmona assisted with the reference list. Drawings were made by Marcos Jimenez and trace fossil specimens were curated by Rodolfo Aredes.

References

Aceñolaza, F.G. 1973: Sobre la presencia de *Oldhamia* sp. en la Formación Puncoviscana de Cuesta Muñano, provincia de Salta, República Argentina. *Revista de la Asociación Geológica Argentina 28*, 56–60.

Aceñolaza, F.G. 1978: El Paleozoico inferior de Argentina según sus trazas fósiles. *Ameghiniana 15*, 15–64.

Aceñolaza, F.G., Aceñolaza, G.F. & Esteban, S. 1999: Bioestratigrafía de la Formación Puncoviscana y unidades equivalentes en el NOA. Relatorio XIV Congreso Geológico Argentino. *Geología del Noroeste Argentino 1*, 91–114.

Aceñolaza, F.G. & Durand, F.R. 1973: Trazas fósiles del basamento cristalino del noroeste argentino. *Boletín de la Asociación Geológica de Córdoba 2*, 45–55.

Aceñolaza, F.G. & Durand, F.R. 1982: El icnogénero *Oldhamia* (traza fósil) en Argentina. Caracteres morfológicos e importancia estratigráfica en formaciones del Cámbrico inferior de Argentina. *5to Congreso Latinoamericano de Geología, Actas 1*, 705–720.

Aceñolaza, F.G. & Durand, F.R. 1984: The trace fossil *Oldhamia*: its interpretation and occurrence in the Lower Cambrian of Argentina. *Neues Jahrbuch fur Geologie und Paläontologie Monatshefte H.12*, 728–740.

Aceñolaza, F.G. & Durand, F.R. 1986: Upper Precambrian–Lower Cambrian biota from the northwest of Argentina. *Geological Magazine 123*, 367–375.

Aceñolaza, F.G., Durand, F.R. & Diaz-Taddei, R. 1976. Geología y contenido paleontológico del basamento metamórfico de la región de Cachi, Provincia de Salta. *6to Congreso Geológico Argentino, Actas 1*, 319–332.

Aceñolaza, F.G. & Toselli, A.J. 1981: Geología del Noroeste Argentino. *Facultad de Ciencias Naturales e Instituto Miguel Lillo, Universidad Nacional de Tucumán, Publicación Especial 1287*, 1–212.

Aceñolaza, G.F. & Aceñolaza, F.G. 2001: Ichnofossils and microbial activity in the Precambrian/Cambrian transition of northwestern Argentina. *Palaeoworld 13*, 241–244.

Aceñolaza, G.F. & Tortello, M.F. 2000: El Alisal: a new locality with trace fossils of the Puncoviscana Formation (Late Precambrian–Early Cambrian) in Salta Province. *In* Aceñolaza, G.F. & Peralta, S. (eds): *Cambrian from the Southern Edge*, 67–69. Instituto Superior de Correlación Geológica Miscelánea 6.

Aceñolaza, G.F. & Tortello, M.F. 2003: El Alisal: a new locality with trace fossils of the Puncoviscana Formation (late Precambrian–early Cambrian) in Salta Province, Argentina. *Geologica Acta 1*, 95–102.

Banks, N.L. 1970: Trace fossils from the late Precambrian and Lower Cambrian of Finmark, Norway. *Geological Journal Special Issue 9*, 19–34.

Becchio, R., Lucassen, F., Franz, G., Viramonte, J. & Wemmer, K. 1999: El basamento paleozoico inferior del noroeste de Argentina (23°–27°S) – Metamorfismo y geocronología. Relatorio XIV Congreso Geológico Argentino. *Geología del Noroeste Argentino 1*, 58–72.

Bromley, R.G. 1996: *Trace Fossils. Biology, Taphonomy and Applications*. Chapman and Hall, London.

Bromley, R.G. & Asgaard, U. 1975: Sediment structures produced by a spatangoid echinoid: a problem of preservation. *Bulletin of the Geological Society of Denmark 24*, 261–281.

Bryant, I.D. & Pickerill, R.K. 1990: Lower Cambrian trace fossils from the Buen Formation of central North Greenland: preliminary observations. *Grønlands Geologiske Undersøgelse, Rapport 147*, 44–62.

Buatois, L.A. & Mángano, M.G. 2002: Ichnology of the Puncoviscana Formation in northwest Argentina: anactualistic ecosystems and the Precambrian–Cambrian transition. *First International Palaeontological Congress (IPC 2002), Geological Society of Australia Abstracts 68*, 25–26.

Buatois, L.A. & Mángano, M.G. 2003a: La icnofauna de la Formación Puncoviscana en el noroeste argentino: implicancias en la colonización de fondos oceánicos y reconstrucción de paleoambientes y paleoecosistemas de la transición precámbrica–cámbrica. *Ameghiniana 40*, 103–117.

Buatois, L.A. & Mángano, M.G. 2003b: Early colonization of the deep sea: ichnologic evidence of deep-marine benthic ecology from the Early Cambrian of northwest Argentina. *Palaios 18*, 572–581.

Buatois, L.A., Mángano, M.G., Aceñolaza, F.G. & Esteban, S.B. 2000: The Puncoviscana ichnofauna of northwest Argentina: a glimpse into the ecology of the Precambrian–Cambrian transition. *In* Aceñolaza, G.F. & Peralta, S. (eds): *Cambrian from the Southern Edge*, 82–84. Instituto Superior de Correlación Geológica Miscelánea 6.

Churkin, M. Jr. & Brabb, E.E. 1965: Occurrence and stratigraphical significance of *Oldhamia*, a Cambrian trace fossil, in East-Central Alaska. *United States Geological Survey Professional Paper 525-D*, D120–D124.

Conway Morris, S. 1989: The persistence of Burgess Shale-type faunas: implications for the evolution of deeper-water faunas. *Transactions of the Royal Society of Edinburgh: Earth Sciences 80*, 271–283.

Crimes, T.P. 1976: Trace fossils from the Bray Group (Cambrian) at Howth, Co. Dublin. *Geological Survey of Ireland Bulletin 2*, 53–67.

Crimes, T.P. 1992: The record of trace fossils across the Proterozoic–Cambrian boundary. *In* Lipps, J. & Signor, P.W. (eds): *Origin and Early Evolution of the Metazoa*, 177–202. Plenum Press, New York.

Crimes, T.P. 2001: Evolution of the deep-water benthic community. *In* Zhuravlev, A.Y. & Riding, R. (eds): *The Ecology of the Cambrian Radiation*, 275–290. Columbia University Press, New York.

Crimes, T.P. & Anderson, M.M. 1985: Trace fossils from Late Precambrian–Early Cambrian strata of southeastern Newfoundland (Canada): temporal and environmental implications. *Journal of Paleontology 59*, 310–343.

Crimes, T.P. & Crossley, J.D. 1968: The stratigraphy, sedimentology, ichnology and structure of the Lower Paleozoic rocks of part of northeastern Co. Wexford. *Proceedings of the Royal Irish Academy 67B*, 185–215.

Crimes, T.P. & Fedonkin, M.A. 1994: Evolution and dispersal of deep sea traces. *Palaios 9*, 74–83.

Crimes, T.P., García Hidalgo, J.F. & Poiré, D.G. 1992: Trace fossils from Arenig flysch sediments of Eire and their bearing on the early colonisation of the deep seas. *Ichnos 2*, 61–77.

Crimes, T.P. & Germs, G.J.B. 1982: Trace fossils from the Nama Group (Precambrian–Cambrian) of southwest Africa (Namibia). *Journal of Paleontology 56*, 890–907.

Crimes, T.P. & Jiang, Z. 1986: Trace fossils from the Precambrian–Cambrian boundary candidate at Meishucun, Jinning, Yunnan, China. *Geological Magazine 123*, 641–649.

Crimes, T.P. & McIlroy, D. 1999: An Ediacaran biota from lower Cambrian strata on the Digermul Peninsula, Arctic Norway. *Geological Magazine 36*, 633–642.

Dhonau, N.B. & Holland, C.H. 1974: The Cambrian of Ireland. *In* Holland, C.H. (ed.): *Cambrian of the British Isles, Norden, and Spitsbergen,* 157–176. John Wiley & Sons, London.

Durand, F.R. 1990: Los Conglomerados del Ciclo Pampeano en el Noroeste Argentino. *Serie Correlación Geológica 4,* 61–69.

Durand, F.R. & Aceñolaza, F.G. 1990a: Caracteres biofaunísticos, paleocológicos y paleogeográficos de la Formación Puncoviscana (Precámbrico Superior–Cámbrico Inferior) del Noroeste Argentino. *Serie Correlación Geológica 4,* 71–112.

Durand, F.R. & Aceñolaza, F.G. 1990b: Presencia de un artrópodo primitivo en la Formación Puncoviscana (Precámbrico Superior–Cámbrico Inferior) de la Quebrada del Toro, Provincia de Salta, Argentina. *Actas 5to Congreso Argentino de Paleontología y Bioestratigrafía 1,* 13–17.

Durand, F.R. & Aceñolaza, F.G. 1992: La presencia de *Multipodichnus* holmi (Traza Fósil) en el basamento del Noroeste Argentino. *Serie Correlación Geológica 9,* 1–15.

Durand, F.R., Lech, R.R. & Tortello, M.F. 1994: Nuevas evidencias paleontológicas en el basamento precámbrico–cámbrico del noroeste argentino. *Acta Geológica Leopoldensia 39,* 691–701.

Durand, F.R. & Spalletti, L.A. 1986: Las facies turbidíticas del Precámbrico superior–Cámbrico inferior en la zona de Corralito, provincia de Salta. *1ra Reunión Argentina de Sedimentología, Resúmenes Expandidos,* 113–116.

Droser, M.L., Jensen, S., Gehling, J.G., Myrow, P.M. & Narbonne, G.M. 2002: Lowermost Cambrian ichnofabrics from the Chapel Island Formation, Newfoundland: implications for Cambrian substrates. *Palaios 17,* 3–15.

Droser, M.L. & Li, X. 2001: The Cambrian radiation and the diversification of sedimentary fabrics. *In* Zhuravlev, A.Y. & Riding, R. (eds): *The Ecology of the Cambrian Radiation,* 137–164. Columbia University Press, New York.

Fedonkin, M., Liñán, E. & Perejón, A. 1983: Icnofósiles de las rocas precámbrico–cámbricas de la Sierra de Córoba. España. *Boletín de la Real Sociedad Española de Historia Natural (Geología) 81,* 125–138.

Fritz, W.H. & Crimes, T.P. 1985: Lithology, trace fossils, and correlation of Precambrian–Cambrian boundary beds, Cassiar Mountains, north-central British Columbia. *Geological Survey of Canada Paper 83-13,* 1–24.

García Hidalgo, J.F. 1993: Las pistas fósiles del Alcudiense superior en el anticlinal de Ibor. Consideraciones cronoestratigráficas. *Geogaceta 13,* 33–35.

Gehling, J.G. 1999: Microbial mats in terminal Proterozoic Siliciclastic Ediacaran masks. *Palaios 14,* 40–57.

Gehling, J.G., Droser, M., Jensen, S. & Runnegar, B. 1998: Similar cycles – different strokes: closing a taphonomic window across the Precambrian–Cambrian boundary. *Inaugural Sprigg Symposium: The Ediacaran Revolution, Abstracts & Programme,* 20–21.

Germs, G.J.B. 1972: New shelly fossils from Nama Group, South West Africa. *American Journal of Sciences 272,* 752–761.

Geyer, G. & Uchman, A. 1995: Ichnofossil assemblages from the Nama Group (Neoproterozoic–Lower Cambrian) in Namibia and the Proterozoic–Cambrian boundary problem revisited. *Beringeria Special Issue 2,* 175–202.

Gibson, G.G. 1989: Trace fossil from late Precambrian Carolina Slatte Belt, South-Central North Carolina. *Journal of Paleontology 63,* 1–10.

Glaessner, M.F. 1969: Trace fossils from the Precambrian and basal Cambrian. *Lethaia 2,* 369–393.

Glaessner, M.F. 1984: *The Dawn of Animal Life: A Biohistorical Study.* Cambridge University Press, Cambridge.

Goldring, R. & Jensen, S. 1996: Trace fossils and biofabrics at the Precambrian–Cambrian boundary interval in western Mongolia. *Geological Magazine 133,* 403–415.

Hagadorn, J.W., Schellenberg, S.A. & Bottjer, D.J. 2000: Palaecology of a large Early Cambrian bioturbator. *Lethaia 33,* 142–156.

Hofmann, H.J. & Cecile, M.P. 1981: Occurrence of *Oldhamia* and other trace fossils in Lower Cambrian(?) argillites, Selwyn Mountains, Yukon. *Geological Survey of Canada, Paper 81-1A,* 281–289.

Hofmann, H.J., Cecile, M.P. & Lane, L.S. 1994: New occurrences of *Oldhamia* and other trace fossils in the Cambrian of the Yukon and Ellesmere Island, Arctic Canada. *Canadian Journal of Earth Sciences 31,* 767–782.

Hofmann, H.J. & Patel, I.M. 1989: Trace fossils from the type 'Etcheminian Series' (Lower Cambrian Ratcliffe Brook Formation), Saint John area, New Brunswick, Canada. *Geological Magazine 126,* 139–157.

Holland, C.H. 2001: Cambrian of Leinster. *In* Holland, C.H. (ed.): *The Geology of Ireland,* 73–81. Dunedin Academic Press, Edinburgh.

Hongn, F.D. 1996: La estructura pre-Grupo Mesón (Cámbrico) del basamento del Valle de Lerma, Provincia de Salta. *13° Congreso Geológico Argentino & 3° Congreso de Exploración de Hidrocarburos, Actas 2,* 137–145.

Hongn, F.D., Tubía, J.M., Aranguren, A. & Mon, R. 2001: El batolito de Tastil (Salta, Argentina): un caso de magmatismo poliorogénico en el basamento andino. *Boletín Geológico y Minero 112,* 113–124.

Jensen, S. 2003: The Proterozoic and Earliest Cambrian trace fossil record: patterns, problems and perspectives. *Integrative and Comparative Biology 43,* 219–228.

Jensen, S., Droser, M.L. & Heim, N.A. 2002: Trace fossils and ichnofabrics of the Lower Cambrian Wood Canyon Formation, Southwest Death Valley Area. *In* Corsetti, F.A. (ed.): *Proterozoic–Cambrian of the Great Basin and Beyond,* 123–135. Volume and Guidebook. The Pacific Section SEPM (Society for Sedimentary Geology), Fullerton, California.

Jensen, S., Gehling, J.G. & Droser, M.L. 1998: Ediacara-type fossils in Cambrian sediments. *Nature 393,* 567–569.

Jensen, S. & Grant, S.W.F. 1998: Trace fossils from the Dividalen Group, northern Sweden: implications for Early Cambrian biostratigraphy of Baltica. *Norsk Geologisk Tidskrift 78,* 305–317.

Jensen, S. & Mens, K. 1999: A Lower Cambrian shallow-water occurrence of the branching "deep water" type trace fossil *Dendrorhaphe* from the Lontova Formation, eastern Latvia. *Paläontologische Zeitschrift 73,* 187–193.

Ježek, P. 1990: Análisis sedimentológico de la Formación Puncoviscana entre Tucumán y Salta. *Serie Correlación Geológica 4,* 9–36.

Li, R., Yang, S. & Li, W. 1997: *Trace Fossils from the Sinian–Cambrian Boundary Strata in China.* Geological Publishing House, Beijing.

Lindholm, R.M. & Casey, J.F. 1990: The distribution and possible biostratigraphic significance of the ichnogenus *Oldhamia* in the shales of the Blow Me Down Brook Formation, western Newfoundland. *Canadian Journal of Earth Sciences 27,* 1270–1287.

MacNaughton, R.B. & Narbonne, G.M. 1999: Evolution and ecology of Neoproterozoic–Lower Cambrian trace fossils, NW Canada. *Palaios 14,* 97–115.

MacNaughton, R.B., Narbonne, G.M. & Dalrymple, R.W. 2000: Neoproterozoic slope deposits, Mackenzie Mountains, northwestern Canada: implications for passive-margin development and Ediacaran faunal ecology. *Canadian Journal of Earth Sciences 37,* 997–1020.

Mángano, M.G. & Buatois, L.A. 2002: Estructura de escalonamiento y explotación del ecoespacio infaunal en planicies mareales cámbricas, Formación Campanario, noroeste argentino. *Resúmenes VIII Congreso Argentino de Paleontología y Bioestratigrafía,* 107–108.

Mángano, M.G. & Buatois, L.A. 2004: Reconstructing Early Phanerozoic intertidal ecosystems: ichnology of the Cambrian Campanario Formation in northwest Argentina. *Fossils and Strata 51*, 17–38.

Mángano, M.G. & Buatois, L.A. in press: Integración de estratigrafía secuencial, sedimentología e icnología para un análisis cronoestratigráfico del Paleozoico inferior del noroeste argentino. *Revista de la Asociación Geológica Argentina.*

Mángano, M.G., Buatois, L.A., Aceñolaza, F.G. & Esteban, S.B. 2000: La icnofauna de la Formación Puncoviscana (Proterozoico Tardío-Cámbrico Temprano): implicancias en la colonización de los fondos oceánicos. *Actas Segundo Congreso Latinoamericano de Sedimentología 1*, 107.

Mángano, M.G. & Droser, M. 2004: The ichnologic record of the Ordovician radiation. *In* Webby, B.D., Droser, M., Paris, F. & Percival, I.G. (eds): *The Great Ordovician Biodiversification Event*, 369–379. Columbia University Press, New York.

McIlroy, D. & Logan, G.A. 1999: The impact of bioturbation on infaunal ecology and evolution during the Proterozoic–Cambrian transition. *Palaios 14*, 58–72.

Mirré, J.C. & Aceñolaza, F.G. 1972: El hallazgo de *Oldhamia* sp. (traza fósil) y su valor como evidencia de edad cámbrica para el supuesto Precámbrico del borde occidental del Aconquija, Prov. de Catamarca. *Ameghiniana 9*, 72–78.

Moczydłowska, M., Jensen, S., Ebbestad, J.O.R., Budd, G.E. & Martí-Mus, M. 2001: Biochronology of the autochthonus Lower Cambrian in the Laisvall-Storuman area, Sweedish Caledonides. *Geological Magazine 138*, 435–453.

Mon, R. & Hongn, F. 1988: Caracterización estructural de la Formación Puncoviscana dentro del basamento del Norte Argentino. *Revista de la Asociación Geológica Argentina 43*, 124–127.

Mon, R. & Hongn, F. 1991: The structure of the Precambrian and Lower Paleozoic Basement of the Central Andes between 22° and 32° S. Lat. *Geologische Rundschau 83*, 745–758.

Moya, M.C. 1998: El Paleozoico inferior en la sierra de Mojotoro, Salta - Jujuy. *Revista de la Asociación Geológica Argentina 53*, 219–238.

Neuman, R.B. 1962: The Grand Pitch Formation: a new name for Grand Falls Formation (Cambrian?) in northeastern Maine. *American Journal of Science 260*, 794–797.

Narbonne, G.M. & Aitken, J.D. 1990: Ediacaran fossils from the Sekwi Brook area, Mackenzie mountains, northwestern Canada. *Palaeontology 33*, 945–980.

Narbonne, G.M. & Gehling, J.G. 2003: Life after snowball: the oldest complex Ediacaran fossils. *Geology 31*, 27–30.

Omarini, R.H. & Baldis, B.A.J. 1984: Sedimentología y mecanismos depositacionales de la Formación Puncoviscana (Grupo Lerma, Precámbrico-Cámbrico) del noroeste argentino. *Actas Noveno Congreso Geológico Argentino 1*, 384–398.

Omarini, R.H., Sureda, R.J., Götze, H.-J., Seilacher, A. & Pflüger, F. 1999: Puncoviscana folded belt in northwestern Argentina: testimony of Late Proterozoic Rodinia fragmentation and pre-Gondwana collisional episodes. *International Journal of Earth Sciences 88*, 76–97.

Orłowski, S. 1989: Trace fossils in the Lower Cambrian sequence in the Swietokrzyskie Mountains, Central Poland. *Acta Palaeontologica Polonica 34*, 211–231.

Orłowski, S. & Żylińska, A. 2002: Lower Cambrian trace fossils from the Holy Cross Mountains, Poland. *Geological Quarterly 46*, 135–146.

Orr, P.J. 1996: The ichnofauna of the Skiddaw Group (Early Ordovician) of the Lake District, England. *Geological Magazine 133*, 193–216.

Orr, P.J. 2001: Colonization of the deep-marine environment during the early Phanerozoic: the ichnofaunal record. *Geological Journal 36*, 265–278.

Pacześna, J. 1986: Upper Vendian and Lower Cambrian ichnocoenoses of the Lublin region, *Biulety Instytutu Geologicznego 355*, 32–47.

Pacześna, J. 1996: The Vendian and Cambrian ichnocoenoses from the Polish part of the East-European Platform. *Prace Państwowego Instytutu Geologicznego CL11*, 1–77.

Pemberton, S.G., Spila, M., Pulham, A.J., Saunders, T., MacEachern, J.A., Robbins, D. & Sinclair, I.K. 2001: Ichnology and sedimentology of shallow to marginal marine systems. Ben Nevis and Avalon Reservoirs, Jeanne d'Arc Basin. *Geological Association of Canada, Short Course Notes 15*, 1–343.

Pickerill, R.K. & Peel, J.S. 1990: Trace fossils from the Lower Cambrian Bastion Formation of north-east Greenland. *Grønlands Geologiske Undersøgelse, Rapport 147*, 5–43.

Plaziat, J.-C. & Mahmoudi, M. 1988: Trace fossils attributed to burrowing echinoids: a revision including new ichnogenus and ichnospecies. *Géobios 21*, 209–233.

Ramos, V.R. 2000: The Southern Central Andes. *In* Cordani, U.G., Milani, E.J., Thomaz Filho, A. & Campos, D.A. (eds): *Tectonic Evolution of South America*, 561–604. 31st International Geological Congress, Rio de Janeiro.

Ruedemann, R. 1942: *Oldhamia* and the Rensselaer Grit problem. *New York State Museum, Bulletin 327*, 5–17.

Runnegar, B. 1992: Evolution of the earliest animals. *In* Schopf, J.W. (ed.): *Major Events in the History of Life*, 999–1007. Jones and Bartlett, Boston.

Seilacher, A. 1955: Spuren und Fazies im Unterkambrium. *In* Schindewolf, O.H. & Seilacher, A. (eds): Beitrage zur Kenntnis des Kambriums in der Salt Range (Pakistan). *Akademiee der Wissenschaften und der Literatur zu Mainz, mathematisch-naturwissenschaftliche Klasse, Abhandlungen 10*, 117–147.

Seilacher, A. 1989: Vendozoa: organismic construction in the Proterozoic biosphere. *Lethaia 22*, 229–239.

Seilacher, A. 1992: Vendobionta and Psammocorallia: lost constructions of Precambrian evolution. *Journal of the Geological Society, London 149*, 607–613.

Seilacher, A. 1997: *Fossil Art*. The Royal Tyrrell Museum of Paleontology, Drumheller, Alberta.

Seilacher, A. 1999: Biomat-related lifestyles in the Precambrian. *Palaios 14*, 86–93.

Seilacher, A., Grazhdankin, D. & Legouta, A. 2003: Ediacaran biota: the dawn of animal life in the shadow of giant protists. *Paleontological Research 7*, 43–54.

Seilacher, A. & Pflüger, F. 1992: Trace fossils from the Late Proterozoic of North Carolina: early conquest of deep-sea bottoms: 5th North American Paleontological Convention, Abstracts and Program. *Paleontological Society, Special Publication 6*, 265.

Smith, A.B. & Crimes, P.T. 1983: Trace fossils formed by heart urchins – a study of *Scolicia* and related traces. *Lethaia 16*, 79–92.

Sweet, N.L. & Narbonne, G.M. 1993: Occurrence of the Cambrian trace fossil *Oldhamia* in southern Québec. *Palaios 29*, 69–73.

Turner, J.C.M. 1960: Estratigrafía de la Sierra de Santa Victoria y adyacencias. *Boletín de la Academia Nacional de Ciencias de Córdoba 41*, 163–196.

Turner, J.C.M. 1972: Puna. *In* Leanza, A. (ed.): *I Simposio de Geología Regional Argentina*, 91–116. Academia Nacional de Ciencias, Córdoba.

Uchman, A. 1995: Taxonomy and paleoecology of flysch trace fossils: the Marnoso-arenacea Formation and associated facies (Miocene, Northern Apennines, Italy). *Beringeria 15*, 1–115.

Van Staden, A. & Zimmermann, U. 2003: Tillites or ordinary conglomerates? Provenance studies on diamictites of the Neoproterozoic Puncoviscana in NW Argentina. *Abstracts of the 3rd Latinoamerican Congress of Sedimentology*, 74–75.

Verniers, J., Herbosch, A., Vanguestaine, M., Geukens, F., Delcambre, B., Pingot, J.L., Belanger, I., Hennebert, M., Debacker, T., Sintubin, M. & de Vos, W. 2001: Cambrian–Ordovician–Silurian lithostratigraphic units (Belgium). *Geologica Belgica 4*, 5–38.

Vidal, G., Jensen, S. & Palacios, T. 1994: Neoproterozoic (Vendian) ichnofossils from Lower Alcudian strata in central Spain. *Geological Magazine 131*, 169–179.

Waggoner, B. 1998: Interpreting the earliest metazoan fossils: what can we learn? *American Zoologist 38*, 975–982.

Walter, M.R., Elphinstone, R. & Heys, G.R. 1989: Proterozoic and Early Cambrian trace fossils from the Amadeus and Georgina Basins, central Australia. *Alcheringa 13*, 209–256.

Young, F.G. 1972: Early Cambrian and older trace fossils from the Southern Cordillera of Canada. *Canadian Journal of Earth Sciences 9*, 1–17.

Zhu, M. 1997: Precambrian–Cambrian trace fossils from eastern Yunnan, China: implications for Cambrian explosion. *Bulletin of National Museum of Natural Science 10*, 275–312.

Reconstructing Early Phanerozoic intertidal ecosystems: ichnology of the Cambrian Campanario Formation in northwest Argentina

M. GABRIELA MÁNGANO & LUIS A. BUATOIS

Mángano, M.G. & Buatois, L.A. **2004 10 25**: Reconstructing Early Phanerozoic intertidal ecosystems: ichnology of the Cambrian Campanario Formation in northwest Argentina. *Fossils and Strata*, No. 51, pp. 17–38. Canada. ISSN 0300-9491.

The Campanario Formation is the middle unit of the upper Lower to Middle Cambrian Mesón Group of northwest Argentina. This formation is interpreted as having accumulated in macrotidal shallow-marine environments with extensive tidal-flat areas flanked seawards by subtidal sandbar complexes. Shallow subtidal and intertidal sand-flat deposits are dominated by vertical domiciles of suspension feeders and passive predators of the *Skolithos* ichnofacies (*Skolithos*, *Arenicolites*, *Diplocraterion*). The ichnogenus *Syringomorpha* also occurs in sand-flat facies, commonly forming high-density, monospecific assemblages. Mixed-flat deposits contain horizontal feeding, locomotion and resting traces as diagnostic components. A relatively low-diversity *Cruziana* ichnofacies is present in these lower-energy deposits. This ichnofauna includes presumed trilobite trace fossils (*Cruziana*, *Rusophycus*, *Diplichnites*), structures produced by sessile cnidarians (*Bergaueria* cf. *B. perata*) and shallow burrows and trails of vermiform organisms (*Planolites*, *Palaeophycus*, *Helminthoidichnites*). Tiering in these tidal-flat deposits is relatively simple. Six ichnoguilds (*Cruziana problematica*, *Palaeophycus*, *Bergaueria*, *Rusophycus leifeirikssoni*, *Syringomorpha* and *Skolithos*) have been defined. These ichnoguilds show a preferential palaeoenvironmental distribution following proximal–distal trends. Modern tidal flats are characterised by an abundant food supply derived from multiple sources, including nutrients brought in by the sea, terrestrially derived organic detritus and autochthonous food production (e.g. primary production, faecal pellets). The inhabitants of modern intertidal areas are exposed to a double set of predators. During submergence they are preyed on by marine organisms and during emergence they are visited by enemies from the land and air. Contrastingly, the Cambrian intertidal environments functioned as refugia in the absence of continental (i.e. air and land) predators. Despite the physical stress, Cambrian tidal flats must have been protected areas where abundant food was available at almost no risk. Due to the paucity of land vegetation (and land-derived detritus), Cambrian intertidal trophic webs were almost entirely based on the organically rich marine source and most likely a significant autochthonous production. Although the picture that emerges from the Cambrian is qualitatively different, Cambrian tidal flats may have resembled modern ones in their ecological role as sites of reproduction and protection. The depth and extent of bioturbation reveal colonisation of a relatively deep infaunal ecospace by coelomate metazoans. Arthropod incursions in Cambrian tidal flats support an early colonisation of intertidal environments and indicate that representatives of the Cambrian evolutionary fauna were able to colonise very shallow-water environments, thus suggesting a significant landward expansion of the Cambrian explosion and the "Agronomic Revolution".

Key words: Cambrian; Ichnology; tidal flats; evolution; ecology; Argentina.

M. Gabriela Mángano [gabriela.mangano@usask.ca] & Luis A. Buatois [luis.buatois@ usask.ca], Department of Geological Sciences, University of Saskatchewan, 114 Science Place, Saskatoon, Canada S7N 5E2

Introduction

Intertidal regions are complex ecosystems where animal communities have developed under rigorous environmental conditions. Numerous studies have documented the ecology of intertidal benthic communities on modern coasts (e.g. Schäfer 1972; Swinbanks & Murray 1981; Reise 1985; Raffaelli & Hawkins 1996; Hild & Günther 1999; Dittman 1999; Dittmann *et al.* 1999; Bertness 1999; Little 2000). However, very little is known about how these ecosystems changed through geological time. In particular, animal–substrate interactions in Early Phanerozoic intertidal ecosystems remain mostly unexplored.

The Cambrian Campanario Formation of the Mesón Group is superbly exposed in northwest Argentina. This formation records deposition in extensive intertidal areas flanked seaward by subtidal sandbar complexes. Trace fossils, representing various forms of behaviour of the benthic fauna, are locally abundant in this unit (Alonso & Marquillas 1981; Manca 1986; Buatois & Mángano 2001; Mángano & Buatois 2003a). Integrated sedimentological and ichnological studies allow us to explore some aspects of the ecology of the Cambrian intertidal ecosystems.

Geological and stratigraphic setting

Outcrops of the Mesón Group extend from the southern Bolivia and northern Jujuy Province, northwest Argentina in the north to the northern Tucumán Province in the south, covering the eastern Cordilleran geological province (Fig. 1). The extent of the basin towards the east is uncertain because no outcrops are available. Subsurface information suggests possible equivalents of the Mesón Group in Formosa and Santiago del Estero provinces (Mingramm *et al.* 1979; Aceñolaza *et al.* 1982). These data need to be re-evaluated because distinction between the Mesón Group and the quartzose sandstones of the lower part of the Santa Rosita Formation (Tilcara Member and coeval strata) is complicated. The Mesón Group unconformably overlies the Neoproterozoic to Nemakit-Daldynian (earliest Cambrian) metasedimentary rocks of the Puncoviscana Formation. An angular unconformity, resulting from the tectonic movements associated with the Tilcaric Orogeny, separates both units (Turner 1979; Aceñolaza *et al.* 1999). The Mesón Group is overlain by the Upper Cambrian to Tremadocian Santa Rosita Formation of the Santa Victoria Group, and is included within the Famatinian Cycle of Aceñolaza & Toselli (1981). The Lower Palaeozoic Basin developed along the active margin of Western Gondwana.

The Mesón Group is divided, from base to top, into the Lizoite, Campanario, and Chalhualmayoc Formations (Turner 1960, 1963). The Lizoite and Chalhualmayoc Formations consist of thick-bedded, large-scale, planar and trough cross-bedded quartzites, while the Campanario Formation is dominated by bioturbated, planar cross-bedded, and ripple cross-laminated sandstones, thinly interbedded sandstones and mudstones, and red mudstones. Hummocky cross-stratified sandstones are locally present in this formation. Deposits of the Mesón

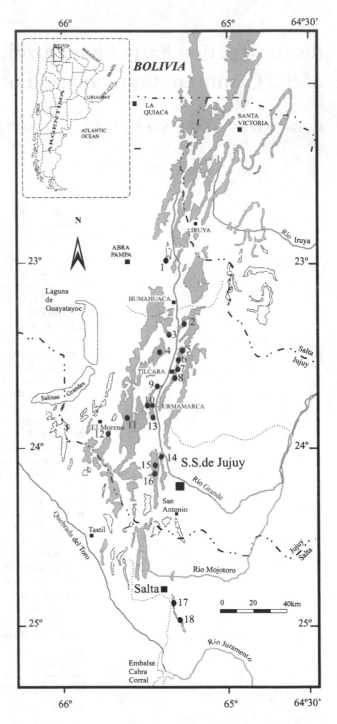

Fig. 1. Outcrop map of the Mesón Group in Jujuy and Salta provinces of northwest Argentina (after Sánchez & Salfity 1999), showing the location of trace fossil localities. 1 = Hornadita. 2 = Quebrada de Moya. 3 = Angosto del Morro de Chucalezna. 4 = Huacalera. 5 = Quebrada de la Huerta. 6 = Angosto de Perchel. 7 = Cordón de Alfarcito. 8 = Maimará. 9 = Incahuasi. 10 = Purmamarca. 11 = Huachichocana. 12 = El Moreno. 13 = Tumbaya. 14 = Leon. 15 = Yala. 16 = Reyes. 17 = Villa Floresta. 18 = Cuesta de la Pedrera.

Group accumulated in a macrotidal shallow-marine seaway. In particular, the Lizoite and Chalhualmayoc Formations represent deposition in large subtidal sandbar complexes and the Campanario Formation mostly records sedimentation in extensive intertidal areas and, to a lesser extent, shallow subtidal regions (Sánchez & Salfity 1990, 1999; Mángano & Buatois 2000; Buatois & Mángano 2001).

The Mesón Group has historically been considered as Middle to Late Cambrian owing to three different lines of evidence: stratigraphic relations, body fossils, and trace fossils. The evidence has been critically reassessed by Mángano & Buatois (in press). The body fossil content of the Mesón Group is meagre, consisting mostly of localised, poorly preserved concentrations of lingulids, including the inarticulate brachiopod *Lingulepis* sp. (Sánchez & Herrera 1994). The trilobite *Asaphiscus* was recorded in strata supposedly of the Lizoite Formation in the Puna geological province, suggesting a late Middle Cambrian age (Aceñolaza 1973; Aceñolaza & Bordonaro 1990). However, recent fieldwork indicates that a lateral equivalent of the Santa Rosita Formation, rather than the Mesón Group, is represented in that area (Buatois & Mángano 2003a). Additionally, the trilobite specimen has been recently reassigned to *Leiostegium douglasi*, a Tremadocian taxon (Vaccari & Waisfeld 2000). The supposed presence of *Cruziana semiplicata* in an outcrop of the Mesón Group has been taken as evidence of a Late Cambrian age. However, the finding is controversial. Originally, Manca (1981) mentioned *C. semiplicata* as float from the Angosto de Perchel section. Study of this specimen fails to reveal any of the diagnostic characteristics of *Cruziana semiplicata*. In a subsequent paper on the Campanario ichnofauna, Manca (1986) did not mention this specimen, but described and illustrated another one as coming from a different outcrop, Huacalera. Re-examination of the specimen confirms the taxonomic assignment, but raises serious doubts regarding its stratigraphic provenance. In fact, the specimen may come from the overlying strata of the Santa Rosita Formation, where this ichnotaxon is widespread. Mángano & Buatois (2003a) recently documented the presence of *Rusophycus leifeirikssoni* in the Campanario Formation. This ichnospecies is known from Upper Cambrian–Tremadocian strata in Newfoundland (Bergström 1976; Fillion & Pickerill 1990). However, the Campanario Formation representative is significantly smaller in size range and records some distinctive morphological differences, suggesting a potential new ichnosubspecies. Therefore, its biostratigraphic meaning remains uncertain. Alonso & Marquillas (1981) recorded the ichnogenus *Syringomorpha* in the Campanario Formation. This identification was questioned by Manca (1989), who collected additional specimens and reassigned them to *Daedalus*. Re-examination of the specimens and additional collections by Mángano & Buatois (2000, 2001) support the original identification by Alonso & Marquillas (1981). *Syringomorpha* is known only from the Lower Cambrian (Mángano & Buatois 2001, in press). Available evidence

suggests that the Mesón Group may range from the upper Lower to Middle Cambrian.

Sedimentary facies, trace fossil distribution and depositional model

Five main sedimentary facies have been recognised in the Campanario Formation. The degree of bioturbation is assessed following the bioturbation index (BI) of Taylor & Goldring (1993).

Large-scale, planar cross-bedded sandstone with superimposed ripples

This facies consists of light grey to light pink, 5–40 cm thick, erosive-based, laterally extensive, well-sorted, medium- to fine-grained quartzite. Reactivation surfaces and herringbone cross-stratification are relatively common. Mud drapes are locally present in foresets and may contain synaeresis cracks (Fig. 2A, B). Bedding planes display a wide variety of structures, including ripple patches, interference ripples, and flat-topped ripples. Ripples are out of phase with respect to the internal structures of the bed. BI ranges from 1 to 5. Trace fossils are rather abundant but poorly diverse, including *Skolithos linearis*, *Arenicolites* isp., and *Diplocraterion parallelum*. Many beds present monospecific or paucispecific assemblages. This facies is common at the transition between the Lizoite and Campanario Formations and is also locally present as a secondary component through the whole Campanario Formation. This facies is interpreted as sandbar deposits. The presence of different types of superimposed ripples (e.g. interference ripples, flat-topped ripples) indicates a shallow-water setting, most likely across the subtidal to lower intertidal transition.

Bioturbated ripple cross-laminated sandstone

This facies comprises light green to red, 3–24 cm thick, either erosive-, sharp- or gradational-based, laterally extensive, medium- to very fine-grained quartzose sandstone. The sand/mud ratio is high (> 1) to infinite (100% sand). Gutter and pot casts are locally present. Ripple cross-lamination is the dominant internal structure. Flaser bedding and mud drapes are common. Wavy bedding is rare. Synaeresis cracks displaying different morphologies (e.g. sinusoidal, spiral) are common in mud drapes and sandstone tops. Bed tops are undulatory, showing asymmetric ripples. A wide variety of bedforms

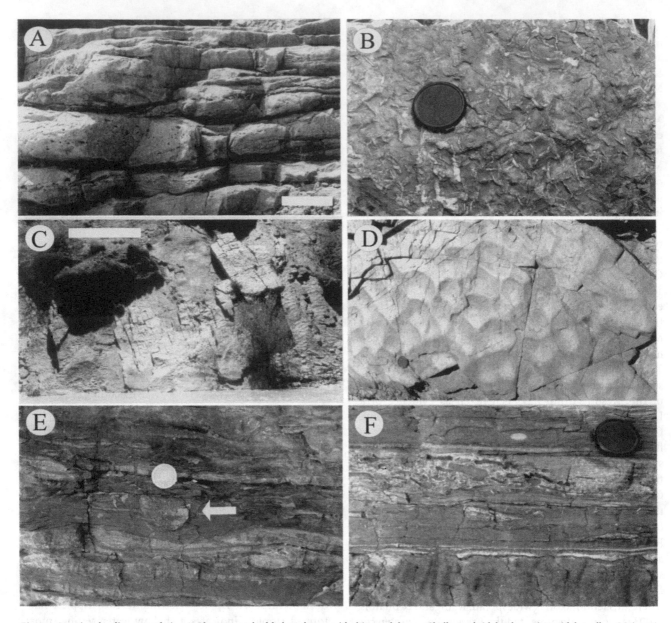

Fig. 2.   Associated sedimentary facies. A: Planar cross-bedded sandstone with thin mud drapes. Shallow subtidal to lower intertidal sandbar. Maimará. Scale bar = 1 m. B: Bedding plane view of mud layers in (A) showing widespread development of synaeresis cracks. Shallow subtidal to lower intertidal sandbar. Maimará. Lens cap diameter = 5.5 cm. C: General view of a bedding plane showing ripple patches. Intertidal sand flat. Angosto de Perchel. Scale bar = 5 m. D: Interference ripples. Intertidal sand flat. Quebrada de Moya. Lens cap diameter (lower left) = 5.5 cm. E: General view of interbedded very fine-grained sandstone and mudstone. Note the preservation of *Rusophycus leifeirikssoni* as a full relief structure (arrow) and associated synaeresis cracks. Intertidal mixed flat. Angosto del Morro de Chucalezna. Coin diameter = 2.3 cm. F: Flat mud pebble conglomerate in a mud-dominated heterolithic interval. Intertidal mixed flat. Angosto del Morro de Chucalezna. Lens cap diameter = 5.5 cm.

are preserved on bedding surfaces, including ripple patches (Fig. 2C), wrinkle marks, interference ripples (Fig. 2D), and flat-topped ripples. Evidence of bidirectionality is pervasive. This facies is usually intensely bioturbated (BI = 3–5). *Skolithos linearis*, *Syringomorpha nilssoni*, *Syringomorpha* isp., *Arenicolites* isp., and *Diplocraterion parallelum* are the dominant elements. Palimpsestic surfaces, recording successive colonisation events, are very common. Clusters of *Rusophycus leifeirikssoni* are locally present. This facies is interpreted

as recording migration of current ripples in the lower intertidal sand flat. The high BI indicates frequent colonisation windows open to the settlement of larvae, particularly of suspension feeders.

Thinly interbedded sandstone and mudstone

This facies consists of light pink, laterally extensive, fine- to very fine-grained sandstone and red siltstone. The sand/mud ratio is about 1:1. Individual beds are 1–11 cm thick. Ripple cross-lamination is the dominant internal structure. Wavy bedding is the dominant bedding type, but flaser bedding is also present. A wide variety of synaeresis cracks occurs (Fig. 2E). Soft sediment deformation structures include load casts and ball and pillow. Bed tops are undulated, showing either symmetric or asymmetric ripples. Thin (3–4 cm thick), erosive-based, graded beds of coarse- to very fine-grained sandstones having flat mud pebble clasts (Fig. 2F) and symmetrical ripples are also present locally. In places this heterolithic facies is intercalated with similar deposits having inclined heterolithic stratification. The degree of bioturbation is highly variable (BI = 0 − 4). Although vertical burrows (*Skolithos linearis*, *Syringomorpha nilssoni*) are present, the dominant form is *Rusophycus leifeirikssoni* (Fig. 2E). Other ichnotaxa include *Cruziana problematica*, *Rusophycus carbonarius*, large *Rusophycus* isp., *Diplichnites* isp., *Planolites* isp., *Palaeophycus tubularis*, *Helminthoidichnites tenuis* and *Bergaueria* cf. *B. perata*. This facies essentially represents the alternation of tidal flood and ebb with slack-water periods in a middle intertidal, mixed-flat environment. Intercalated graded beds and flat pebble conglomeratic lenses are interpreted as storm deposits. Beds displaying inclined heterolithic stratification accumulated in intertidal run-off channels (Thomas *et al.* 1987).

Mudstone

This facies is composed of red, laterally extensive, massive- to parallel-laminated mudstone interbedded with thin (0.5–3.0 cm thick) very fine-grained sandstone. The sand/mud ratio is low (<1). Lenticular bedding is the dominant bedding type. Synaeresis cracks are abundant. Desiccation cracks are conspicuous features, although not extremely common. Thin (3–6 cm thick), erosive-based, graded beds of medium- to very fine-grained sandstones, having mudstone clasts and symmetrical ripples, are present locally. Trace fossils are rare (BI = 0 − 1), mostly represented by *Planolites* isp. and indistinct burrow mottlings. Isolated vertical traces (e.g. *Skolithos linearis*) penetrate into the mudstone from overlying sandstone beds and are not integral components of the facies. They commonly represent firm ground suites delineating transgressive surfaces. This facies was most likely deposited in an upper intertidal mud flat. Intercalated graded sandstone beds with mudstone clasts may record storm deposition.

Hummocky cross-stratified sandstone

This facies consists of light pink, erosive-based, fine- to very fine-grained sandstone. Individual beds are either tabular at outcrop scale or tend to pinch out laterally. Scours up to 2 cm deep and filled with mudstone clasts are present locally. Beds are 3–60 cm thick. Internal sedimentary structures follow Dott & Bourgeois' (1982) model of hummocky cross-stratified beds, as modified by Walker *et al.* (1983). Hummocky and cross-laminae divisions are very well developed. Some beds display a lower interval with parallel lamination. Symmetrical to near-symmetrical ripples are present at the top of some beds. Typically, thinner beds in the lower part of the package are separated by thin mudstone partings or layers, 1–50 mm thick, while sandstone bed amalgamation is common at the tops of packages. Bioturbation is relatively rare (BI = 0 − 1) in this facies and consists of opportunistic suites of *Syringomorpha nilssoni* or *Skolithos linearis* colonising hummocky cross-stratified beds. Hummocky cross-stratified beds are thought to result from episodic storm wave activity and wave-generated surges (e.g. Bourgeois 1980; Dott & Bourgeois 1982; Cheel & Leckie 1993). This facies most likely accumulated in a shoreface environment.

Depositional model

The Mesón Group records deposition in a macrotidal shallow-marine environment with extensive tidal-flat areas flanked seaward by subtidal sandbar complexes (Sánchez & Salfity 1990, 1999; Mángano *et al.* 2000; Mángano & Buatois 2000; Buatois & Mángano 2001). In this context, the Lizoite and Chalhualmayoc Formations represent deposition in high-energy, subtidal sandbar complexes dissected by subtidal channels, and the Campanario Formation mostly records deposition in tidal-flat settings. Therefore, the Campanario Formation represents the most proximal unit, while the Lizoite and Chalhualmayoc Formations are the most distal units (Moya 1998). The Campanario Formation facies as described, are composed of stacked, fining and thinning upward parasequences that record tidal-flat progradation (Fig. 3). Subtidal sandbars may develop in two different settings: open shelves (e.g. Smith 1988) and estuaries (e.g. Dalrymple 1984). In the case of the Mesón Group, the absence of an apparent valley incision and the regional extension of these deposits argue against an estuarine setting. The elongated morphology of the Cambrian basin suggests that the sandbar complexes were formed in a northwest–southeast trending seaway. This morphology led to tidal amplification and promoted a macrotidal regime.

Trace fossil distribution patterns and ichnofabrics

Trace fossils are relatively rare in the Lizoite and Chalhualmayoc Formations. Opportunistic assemblages

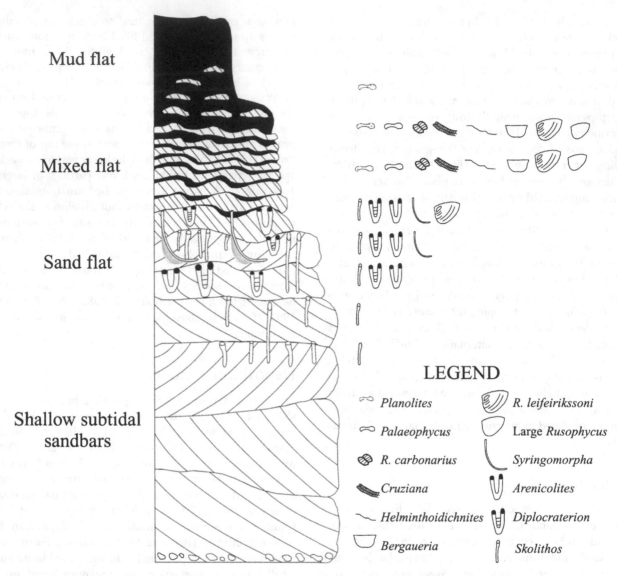

Mud flat

Mixed flat

Sand flat

Shallow subtidal
sandbars

LEGEND

〰	*Planolites*	🐚	*R. leifeirikssoni*
〰	*Palaeophycus*	⌴	Large *Rusophycus*
🐚	*R. carbonarius*	⌐	*Syringomorpha*
🐛	*Cruziana*	U	*Arenicolites*
〜	*Helminthoidichnites*	𝖴	*Diplocraterion*
⌣	*Bergaueria*			*Skolithos*

Fig. 3.   Idealised parasequence of the Campanario Formation showing the distribution of trace fossils and the different subenvironments represented.

of the *Skolithos* ichnofacies are associated with colonisation windows and mostly consist of monospecific suites of *Skolithos linearis*. In contrast, trace fossils are abundant in the Campanario Formation. Shallow subtidal and intertidal sand-flat deposits are dominated by vertical domiciles of suspension feeders and passive carnivores of the *Skolithos* ichnofacies (e.g. *Skolithos linearis*, *Arenicolites* isp., *Diplocraterion parallelum*) (Fig. 4). *Skolithos linearis* commonly forms piperocks. The feeding trace *Syringomorpha* ispp., which is abundant in the sand-flat facies, is also a component of the *Skolithos* ichnofacies (Fig. 5). Elements of the *Cruziana* ichnofacies are rare and are represented by local occurrences of *Rusophycus. leifeirikssoni*. Mángano & Buatois (2001) recognised two main forms of *Syringomorpha*. Some specimens consist of a simple spreite formed by a single, J-shaped causative burrow and are included in *Syringomorpha nilssoni*

(Fig. 5A, B). Other specimens, however, display an apparently more complex structure, recording lateral shifting of the J-shaped causative burrow and are herein referred to as *Syringomorpha* isp. Mángano & Buatois (2001) also analysed the role of *Syringomorpha* as an ichnofabric-forming ichnotaxon. They recognised the high density of individuals forming a composite ichnofabric that records multiple colonisation events in intertidal areas (Fig. 6). The composite nature of this ichnofabric is revealed by complex cross-cutting relationships of specimens. Pervasive bioturbation results from the activity of successive suites of deep infaunal organisms forming palimpsest assemblages. The high degree of bioturbation and preferential preservation of closely spaced, vertical components make this ichnofabric analogous to *Skolithos* piperock. Like *Skolithos* piperock, the *Syringomorpha* ichnofabric occurs in moderate- to high-energy settings, being particularly abundant in the sand flat.

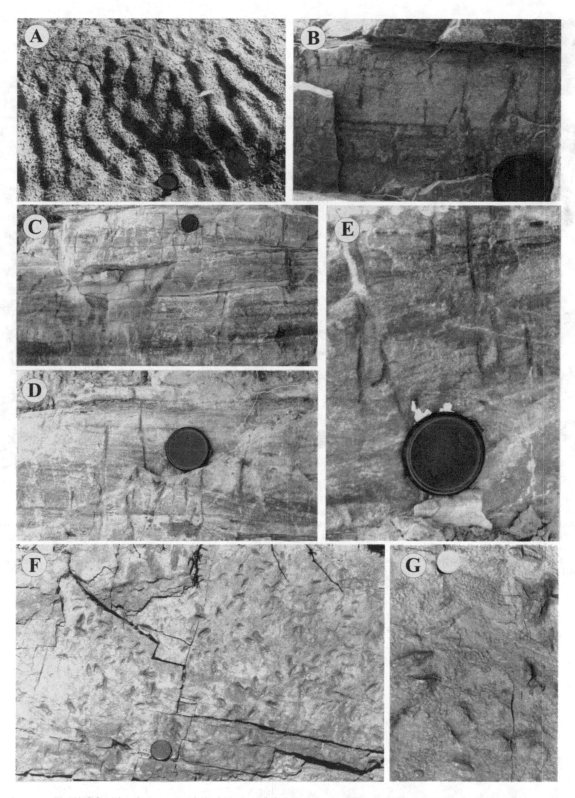

Fig. 4.   Elements of the *Skolithos* ichnofacies. A: Bedding plane view of *Skolithos linearis* occurring in high density at a rippled sandstone surface. Shallow subtidal to lower intertidal sandbar. Angosto de Perchel. Lens cap diamter = 5.5 cm. B: Low density of *Skolithos linearis* in planar cross-bedded sandstone. Shallow subtidal to lower intertidal sandbar. Angosto de Perchel. Lens cap diameter = 5.5 cm. C: General view of a planar cross-bedded sandstone displaying *Arenicolites* isp. at the top. Shallow subtidal to lower intertidal sandbar. Angosto del Perchel. Lens cap diameter = 5.5 cm. D: *Arenicolites* isp. at the top of a planar cross-bedded sandstone. Shallow subtidal to lower intertidal sandbar. Angosto del Perchel. Lens cap diameter = 5.5 cm. E: Close-up showing long, deep specimens of *Arenicolites* isp. Shallow subtidal to lower intertidal sandbar. Angosto del Perchel. Lens cap diameter = 5.5 cm. F: General view of the top of a rippled sandstone showing high density of *Diplocraterion parallelum*. Intertidal sand flat. Quebrada de Moya. Lens cap diameter = 5.5 cm. G: Close-up showing *Diplocraterion parallelum*. Intertidal sand flat. Quebrada de Moya. Coin diameter = 1.8 cm.

Fig. 5.   The ichnogenus *Syringomorpha* in intertidal sand-flat deposits. A: *Syringomorpha nilssoni* showing oblique causative burrow and associated spreite. Cuesta de la Pedrera. Scale bar = 1 cm. B: *Syringomorpha nilssoni* with subvertical causative burrow and associated spreite. Cordon de Alfarcito. Coin diameter = 1.8 cm. C: General view of a bioturbated sandstone hosting a *Syringomorpha* ichnofabric. The sandstone represents a *Syringomorpha* piperock but the closely spaced vertical burrows may be confused with a *Skolithos* piperock. Angosto del Morro de Chucalezna. Lens cap diameter = 5.5 cm. D: Close-up of the bioturbated sandstone showing the development of typical *Syringomorpha* spreiten towards the base of the unit. Angosto del Morro de Chucalezna. Coin diameter = 1.8 cm.

Syringomorpha Ichnofabrics

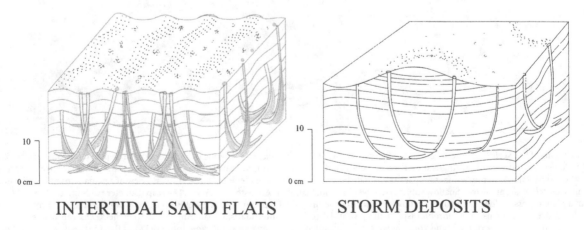

INTERTIDAL SAND FLATS   STORM DEPOSITS

Fig. 6.   Schematic reconstructions of the *Syringomorpha* ichnofabrics in intertidal sand-flat and storm deposits.

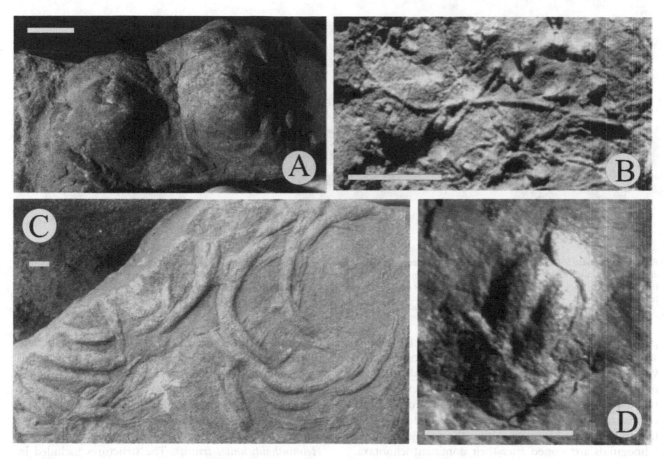

Fig. 7. Elements of the *Cruziana* ichnofacies. A: *Bergaueria* cf. *B. perata.* Huachichocana. B: *Helminthoidichnites tenuis.* Angosto de Perchel. C: *Cruziana problematica* Purmamarca. D: *Rusophycus carbonarius.* Villa Floresta. All scale bars = 1 cm.

Mixed-flat deposits contain elements of both the *Skolithos* and *Cruziana* ichnofacies. Horizontal feeding, locomotion, and resting traces of the *Cruziana* ichnofacies typically occur on soles of the sandstone interbeds (Fig. 7). Vertical trace fossils of the *Skolithos* ichnofacies occur locally, commonly recording opportunistic colonisation of interbedded storm deposits. The *Cruziana* ichnofacies is represented by *Rusophycus leifeirikssoni, Cruziana problematica, Rusophycus carbonarius,* large *Rusophycus* isp., *Diplichnites* isp., *Planolites* isp., *Palaeophycus tubularis, Helminthoidichnites tenuis* and *Bergaueria* cf. *B. perata. Rusophycus leifeirikssoni* is the most conspicuous ichnotaxon (Fig. 8) and occurs commonly forming clusters (Fig. 9) (Mángano & Buatois 2003a).

Mud-flat deposits are only sparsely bioturbated. Trace fossils are rare, mostly represented by isolated *Planolites* isp. Burrow mottlings occur locally.

Tide-dominated deposits are locally interbedded with storm-dominated facies. Bioturbation is sparse in these deposits. *Syringomorpha nilssoni* occurs in moderate to low densities in hummocky cross-stratified sandstones (Mángano & Buatois 2001). This ichnofauna records opportunistic colonisation after storms (Fig. 6). Causative burrows extend from the top of the tempestite, developing

a wide spreite structure at the lower part of the storm bed. In contrast to tidal-flat examples, this ichnofabric is simple and represents a single bioturbation event following episodic sedimentation.

Tiering structure and ichnoguilds

Tiering is the vertical partitioning of a community. Sediments are vertically zoned in terms of physical, chemical and biological factors. Thus, elements of the endobenthic communities live at a certain depth with respect to the sediment–water interface (Ausich & Bottjer 1982; Bromley 1990, 1996). Analysis of the tiering structure suggests that organisms tend to group together to exploit the same tiers in similar ways. Accordingly, Bromley (1990, 1996) proposed the ichnoguild concept, following utilisation of the term "guild" for the analysis of body fossils by Bambach (1983). According to Bromley (1990, 1996), an ichnoguild reflects three parameters: bauplan, food source and use of space. In terms of bauplan, biogenic structures are categorised as

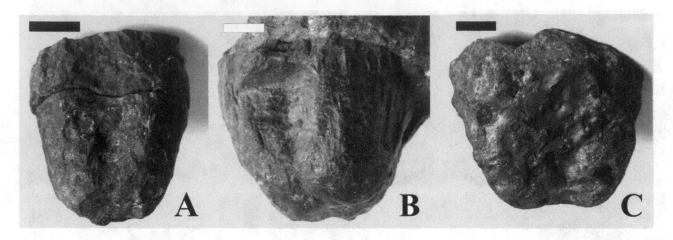

Fig. 8.   Rusophycus leifeirikssoni. A: Posterior view of a deep specimen displaying typical general morphology with subvertical posterior lobes. Angosto de Perchel. B: Posterior view showing a wide axial groove with poorly preserved perpendicular ridges and divergent lobes covered with longitudinal (exopodal?) scratches. Huachichocana. C: Posterior view of a poorly preserved specimen cross-cut by *Skolithos linearis*. Angosto de Perchel. All scale bars = 1 cm.

permanent to semi-permanent burrows produced by stationary organisms or transitory structures made by vagile animals. Food source is reflected by trophic analysis of trace fossils, including categories such as detritus feeding, deposit feeding, suspension feeding, gardening and chemosymbiosis. Use of space is approximately equivalent to the vertical position within the tiering structure. Ichnoguilds are named after their dominant ichnotaxa. Ichnoguilds provide valuable information for understanding the patterns of ecospace utilisation through geological time, as well as being useful tools for elucidating the adaptive strategies displayed by benthic producers (e.g. Bromley 1994; Buatois *et al.* 1998; Mángano *et al.* 2002; Buatois & Mángano 2003b).

Tiering in these tidal-flat deposits is relatively simple (Fig. 10). Shallow-tier trace fossils include *Cruziana problematica, Rusophycus carbonarius, Diplichnites* isp., *Planolites* isp., *Palaeophycus tubularis* and *Helminthoidichnites tenuis. Rusophycus leifeirikssoni,* large *Rusophycus* isp. and *Bergaueria* cf. *B. perata* are deeper structures, commonly cross-cutting shallow-tier elements, and are grouped as middle-tier ichnotaxa. Deep-tier ichnofossils include *Skolithos linearis, Syringomorpha nilssoni, Syringomorpha* isp., *Arenicolites* isp. and *Diplocraterion parallelum.* Six ichnoguilds (*Cruziana problematica, Palaeophycus, Bergaueria, Rusophycus leifeirikssoni, Syringomorpha* and *Skolithos*) are defined here (Fig. 10).

Cruziana problematica ichnoguild

This ichnoguild includes *Cruziana problematica, Rusophycus carbonarius, Diplichnites* isp., *Helminthoidichnites tenuis* and *Planolites* isp. The occurrence and abundance of these shallow endobenthic and superficial structures is strongly biased by preservational problems. In fact, some structures, such as small *Diplichnites* isp, are very

rare and, where observed, are commonly incomplete undertraces preserved on ripple crests. The *Cruziana problematica* ichnoguild is made up of transitory, very shallow-tier, deposit-feeder structures, produced by vagile small arthropods and vermiform animals. The possibility that some small carnivores are represented cannot be ruled out for these very simple structures (i.e. *Helminthoidichnites tenuis*). The structures included in this ichnoguild are horizontal to subhorizontal trace fossils that reflect exploitation of detritus concentrated in the uppermost millimetres of the substrate. In fact, some specimens of *Cruziana problematica* exhibit the classic circling behavioural pattern (scribbling pattern *sensu* Seilacher 1970) suggestive of a primitive foraging strategy (Fig. 7C). Trace fossils are mostly interfacial and, therefore, do not produce significant disturbance of the primary sedimentary fabric.

Palaeophycus ichnoguild

This is a monospecific ichnoguild, represented by the ichnospecies *Palaeophycus tubularis.* It consists of semi-permanent, shallow-tier, suspension-feeder or predator structures produced by vagile vermiform organisms. These organisms constructed horizontal to subhorizontal burrows that are maintained as open structures to feed on particles in suspension or to actively predate on other worms.

Bergaueria ichnoguild

This is represented by *Bergaueria* cf. *B. perata* and large *Rusophycus* isp. The *Bergaueria* ichnoguild consists of permanent to transitory, middle-tier structures produced by sessile cnidarians and arthropods (large trilobites?). Although the material that can confidently be included in

Fig. 9. Bedding plane view (top) of a concentration of *Rusophycus leifeirikssoni* in intertidal mixed-flat deposits. Angosto del Morro de Chucalezna. A: Note the complex structure formed by multiple imbricated specimens of *Rusophycus leifeirikssoni* (centre right) and the dominance of discrete specimens displaying typical kidney shape. Length of hammer = 33.5 cm. B: Close-up showing associated shrinkage cracks and cross-sectional views of *Skolithos linearis.* Lens cap diameter = 5.5 cm.

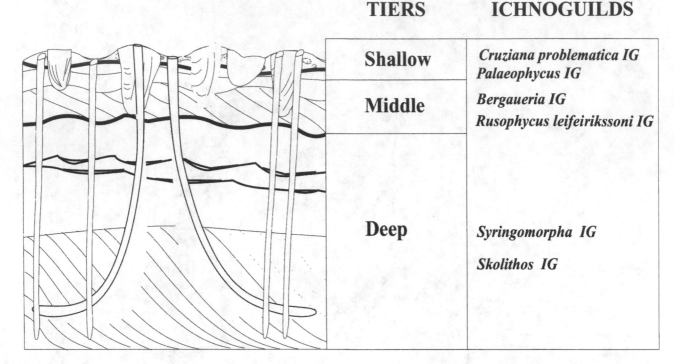

TIERS	ICHNOGUILDS
Shallow	*Cruziana problematica IG* *Palaeophycus IG*
Middle	*Bergaueria IG* *Rusophycus leifeirikssoni IG*
Deep	*Syringomorpha IG* *Skolithos IG*

Fig. 10.   Tiering structure and ichnoguilds of the Campanario ichnofauna.

this ichnoguild is scarce, it suggests that a higher trophic level was probably present in the Campanario food web. It may record activities of primary carnivores, although the participation of higher-level carnivores cannot be confirmed.

Rusophycus leifeirikssoni ichnoguild

This is also a monospecific ichnoguild, and is made up of transitory, middle-tier structures, either related to nesting or endobenthic feeding behaviour of trilobites. The assignment of these structures to trilobites or trilo-bitomorphs is based on fine morphology (see Mángano & Buatois 2003a). The maximum depth recorded is approximately 4 cm at the base of cross-laminated sandstones where *Rusophycus leifeirikssoni* tends to form clusters of multiple individuals. Although a functional analysis of these structures has been attempted (Mángano & Buatois 2003a), the precise ethological meaning remains unclear (see Discussion).

Syringomorpha ichnoguild

This ichnoguild includes *Syringomorpha nilssoni* and *Syringomorpha* isp. It is made up of permanent, deep-tier, probable deposit-feeder (or gardening?) structures exploiting microbial films on sand grains and meiofauna. The maximum penetration depth recorded is 30 cm. Infaunal activity of the *Syringomorpha* producer in

sand-flat deposits may result in pervasive bioturbation and destruction of the primary sedimentary fabric.

Skolithos ichnoguild

This ichnoguild is represented by *Skolithos linearis*, *Arenicolites* isp. and *Diplocraterion parallelum*. It consists of permanent, deep-tier, suspension-feeder or passive predator structures probably produced by polychaetes or phoronids. The ichnoguild records the activity of deep vermiform burrowers that constructed vertical structures to feed on particles suspended in the water column or to passively prey on other organisms. The maximum penetration depth recorded is 40 cm. Intensive biotur-bation commonly results in the production of *Skolithos* piperock.

In terms of ichnofacies, the deep-tier ichnoguilds (*Skolithos* and *Syringomorpha*) are included in the *Skolithos* ichnofacies, while the shallow- to middle-tier ichnoguilds (*Helminthoidichnites*, *Palaeophycus*, *Bergaueria* and *Rusophycus leifeirikssoni*) are collectively included in the *Cruziana* ichnofacies, albeit with reduced diversity.

Taphonomic and ecological controls on trace fossil zonation

Trace fossils and ichnoguilds show a preferential palaeoenvironmental distribution following proximal–

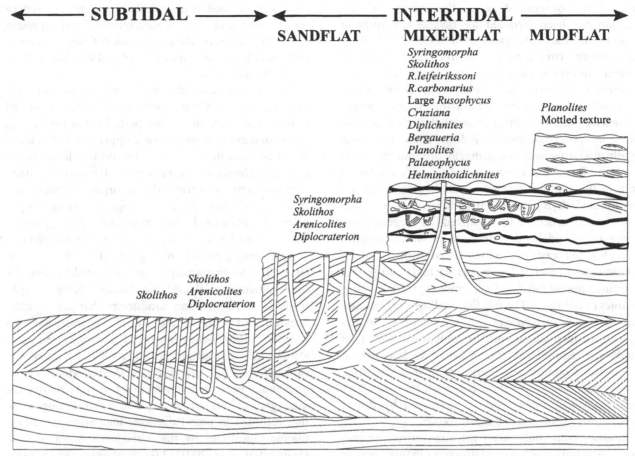

Fig. 11. Trace fossil distribution along a bathymetric gradient. Note that elements of the deep-tier ichnoguilds tend to occur in the high-energy, most distal regions, while trace fossils of the middle- and shallow-tier ichnoguilds are dominant in the low-energy, proximal regions.

distal trends (Fig. 11). Although there is some superimposition, deep-tier ichnoguilds tend to occur in the higher-energy, seaward distal regions, specifically in the shallow subtidal to intertidal transition and in the lower intertidal sand flats. On the other hand, middle- and shallow-tier ichnoguilds are dominant in the moderate- to low-energy, proximal regions, mainly the middle intertidal mixed flat. This resultant pattern of distribution of biogenic structures is shaped by the interplay of key environmental parameters overprinted by a strong taphonomic control. The dominance of endogenic structures in these tidal flats reflects, on the one hand, the fact that the inhabitants of the seashore were marine organisms that needed to avoid severe water loss and high temperatures for survival, and on the other hand, epigenic structures that may have been produced subaqueously during high tide or subaereally recording brief incursions during low tide that were most likely destroyed by wave and tidal processes. Hydrodynamic energy, substrate and food supply are among the most important palaeoenvironmental factors affecting

the distribution of the biogenic structures in these macrotidal shorelines.

Hydrodynamic energy is a major controlling factor in coastal and shallow-marine environments. The high energy of tides and waves controls substrate nature, food availability and distribution, in turn determining zonation of benthic fauna and, therefore, trace fossil distribution along tidal shorelines. High- to moderate-energy zones, represented mostly by the subtidal–intertidal transition and the sand flat, are typically dominated by deep, vertical burrows, such as *Skolithos*, *Diplocraterion* or *Arenicolites* (e.g. Cornish 1986; Simpson 1991; Bromley & Hanken 1991). The construction of deep, vertical burrows serves to buffer the destructive impact of high- energy waves and tidal currents producing significant erosion in shallow subtidal to lower intertidal zones. The *Skolithos* and *Syringomorpha* ichnoguilds are dominant under high- and moderate-energy conditions in the Campanario tidal deposits. Repeated erosion events are also indicated by the presence of palimpsest surfaces and composite ichnofabrics that record amalgamation of repeated colonisation events. A similar situation has been recorded for bivalve structures in ancient tidal flats (Mángano *et al.* 1998, 2002). Even higher-energy conditions are recorded in the subtidal sandbar complexes, mostly represented by the Lizoite and Chalhualmayoc Formations, which are for the most part unbioturbated or have distinct horizons with monospecific assemblages of *Skolithos linearis* reflecting occasional and short-term colonisation windows. High-energy and rapidly migrating bedforms generally preclude the establishment of a mobile epifaunal and/or shallow infaunal biota in the subtidal area where tidal currents reach a maximum (Dalrymple 1992). In contrast, the dominance of horizontal structures in the mixed flat suggests lower energy in the middle intertidal zone. However, the presence of gutter casts and scours filled with intraclasts indicates occasional storm events of high energy that sculptured the tidal-flat surface, removing relatively large volumes of sediment. The fact that shallow-tier, locomotion structures of trilobites (i.e. *Cruziana*) are underrepresented in these deposits in comparison with their middle-tier equivalents (i.e. *Rusophycus leifeirikssoni*) may reflect a taphonomic overprint. Shallow-tier structures have lower preservation potential; they may have been eroded during storms or destroyed by deeper bioturbators (Mángano & Buatois 2003a).

Substrate commonly exerts a strong control, not only on the distribution of biogenic structures in tidal deposits (Gingras *et al.* 2001), but also on the preserved morphology of the trace fossils that reflects different degrees of substrate cohesiveness (Mángano *et al.* 1998, 2002). In the Campanario Formation, trace fossils are usually segregated according to substrate texture and heterogeneity. The *Skolithos* and *Syringomorpha* ichnoguilds tend to be widespread in relatively clean, medium- to very fine-grained sandstone. The *Cruziana problematica*, *Palaeophycus* and *Rusophycus leifeirikssoni* ichnoguilds are dominant in interbedded fine- to silty very fine-grained sandstone and mudstone. Additionally, preservation of horizontal, interfacial trace fossils is favoured by the presence of sandstone/mudstone interfaces because homogeneous lithologies would inhibit preservation and visibility of these biogenic structures (Mángano *et al.* 2002). Additionally, animals are not passive to the physical properties of the sediment, but can actually substantially modify substrate attributes (Bromley 1996). Mobile, mostly deposit- and detritus-feeder infauna and epifauna, whose feeding and defecation activities may provide abundant particles in suspension, destabilise the substrate (Rhoads & Young 1970; Rhoads 1974). However, vagile deposit-feeder structures of the Campanario mixed flats represent a shallow tier and no significant bioturbation is recorded. Sedentary organisms that build mucus-lined tubes within the sediment reduce resuspension and erosion susceptibility, and represent sediment stabilisers. Deep suspension-feeder structures (e.g. *Skolithos linearis*) may have played this role in these Cambrian tidal flats.

Food supply is another important control on trace fossil distribution. Undoubtedly, the *Skolithos* ichnoguild reflects abundance of organic particles that are kept in suspension in the more energetic upper subtidal to lower intertidal areas. In contrast, shallow deposit-feeder structures (e.g. *Helminthoidichnites tenuis*) tend to concentrate in finer-grained substrates rich in organic detritus. The trophic significance of the ichnogenus *Syringomorpha* is less straightforward. The presence of a spreite, particularly the complex structure present in *Syringomorpha* isp., suggests underground mining activities of a deposit feeder. This feeding strategy is not supported, however, by the sedimentological evidence, because *Syringomorpha* tends to occur in clean sandstones that are typically impoverished in organic content. This anomaly has been noted for other feeding traces in Palaeozoic clean sandstones, such as *Daedalus* (Seilacher 2000). An alternative explanation is that the *Syringomorpha* producer fed on epigranular bacteria or meiofauna that were coating the sand grains. This strategy is somewhat analogous to that recorded in post-Palaeozoic high-energy, shallow-marine sandstones by the ichnogenus *Macaronichnus* (Pemberton *et al.* 2001). Like *Syringomorpha*, *Macaronichnus* exploits the deep sand habitat at the "toe of the beach" where an abundant and constantly replenishing food supply is available. This high-energy setting presents two key features: (1) it is a zone of nutrient convergence, with nutrients derived from the land via low tide drainage and from the sea through intense pumping in the innermost surf zone; and (2) it represents a permanent high "oxygen window" that allows deep penetration of the infauna into the sediment (Pemberton *et al.* 2001).

The bathymetric distribution of these environmental parameters together with the associated preservational biases result in an ichnofacies gradient of tide-dominated shorelines that is opposite to that of wave-dominated shoreface to offshore environments. As overall tidal energy increases from supratidal to subtidal settings, the *Skolithos* ichnofacies tends to occur seaward of the *Cruziana* ichnofacies (Mángano *et al.* 2002; Mángano & Buatois in press). This shoreward decrease in energy parallels a decrease in oxygenation, sand content, amount of organic particles in suspension, and mobility of the substrate. This gradient is consistent with information from modern tide-influenced environments (e.g. Bajard 1966; Howard & Dörjes 1972; Beukema 1976; Swinbanks & Murray 1981; Ghare & Badve 1984; Frey *et al.* 1987; Gingras *et al.* 1999).

Cambrian intertidal ecosystems and landward expansion of the "Agronomic Revolution" – a discussion

As stated by Butterfield (2001a), evolutionary palaeo-ecology presents the unique challenge of reconstructing ecosystems occupied largely or entirely by extinct organisms. In the ichnological approach, organism auto-ecology is inferred from the functional analysis of the biogenic structures, but the reconstruction of the syneco-logical picture is a more complex issue that necessarily involves a certain degree of analogy and uniformitarian assumptions. In the case of Cambrian tide-dominated shorelines, the uniformitarian reasoning becomes quite uncertain because Cambrian ecology is still an incomplete puzzle in which many pieces are lacking (e.g. Burzin et al 2001; Butterfield 2001a). However, ichnology may provide the unique opportunity of the study of *in situ* relationships among biogenic structures and between physical and biogenic processes, offering a dynamic picture of intertidal to shallow subtidal ecology.

Despite its dangers, modern tidal flats are highly populated. They are characterised by an abundant food supply derived from multiple sources, including nutrients and plankton that drifted in from the sea, terrestrially derived organic detritus, and autochthonous food production. Many marine and terrestrial species migrate to the tidal flat in search of food or protection. Tidal flats are particularly rich in food because they congregate detritus from the land and from the sea, promoting the development of a complex benthic community. In modern shorelines, the rising tide transports larvae, juvenile individuals and adults to the tidal flat. Some are periodic visitors that feed and return to subtidal areas with the subsequent ebb flow, others become temporary or permanent residents of the intertidal area (Reise 1985; Palmer 1995). Also, the inhabitants of the intertidal area are exposed to a double set of predators significantly affecting community structure. During submergence they are preyed on by other marine organisms and during emergence they are visited by predators from the land and air. In modern shorelines, crustaceans and fishes are the main visiting foragers, whereas birds represent the most significant continental visitors (Reise 1985; Little 2000).

The picture that emerges of a Cambrian intertidal ecosystem is contrastingly different (Fig. 12). The paucity of land vegetation and the absence of land-derived inhabitants and predatory visitors in the Cambrian were major controlling factors in the composition and structure of intertidal communities. The earliest accepted records of land vegetation are from spores in Middle Ordovician rocks of Saudi Arabia (Strother et al. 1996) and spore-containing plant fragments in Upper Ordovician deposits of Oman (Wellman et al. 2003). However, terrestrial micro-organisms are known by the Late Archean (Watanabe et al. 2000) and were probably widespread by the Proterozoic (Horodyski & Knauth 1994; Prave 2002). In fact, spore-like microfossils, referred to as cryptospores, are known by the Middle Cambrian (Strother 2000; Strother & Beck 2000). The Lower Palaeozoic cryptospore record is relatively widespread and may reflect the establishment of a nascent semi-aquatic to subaerial flora of bryophyte grade (Strother 2000). Cryptospores are usually preserved in continental and marginal-marine environments (Strother 2000). In any case, Cambrian intertidal trophic webs were most likely based mainly on the organically rich marine source of phytoplankton and recently introduced mesozo-oplankton (Butterfield 2001a), and abundant autochtho-nous production, with minor contributions from the incipient land vegetation. Energy transfers to higher trophic levels may have been less efficient, inevitably resulting in shorter trophic webs (Budd 2001). Although the post-Nemakit-Daldynian (post-earliest Cambrian) tidal flats may have in many ways resembled recent tidal flats as sites of enriched food resources and protection, Cambrian coastal environments functioned as efficient refugia in the absence of land and air predators. In fact, predation pressure must have been significantly lower in Cambrian shoreline ecosystems.

The presence of multiple trophic guilds and a well-established suspension-feeding infauna provides evidence of a significant change in complexity in post-Nemakit-Daldynian intertidal to shallow subtidal communities, suggesting that the plankton was closely coupled to the benthos (cf. Butterfield 2001a, b). Moczydłowska (2002) noted synchronous replacements of acritarch and trilobite assemblages in the Lower Cambrian of the Swedish Caledonides. These coupled evolutionary events have been interpreted as a bottom-up cascade effect from primary producers (phytoplankton) to consumers (metazoa) (cf. Moczydłowska 2002). However, the key innovation introduced by the evolution of filter-feeding mesozooplakton may have triggered not only the evolution of large metazoans (Butterfield 2001b), but also the agronomic revolution (Seilacher 1999). The effect of mesozooplankton evolution in benthic ecology may have been enormous (Butterfield 2001b). By repacking unicellular phytoplankton as nutrient-rich particles 10–100 times larger, zooplankton produced a more concentrated and exploitable resource for the benthos. This significant increase in the delivery of labile, nutrient-rich particles into the sediment may be responsible for the most significant change in the history of benthic ecology: the shift from matgrounds to mixgrounds (cf. Seilacher 1999). In fact, as recorded

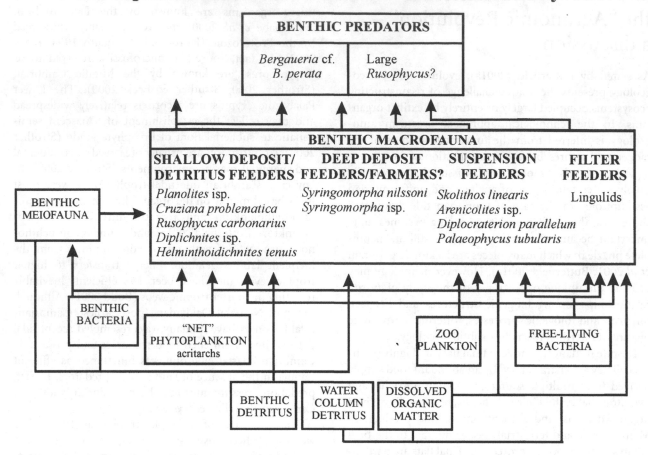

Fig. 12. Hypothetical diagram of the Campanario Formation shallow subtidal to intertidal trophic web. Megafaunal trophic levels are based on trace fossils and body fossils (lingulids); other components of the web are inferred from indirect evidence or analogy (inspired by Burzin *et al.* 2001).

by the ichnological evidence, the presence of metazoa able to exploit the endobenthic environment preceded the establishment of a modern endobenthic ecological structure (i.e. "mixground ecology" *sensu* Seilacher 1999).

As outlined by Budd (2001), arthropods were more important in Cambrian marine ecosystems than they are now; and this is supported by the ichnological record in the Campanario Formation. The significance of trace fossils attributable to trilobites in tidal-flat facies remains a matter of debate (Mángano & Buatois 2003a). Four main alternatives can be invoked to explain trilobite incursions to intertidal areas: the "trilobite pirouette", "hunting burrow", "microbial garden" and "trilobite nursery" hypotheses. The former three are related to feeding activities, while the latter reflects reproductive behaviour. The "trilobite pirouette" hypothesis (after Seilacher 1997: 34) is based on the fact that some ichnospecies of *Cruziana* (e.g. *C. semiplicata* and *C. rugosa*) attributed to trilobites tend to form circular,

scribbling trails that efficiently cover bedding surfaces, suggesting feeding strategies, namely, browsing and detritus feeding. These structures may record combined locomotion and feeding behaviour and are, therefore, regarded as grazing traces (Seilacher 1970; Bergström 1976). *Cruziana semiplicata* occurs in association with desiccation cracks in the Upper Cambrian of Oman (Fortey 2001, pers. comm.), while *Cruziana rugosa* forms scribbling patterns in Lower Ordovician strata of Portugal (Seilacher 2003). These occurrences of *Cruziana* are best regarded as feeding incursions into the intertidal area.

The "hunting burrow" hypothesis proposes that trilobite trace fossils (deep burrowing *Rusophycus dispar*) from the Lower Cambrian Mickwitzia Sandstone of Sweden record predation on worms (Bergström 1973; Jensen 1990). In the *Rusophycus dispar* examples, the axis of the trilobite trace fossils is nearly parallel to the worm burrows. Also, the worm burrows closely follow the curvature of the *Rusophycus dispar* trace and are commonly in contact with only one of its lobes. Although

Brandt *et al.* (1995) made a similar suggestion for a trilobite trace in the Upper Ordovician of Ohio, the interpretation remains controversial. The association of *Gyrolithes saxonicus* and *Rusophycus leifeirikssoni* from the Upper Cambrian–Lower Ordovician Beach Formation of Newfoundland led Fillion & Pickerill (1990) to suggest that this *Rusophycus* could represent a predation structure. However, the predation interpretation of *Rusophycus dispar* has been questioned recently by Rydell *et al.* (2001).

In the "microbial garden" hypothesis, trilobites visit the shoreline to feed from the enriched food resources that result from the infaunal activity of other organisms (Mángano & Buatois 2003a). This kind of relationship is based on the notion of "promotion" as proposed by Reise (1985). Animal activity may oxidise and fertilise the surroundings, provoking an improvement in the environment and positive effects on other organisms. As this sort of interaction has no apparent reciprocal advantageous or detrimental effects it is not considered an ecological interaction *sensu strictum* (cf. Reise 1985). In modern coastal regions, benthic megafaunal activity, particularly of polychaetes and tellinid bivalves, results in an increase in nutrients, and micro- and meiofaunal content. A well-documented example is *Macoma baltica*, an infaunal detritus feeder that employs an inhalant siphon that extends up to the sediment surface (Schäfer 1972: fig. 156). *Macoma balthica* is thought to be responsible for an upward diffusion of nutrients. Faecal pellets released from the exhalent siphon provide an organic substrate for heterotrophs. Additionally, metabolites excreted by *Macoma balthica* provide important nutrients. Together, these components probably enhance microbial growth, which in turn, increases the abundance of bacteria-feeding platyhelminths and polychaetes. Their densities are high enough to attract predatory platyhelminths (Reise 1985). In short, as documented by Reise (1985), three levels of the trophic web are maintained by *Macoma balthica* living imperceptibly below the sediment surface. The intense endobenthic activity in Cambrian tidal flats, particularly by pervasive *Syringomorpha* producers, may have resulted from an unusual concentration of organic detritus, microfauna and meiofauna, ultimately causing the immigration of trilobites to the shore. The plausibility of this scenario is increased by the well-established mixground ecology in the tidal flats of the Campanario Formation.

Finally, the "trilobite nursery" hypothesis invokes reproductive behaviour to explain the occurrence of clusters of *Rusophycus* in tidal-flat deposits. Recent tidal flats are important nurseries in which eggs and juveniles are safe from their marine predators (Reise 1985). Numerous organisms nest in tidal flats (e.g. limulids, fishes, turtles) only to return to subtidal settings after birth. In their pioneering paper, Fenton & Fenton (1937) interpreted deep trilobite burrows (*R. jenningsi*) as structures dug for the reception of eggs, and compared these traces with those produced by the modern *Limulus*. Behavioural studies of limulids may shed some light on occurrences of trilobite traces in intertidal deposits. Studies of basic patterns of horseshoe crab breeding behaviour show that the male and female couple offshore and then approach the beach. During the rising tide, the female digs burrows into the sand in which the eggs are laid (Sekiguchi 1988). The burrowing activity is accomplished with the appendages and also with the anterior margin of the prosoma, which is plunged into the sand (Eldredge 1970; Sekiguchi 1988). Mángano *et al.* (1996) compared these modern analogues to the clusters of *Rusophycus latus* in Upper Cambrian intertidal deposits of the Tilcara Member (Santa Rosita Formation, northwest Argentina) and suggested that these structures probably record nesting behaviour. Other evidence of reproductive behaviour in littoral environments was proposed by Braddy (2001). Based on his study of Late Silurian concentrations of eurypterid exuviae, he suggested that these arthropods may have migrated *en masse* into paralic environments to moult and mate.

Of these four alternative explanations, the "trilobite pirouette" hypothesis, can only be applied to some specimens of small *Cruziana problematica* that display the typical circling pattern. These simple *Cruziana*, however, can reflect the primitive foraging behaviour of different arthropods other than trilobites feeding in the intertidal deposits of the Campanario Formation. In fact, as previously discussed, because of a taphonomic filter, specimens of *Cruziana* are relatively rare in the unit. Although *Rusophycus leifeirikssoni* is directly associated with vertical worm burrows (e.g. *Skolithos linearis*, *Syringomorpha* ispp.), it is difficult to find strong support for *Rusophycus leifeirikssoni* producers preying on worms (Mángano & Buatois 2003a). Furthermore, the relatively small size of *Rusophycus leifeirikssoni* with respect to the worm burrows also makes the "hunting burrow" hypothesis unlikely. In contrast, both the "microbial garden" and "trilobite nursery" hypotheses may better explain the occurrences of clusters of *Rusophycus leifeirikssoni* in the Campanario tidal flats. In particular, the wide spreite of *Syringomorpha* may have hosted a food source attractive to the trilobite fauna. Rich mucus-lined, closely packed *Syringomorpha* may have promoted the development and high concentration of microbes and meiofauna (Mángano & Buatois 2003a). High-density assemblages of *Rusophycus leifeirikssoni* in intertidal deposits may record trilobite incursions to feed in the rich coastal areas. Although this ecological scenario is plausible, the complex cross-cutting relationships between *Rusophycus leifeirikssoni* and *Syringomorpha* ispp., however, make it difficult to corroborate the "microbial garden" hypothesis. Alternatively, *Rusophycus leifeirikssoni* may represent a nesting structure. Cambrian intertidal ecosystems may have acted as in modern ones, as nurseries. The localised abundance of *Rusophycus leifeirikssoni* in the Campanario tidal-flat deposits, the tendency to form clusters and the overall morphology, support the nesting behaviour (Mángano & Buatois 2003a).

Regardless of their precise ethological significance, the presence of presumed trilobite trace fossils in intertidal deposits of the Campanario Formation indicates early incursions in the intertidal zone and that representatives of the "Cambrian evolutionary fauna" (Sepkoski 1997) were able to colonise very shallow-water and marginal environments. Although the Cambrian explosion is commonly associated with open-marine, subtidal settings, ichnological data suggest a significant landward expansion of this evolutionary radiation. Additionally, the high density of vertical burrows in tidal-flat facies reveals that the "Agronomic Revolution" of Seilacher (1999) was not restricted to open-marine, shelfal environments. The depth and extent of bioturbation reveal colonisation of a relatively deep infaunal ecospace mostly by suspension feeders and some unorthodox deposit feeders feeding on clean sand sediment rich in epigranular microbes and meiofauna. *Skolithos* and *Syringomorpha* ichnofabrics dominated by vertical structures record the advent of deep burrowing by coelomate metazoans during the Cambrian radiation (cf. Droser & Li 2001). Assemblages of the *Cruziana* ichnofacies are exceptionally preserved, recording the activities of very shallow vermiform animals, and small epibenthic and shallow endobenthic arthropods. Deep rusophycids are larger and deeper than other shallow trace fossils, leading to increased disturbance of the original stratification. The tiering structure, however, remains relatively simple compared with Ordovician counterparts (Mángano & Droser 2004).

The presence of trace fossils attributed to trilobites in intertidal deposits of the Campanario Formation is by no means a curiosity. In fact, the comparative analysis of ichnofaunas from Lower Palaeozoic shallow-marine clastic successions elsewhere indicates that trilobite trace fossils were common, not only in lower shoreface to offshore settings, but also in the tidal-flat environments (see Mángano *et al.* 2002). Crimes *et al.* (1977) discussed the ichnology of Precambrian–Cambrian shallow-water successions in Spain, documenting the presence of abundant trilobite structures, such as *Cruziana*, *Rusophycus*, and *Diplichnites*, in low-energy, thinly bedded heterolithic, intertidal facies. Astini *et al.* (2000) recorded *Cruziana* in Lower Cambrian intertidal strata of the Precordillera, western Argentina. The trilobite trace fossils are associated with a set of physical sedimentary structures indicative of very shallow water and periodic subaerial emergence, including flat-topped ripples and desiccation cracks. The presence of halite pseudomorphs supports a stressed, hypersaline environment. Mikuláš (1995) described an arthropod ichnofauna (*Monomorphichnus*, *Diplichnites*, ?*Dimorphichnus*, ?*Rusophycus*) from Lower Cambrian marginal-marine deposits of the Czech Republic. This ichnofauna represents one of the earliest pieces of evidence for colonisation in marginal-marine or paralic settings. Selley (1970) described trilobite trace fossils from supposedly abandoned braided channel deposits in the Cambrian of Jordan. These deposits were subsequently reinterpreted as having formed during an Early Cambrian transgression that led to the establishment of estuarine conditions (Amireh *et al.* 1994). Legg (1985) documented sedimentary facies and ichnofaunas from a Middle Cambrian tide-influenced delta system in Spain. He recorded a variety of trilobite traces (*Cruziana*, *Rusophycus*) in low-energy heterolithic facies of intertidal origin. Seilacher (1977) illustrated specimens of *Cruziana* cross-cut by desiccation cracks and radular traces of gastropod-like organisms in sandstones of probable Late Cambrian age in Saudi Arabia. Mángano & Buatois (2003b) analysed Upper Cambrian tidal siliciclastic rocks of the Tilcara Member (Santa Rosita Formation) in northwest Argentina and noted that tidal-flat assemblages were dominated by worm (*Palaeophycus*) and presumed trilobite trace fossils (*Cruziana*, *Rusophycus*).

Baldwin (1977) documented trilobite trace fossils (*Cruziana*, *Rusophycus*) in tidal-flat deposits of the Cambrian–Ordovician of Spain. He concluded that the dominance of trilobite trace fossils in intertidal areas results from both actual abundance of trace makers and enhanced preservational potential of the structures. Fillion & Pickerill (1990) described in detail the trace fossil content of Cambrian–Ordovician siliciclastic rocks of Newfoundland, Canada. Trilobite traces, such as *Cruziana*, *Rusophycus*, *Monomorphichnus*, *Diplichnites* and *Dimorphichnus*, were recorded in the tidal-flat facies. Additionally, MacNaughton *et al.* (2002) recognised trackways in deposits across the Cambrian–Ordovician transition that they attributed to amphibious euthycarcinoids. They are preserved in coastal eolian dunes, and reveal subaerial incursions by arthropods. Durand (1985) presented an exhaustive sedimentological and ichnological study of Ordovician tidalites of France. He identified various trilobite structures in heterolithic facies of intertidal to upper subtidal origin. Mángano *et al.* (2001) described various ichnospecies of *Cruziana* of the *rugosa* group in Lower to Middle Ordovician tidal-flat deposits of the Mojotoro Formation in northwest Argentina. To conclude, this brief review suggests that, although presumed trilobite trace fossils are usually regarded as indicators of open-marine shelf to near-shore settings, they were also common in restricted, stressed, marginal environments.

Conclusions

1. The Campanario Formation contains ichnofaunas that reflect colonisation of extensive tidal-flat areas flanked seawards by subtidal sandbar complexes. Shallow subtidal and intertidal sand-flat deposits are dominated by vertical domiciles of suspension feeders and passive predators of the *Skolithos* ichnofacies (*Skolithos linearis*, *Arenicolites* isp., *Diplocraterion* isp.) and by the ichnogenus *Syringomorpha*. Mixed-flat deposits contain horizontal feeding, locomotion and resting traces, representing a depauperate *Cruziana* ichnofacies. This ichnofauna includes trilobite trace fossils (*Cruziana problematica*, *Rusophycus carbonarius*, *Rusophycus leifeirikssoni*, large *Rusophycus* isp., *Diplichnites* isp.), structures produced by sessile cnidarians (*Bergaueria* cf. *B. perata*) and shallow burrows and trails of vermiform organisms (*Planolites* isp., *Palaeophycus tubularis*, *Helminthoidichnites tenuis*).

2. Tiering in these tidal-flat deposits is relatively simple. Six ichnoguilds (*Cruziana problematica*, *Palaeophycus*, *Bergaueria*, *Rusophycus leifeirikssoni*, *Syringomorpha* and *Skolithos*) have been defined. These ichnoguilds show a preferential palaeoenvironmental distribution following proximal–distal trends. Deep-tier ichnoguilds tend to occur in the high-energy, most distal regions, specifically in the subtidal to intertidal transition and in the lower intertidal sand flats. On the other hand, middle- and shallow-tier ichnoguilds are dominant in the protected, lower-energy, proximal regions, mainly the middle intertidal mixed flat. The resultant ichnofauna is therefore shaped by the interplay of key environmental parameters, such as energy, substrate nature and food supply, overprinted by taphonomic factors.

3. The presence of the presumed trilobite trace fossil *Rusophycus leifeirikssoni* forming assemblages of clustered individuals in tidal-flat deposits suggests nesting behaviour or feeding in microbial- and meiofaunal-rich sediments. Trilobite incursions to Cambrian tidal flats support an early colonisation of intertidal environments and indicate that representatives of the Cambrian evolutionary fauna were able to cope with physically stressed conditions in very shallow, marginal-marine environments. This environmental shift from the open-marine shelf to the coast indicates an early landward expansion of the Cambrian explosion and the agronomic revolution.

4. In the absence of widespread terrestrial vegetation, Cambrian intertidal trophic webs were mostly based on the organically rich marine source and its autochthonous production. Despite their physical hazards, Cambrian tidal flats provided not only abundant food resources but also represented an efficient site for protection from marine predators in the absence of land and air predators. These favourable conditions may have promoted landward migrations in search for food and sites for moulting and reproductive activities.

Acknowledgements

This project was supported by the Antorchas Foundation, the Percy Sladen Memorial Fund and the Research Council of the University of Tucumán (CIUNT). Several colleagues are thanked for valuable exchanges on Cambrian ecosystems and ichnofaunas, particularly Nick Butterfield, Sören Jensen, Ron Pickerill, Dolf Seilacher and Paul Strother. Detailed reviews by Richard Bromley and Sören Jensen were particularly helpful. Cristina Moya kindly showed us outcrops in Sierra de Mojotoro, around the city of Salta. Drawings were made by Eric Gómez Hasselrod and Marcos Jimenez and trace fossil specimens were curated by Rodolfo Aredes.

References

Aceñolaza, F.G. 1973: Sobre la presencia de trilobites en las cuarcitas del Grupo Mesón en Potrerillos, provincia de Salta. *Revista de la Asociación Geológica Argentina 28*, 309–311.

Aceñolaza, F.G. & Bordonaro, O. 1990: Presencia de *Asaphiscus* (Asaphiscidae–Trilobita) en la Formación Lizoite, Potrerillos, Salta y su significado geológico. *Serie Correlación Geológica 5*, 21–28.

Aceñolaza, F.G., Buatois, L.A., Mángano, M.G., Esteban, S.B., Tortello, F. & Aceñolaza, G.F. 1999: Cámbrico y Ordovícico del noroeste argentino. *Anales del Instituto y Recursos Minerales 29*, 169–187.

Aceñolaza, F.G., Fernández, R.I. & Manca, N. del V. 1982: Caracteres bioestratigráficos y paleoambientales del Grupo Mesón (Cámbrico Medio-Superior), centro-oeste de América del Sur. *Estudios Geológicos 38*, 385–392.

Aceñolaza, F.G. & Toselli, A.J. 1981: Geología del Noroeste Argentino. Facultad de Ciencias Naturales e Instituto Miguel Lillo, Universidad Nacional de Tucumán, Publicación Especial 1287, 1–212. San Miguel de Tucumán.

Alonso, R.N. & Marquillas, R.A. 1981: Trazas fósiles de la Formación Campanario (Grupo Mesón, Cámbrico) en el norte argentino. Consideraciones ambientales y geocronológicas. *Revista del Instituto de Geología y Minería 4*, 95–110.

Amireh, B.S., Schneider, W. & Abed, A.M. 1994: Evolving fluvial-transitional-marine deposition through the Cambrian sequence of Jordan. *Sedimentary Geology 89*, 65–90.

Astini, R.A., Mángano, M.G. & Thomas, W.A. 2000: El icnogénero *Cruziana* en el Cámbrico Temprano de la Precordillera Argentina: El registro más antiguo de Sudamérica. *Revista de la Asociación Geológica Argentina 55*, 111–120.

Ausich, W.I. & Bottjer, D.J. 1982: Tiering in suspension-feeding communities on soft substrata throughout the Phanerozoic. *Science 216*, 173–174.

Bajard, J. 1966: Figures et structures sédimentaires dans la zone intertidale de la partie orientale de la Baie du Mont-Saint-Michel. *Revue de Géographie physique et de Géologie Dynamique 8*, 39–111.

Baldwin, C.T. 1977: The stratigraphy and facies associations of trace fossils in some Cambrian and Ordovician rocks of north western Spain. *Geological Journal Special Issue 9*, 9–40.

Bambach, R.K. 1983: Ecospace utilization and guilds in marine communities through the Phanerozoic. *In* Tevesz, M.J.S. & McCall, P.L. (eds): *Biotic Interactions in Recent and Fossil Benthic Communities*, 719–746. Plenum Press, New York.

Bergström, J. 1973: Organization, life and systematics of trilobites. *Fossils and Strata 2*, 1–69.

Bergström, J. 1976: Lower Palaeozoic trace fossils from eastern Newfoundland. *Canadian Journal of Earth Sciences 13*, 1613–1633.

Bertness, M.D. 1999: *The Ecology of Atlantic Shorelines*. Sinauer Associates, Sunderland, MA.

Beukema, J.J. 1976: Biomass and species richness of the macro-benthic animals living on the tidal flats of the Dutch Wadden Sea. *Netherlands Journal of Sea Research 10*, 236–261.

Bourgeois, J. 1980: A transgressive shelf sequence exhibiting hummocky stratification: the Sebastian Sandstone (Upper Cretaceous), southwestern Oregon. *Journal of Sedimentary Petrology 50*, 681–702.

Braddy, S.J. 2001: Eurypterid palaeoecology: palaeobiological, ichnological and comparative evidence for a 'mass-moult-mate' hypothesis. *Palaeogeography, Palaeoclimatology, Palaeoecology 172*, 115–132.

Brandt, D.S., Meyer, D.L. & Lask, P.B. 1995: *Isotelus* (trilobita) "hunting burrow" from Upper Ordovician strata, Ohio. *Journal of Paleontology 69*, 1079–1083.

Bromley, R.G. 1990: *Trace Fossils. Biology and Taphonomy*. Unwin Hyman, London.

Bromley, R.G. 1994: The palaeoecology of bioerosion. *In* Donovan, S.K. (ed.): *The Palaeobiology of Trace Fossils*, 134–154. John Wiley & Sons, Chichester.

Bromley, R.G. 1996: *Trace Fossils. Biology, Taphonomy and Applications*. Chapman & Hall, London.

Bromley, R.G. & Hanken, N.M. 1991: The growth vector in trace fossils: examples from the Lower Cambrian of Norway. *Ichnos 1*, 261–276.

Buatois, L.A. & Mángano, M.G. 2001: Ichnology, sedimentology and sequence stratigraphy of the Cambrian Mesón Group in northwest Argentina. *In* Buatois, L.A. & Mángano, M.G. (eds): *Ichnology, Sedimentology and Sequence Stratigraphy of Selected Lower Paleozoic, Mesozoic and Cenozoic Units of Northwest Argentina*, 8–16. Fourth Argentinian Ichnologic Meeting and Second Ichnologic Meeting of Mercosur. Field Guide.

Buatois, L.A. & Mángano, M.G. 2003a: Sedimentary facies and depositional evolution of the Upper Cambrian to Lower Ordovician Santa Rosita Formation in northwest Argentina. *Journal of South American Earth Sciences 16*, 343–363.

Buatois, L.A. & Mángano, M.G. 2003b: Early colonization of the deep sea: ichnologic evidence of deep-marine benthic ecology from the Early Cambrian of northwest Argentina. *Palaios 18*, 572–581.

Buatois, L.A., Mángano, M.G., Genise, J.F. & Taylor, T.N. 1998: The ichnologic record of the invertebrate invasion of nonmarine ecosystems: evolutionary trends in ecospace utilization, environmental expansion, and behavioral complexity. *Palaios 13*, 217–240.

Budd, G.E. 2001: Ecology of nontrilobite arthropods and lobopods in the Cambrian. *In* Zhuravlev, A.Y. & Riding, R. (eds): *The Ecology of the Cambrian Radiation*, 404–427. Columbia University Press, New York.

Burzin, M.B., Debrenne, F. & Zhuravlev, A.Y. 2001: Evolution of shallow-water level-bottom communities. *In* Zhuravlev, A.Y. & Riding, R. (eds): *The Ecology of the Cambrian Radiation*, 200–216. Columbia University Press, New York.

Butterfield, N.J. 2001a: Ecology and evolution of Cambrian plankton. *In* Zhuravlev, A.Y. & Riding, R. (eds): *The Ecology of the Cambrian Radiation*, 217–237. Columbia University Press, New York.

Butterfield, N.J. 2001b: Cambrian food webs. *In* Briggs, D.E.G. & Crowther, P.R. (eds): *Palaeobiology II*, 40–43. Blackwell Science, Oxford.

Cheel, R.J. & Leckie, D.A. 1993: Hummocky cross-stratification. *Sedimentology Review 1*, 103–122.

Cornish, F.G. 1986: The trace fossil *Diplocraterion*: evidence of animal–sediment interactions in Cambrian tidal deposits. *Palaios 1*, 478–491.

Crimes, T.P., Legg, I., Marcos, A. & Arboleya, M. 1977: Late Precambrian–low Lower Cambrian trace fossils from Spain. *Geological Journal Special Issue 9*, 91–138.

Dalrymple, R.W. 1984: Morphology and internal structure of sandwaves in the Bay of Fundy. *Sedimentology 31*, 365–382.

Dalrymple, R.W. 1992: Tidal depositional systems. *In* Walker, R.G. & James, N.P. (ed.): *Facies Models: Response to Sea Level Change*, 195–218. Geological Association of Canada, Ontario.

Dittmann, S. 1999: Biotic interactions in a *Lanice conchilega*-dominated tidal flat. *In* Dittmann, S. (ed.): *The Wadden Sea Ecosystem, Stability Properties and Mechanisms*, 153–162. Springer, Berlin.

Dittmann, S., Günther, C.-P. & Schleier, U. 1999: Recolonization of tidal flats after disturbance. *In* Dittmann, S. (ed.): *The Wadden Sea Ecosystem, Stability Properties and Mechanisms*, 175–192. Springer, Berlin.

Dott, R.H. Jr & Bourgeois, J. 1982: Hummocky stratification: significance of its variable bedding sequences. *Geological Society of America, Bulletin 93*, 663–680.

Droser, M.L. & Li, X. 2001: The Cambrian radiation and the diversification of sedimentary fabrics. *In* Zhuravlev, A.Y. & Riding, R. (eds): *The Ecology of the Cambrian Radiation*, 137–169. Columbia University Press, New York.

Durand, J. 1985: Le Gres Armoricain. Sédimentologie-Traces fossiles. Milieux de dépôt. *Centre Armoricain d'Etude structurale des Socles, Mémoires et Documents 3*, 1–150.

Eldredge, N. 1970: Observations on burrowing behavior in *Limulus polyphemus* (Chelicerata, Merostomata), with implications on the functional anatomy of trilobites. *American Museum Novitates 2436*, 1–17.

Fenton, C.L. & Fenton, M.A. 1937: Trilobite "nests" and feeding burrows. *American Midland Naturalist 18*, 446–451.

Fillion, D. & Pickerill, R.K. 1990: Ichnology of the Lower Ordovician Bell Island and Wabana Groups of eastern Newfoundland. *Palaeontographica Canadiana 7*, 1–119.

Frey, R.W., Howard, J.D. & Hong, J.S. 1987: Prevalent Lebensspuren on a modern macrotidal flat, Inchon, Korea: ethological and environmental significance. *Palaios 2*, 571–593.

Ghare, M.A. & Badve, R.M. 1984: Observations on ichnoactivity from the intertidal environment, west coast of Raigad District, Maharashtra. *Biovigyanam 10*, 173–178.

Gingras, M.K., Pemberton, S.G. & Saunders, T. 2001: Bathymetry, sediment texture, and substrate cohesiveness: their impact on modern *Glossifungites* trace assemblages at Willapa Bay, Washington. *Palaeogeography, Palaeoclimatology, Palaeoecology 169*, 1–21.

Gingras, M.K., Pemberton, S.G., Saunders, T. & Clifton, H.E. 1999: The ichnology of Modern and Pleistocene brackish-water deposits at Willapa Bay, Washington: variability in estuarine settings. *Palaios 14*, 352–374.

Hild, A. & Günther, C.-P. 1999: Ecosystem engineers: *Mytilus edulis* and *Lanice conchilega*. *In* Dittmann, S. (ed.): *The Wadden Sea Ecosystem, Stability Properties and Mechanisms*, 43–49. Springer, Berlin.

Horodyski, R.J. & Knauth, L.P. 1994: Life on land in the Precambrian. *Science 263*, 494–498.

Howard, J.D. & Dörjes, J. 1972: Animal–sediment relationships in two beach-related tidal flats; Sapelo Island, Georgia. *Journal of Sedimentary Petrology 42*, 608–623.

Jensen, S. 1990: Predation by early Cambrian trilobites on infaunal worms – evidence from the Swedish Mickwitzia Sandstone. *Lethaia 23*, 29–42.

Legg, I.C. 1985: Trace fossils from a Middle Cambrian deltaic sequence, North Spain. *Society of Economic Paleontologists and Mineralogists Special Publication 35*, 151–165.

Little, C. 2000: *The Biology of Soft Shores and Estuaries*. Oxford University Press, New York.

MacNaughton, R.B., Cole, J.M., Dalrymple, R.W., Braddy, S.J., Briggs, D.E.G. & Lukie, T.D. 2002: First steps on land: arthropod trackways in Cambrian–Ordovician eolian sandstone, southeastern Ontario, Canada. *Geology 5*, 391–394.

Manca, N. 1981: *Contribución al conocimiento geológico de la zona Angosto de Perchel*. Facultad de Ciencias Naturales e Instituto Miguel Lillo, Universidad Nacional de Tucumán, Trabajo de Seminario.

Manca, N. 1986: Caracteres icnológicos de la Formación Campanario (Cámbrico superior) en Salta y Jujuy. *Ameghiniana 23*, 75–87.

Manca, N. 1989: La presencia de *Daedalus labechei* (traza fósil) en la Formación Campanario (Cámbrico Superior) de la provincia de Jujuy. *Actas IV Congreso Argentino de Paleontología y Bioestratigrafía 4*, 131–138.

Mángano, M.G. & Buatois, L.A. 2000: Ichnology, sedimentary dynamics, and sequence stratigraphy of the Mesón Group: a Cambrian macrotidal shallow-marine depositional system in northwest Argentina. *Instituto Superior de Correlación Geológica, Miscelánea 6*, 109–110.

Mángano, M.G. & Buatois, L.A. 2001: The *Syringomorpha* ichnofabric: pervasive bioturbation and the Cambrian explosion. *Abstracts 6th International Ichnofabric Workshop*, Porlamar-Puerto La Cruz, 2001.

Mángano, M.G. & Buatois, L.A. 2003: *Rusophycus leifeirikssoni* en la Formación Campanario: Implicancias paleobiológicas, paleoecológicas y paleoambientales. In: Buatois, L.A. & Mángano, M.G. (eds.): Icnología: Hacia una convergencia entre geología y biología. *Publicación Especial de la Asociación Paleontológica Argentina 9*, 65–84.

Mángano, M.G. & Buatois, L.A. 2003b: Trace fossils. *In* Benedetto, J.L. (ed.): *Ordovician Fossils of Argentina*, 507–553. Universidad Nacional de Córdoba, Secretaría de Ciencia y Tecnología.

Mángano, M.G. & Buatois, L.A. in press a: Integración de estratigrafía secuencial, sedimentología e icnología para un análisis cronoestratigráfico del Paleozoico inferior del noroeste argentino. *Revista de la Asociación Geológica Argentina.*

Mángano, M.G. & Buatois, L.A. 2004: Ichnology of Carboniferous tide-influenced environments and tidal flat variability in the North American Midcontinent. In: McIlroy, D. (Ed.), The application of ichnology to palaeoenvironmental and stratigraphic analysis. *Geological Society, London, Special Publication 228*, 157–178.

Mángano, M.G., Buatois, L.A. & Aceñolaza, G.F. 1996: Trace fossils and sedimentary facies from an Early Ordovician tide-dominated shelf (Santa Rosita Formation, northwest Argentina): implications for ichnofacies models of shallow marine successions. *Ichnos 5*, 53–88.

Mángano, M.G., Buatois, L.A. & Esteban, S.B. 2000: Ichnology of subtidal sandwave complexes and intertidal flats: the Cambrian Mesón Group, northwest Argentina. *Abstracts 31st International Geological Congress*. CD-Rom.

Mángano, M.G., Buatois, L.A. & Moya, M.C. 2001: Trazas fósiles de trilobites de la Formación Mojotoro (Ordovícico inferior-medio de Salta, Argentina): Implicancias paleoecológicas, paleobiológicas y bioestratigráficas. *Revista Española de Paleontología 16*, 9–28.

Mángano, M.G., Buatois, L.A., West, R.R. & Maples, C.G. 1998: Contrasting behavioral and feeding strategies recorded by tidal-flat bivalve trace fossils from the Upper Carboniferous of eastern Kansas. *Palaios 13*, 335–351.

Mángano, M.G., Buatois, L.A., West, R.R. & Maples, C.G. 2002: Ichnology of an equatorial tidal flat: the Stull Shale Member at Waverly, eastern Kansas. *Bulletin of the Kansas Geological Survey 245*, 1–130.

Mángano, M.G. & Droser, M. 2004: The ichnologic record of the Ordovician radiation. *In* Webby, B.D., Droser, M.L., Paris, F. & Percival, I.G. (eds): *The Great Ordovician Biodiversification Event*, 369–379. Columbia University Press, New York.

Mikuláš, R. 1995: Trace fossils from the Paseky Shale (Early Cambrian, Czech Republic). *Journal of the Czech Geological Society 40*, 37–45.

Mingramm, A., Russo, A., Pozzo, A. & Cazau, L. 1979: Sierras Subandinas. *Academia Nacional de Ciencias de Córdoba 1*, 95–137.

Moczydłowska, M. 2002: Early Cambrian phytoplankton diversification and appearance of trilobites in the Swedish Caledonides with implications for coupled evolutionary events between primary producers and consumers. *Lethaia 35*, 191–214.

Moya, M.C. 1998: El Paleozoico inferior en la sierra de Mojotoro, Salta - Jujuy. *Revista de la Asociación Geológica Argentina 53*, 219–238.

Palmer, J.D. 1995: *The Biological Rhythms and Clocks of Intertidal Animals*. Oxford University Press, New York.

Pemberton, S.G., Spila, M., Pulham, A.J., Saunders, T., MacEachern, J.A., Robbins, D. & Sinclair, I.K. 2001: Ichnology and sedimentology of shallow to marginal marine systems. Ben Nevis and Avalon Reservoirs, Jeanne d'Arc Basin. *Geological Association of Canada, Short Course Notes 15*, 1–343.

Prave, A.R. 2002: Life on land in the Proterozoic: evidence from the Torridonian rocks of northwest Scotland. *Geology 30*, 811–814.

Raffaelli, D. & Hawkins, S. 1996: *Intertidal Ecology*. Chapman & Hall, London.

Reise, K. 1985: Tidal flat ecology. An experimental approach to species interactions. *Ecological Studies 54*, 1–191.

Rhoads, D.C. 1974: Organism–sediment relations in the muddy sea floor. *Oceanography and Marine Biology Annual Review 12*, 263–300.

Rhoads, D.C. & Young, D.K. 1970: The influence of deposit-feeding organisms on sediment stability and community trophic structures. *Journal of Marine Research 28*, 150–178.

Rydell, J., Hammarlund, J. & Seilacher, A. 2001: Trace fossil associations in the Swedish Mickwitzia sandstone (Lower Cambrian): did trilobites really hunt for worms? *Geologiska Föreningens i Stockholm Förhandlingar 123*, 247–250.

Sánchez, M.C. & Herrera, Z. 1994: Braquiópodos inarticulados cámbricos en la Formación Campanario (Grupo Mesón), Rio Reyes, Provincia de Jujuy. *Resúmenes VI Congreso Argentino de Paleontología y Bioestratigrafía*, 69.

Sánchez, M.C. & Salfity, J.A. 1990: Litofacies del Grupo Mesón (Cámbrico) en el oeste del Valle de Lerma (Cordillera Oriental argentina). *Actas X Congreso Geológico Argentino 2*, 129–192.

Sánchez, M.C. & Salfity, J.A. 1999: La cuenca cámbrica del Grupo Mesón en el Noroeste Argentino: desarrollo estratigráfico y paleogeográfico. *Acta Geológica Hispánica 34*, 123–139.

Schäfer, W. 1972: *Ecology and Palaeoecology of Marine Environments*. University of Chicago Press, Chicago.

Seilacher, A. 1970: *Cruziana* stratigraphy of "non-fossiliferous" Palaeozoic sandstones. *Geological Journal Special Issue 3*, 447–476.

Seilacher, A. 1977: Evolution of trace fossil communities. *In* Hallam, A. (ed.): *Patterns of Evolution*, 359–376. Elsevier Science, Amsterdam.

Seilacher, A. 1997: *Fossil Art*. The Royal Tyrrell Museum of Paleontology, Drumheller, Alberta.

Seilacher, A. 1999: Biomat-related lifestyles in the Precambrian. *Palaios 14*, 86–93.

Seilacher, A. 2000: Ordovician and Silurian Arthrophycid ichnostratigraphy. *In* Sola, M.A. & Worsley, D. (eds): *Geological Exploration in Murzuk Basin*, 237–258. Elsevier Science, Amsterdam.

Seilacher, A. 2003: Arte Fóssil. Divulgaçoes do Museu de Ciências e Tecnologia. *UBEA/PUCRS Publicação Especial 1*, 1–86.

Sekiguchi, K. 1988: Ecology. *In* Sekiguchi, K. (ed.): *Biology of Horseshoe Crabs*, 50–68. Science House, Tokyo.

Selley, R.C. 1970: Ichnology of Paleozoic sandstones in the Southern Desert of Jordan: a study of trace fossils in their sedimentologic context. *Geological Journal Special Issue 3*, 477–488.

Sepkoski, J.J. Jr 1997: Biodiversity: past, present, and future. *Journal of Paleontology 71*, 533–539.

Simpson, E.L. 1991: An exhumed, Lower Cambrian tidal flat: the Antietam Formation, central Virginia, U.S.A. *Canadian Society of Petroleum Geologists, Memoir 16*, 123–134.

Smith, D.B. 1988: Bypassing of sand over sand waves and through a sand wave field in the central region of the southern North Sea. *In* De Boer, P.L., van Gelder, A. & Nio, S.D. (eds): *Tide-influenced Sedimentary Environments and Facies*, 51–64. D. Reidel, Boston.

Strother, P.K. 2000: Cryptospores: the origin and early evolution of the terrestrial flora. *The Paleontological Society Papers 6*, 3–20.

Strother, P.K., Al-Hajri, S. & Traverse, A. 1996: New evidence of land plants from the lower Middle Ordovician of Saudi Arabia. *Geology 24*, 55–58.

Strother, W.A. & Beck, J.H. 2000: Spore-like microfossils from Middle Cambrian strata: expanding the meaning of the term cryptospore. *In* Harley, M.M., Morton, C.M. & Blackmore, S. (eds): *Pollen and Spores: Morphology and Biology*, 413–424. Royal Botanic Gardens, Kew.

Swinbanks, D.D. & Murray, J.W. 1981: Biosedimentological zonation of Boundary Bay tidal flats, Fraser River Delta, British Columbia. *Sedimentology 28*, 201–237.

Taylor, A.M. & Goldring, R. 1993: Description and analysis of bioturbation and ichnofabric. *Journal of the Geological Society, London 150*, 141–148.

Thomas, R.G., Smith, D.G., Wood, J.M., Visser, J., Caverley-Range, E.A. & Koster, E.H. 1987: Inclined heterolithic stratification – terminology, description, interpretation and significance. *Sedimentary Geology 53*, 123–179.

Turner, J.C.M. 1960: Estratigrafía de la Sierra de Santa Victoria y adyacencias. *Boletín de la Academia Nacional de Ciencias de Córdoba 41*, 163–196.

Turner, J.C.M. 1963: The Cambrian of northern Argentina. Symposium of Petroleum Geology of South America. *Geological Society Digest 31*, 193–211.

Turner, J.C. 1979: Paleozoico Inferior de América del Sur. *Anales de la Academia de Ciencias Exactas, Físicas y Naturales 31*, 25–71.

Vaccari, N.E. & Waisfeld, B.G. 2000: Trilobites tremadocianos de la Formación Las Vicuñas, Puna Occidental, provincia de Salta, Argentina. *Actas XIV Congreso Geológico Boliviano*. CD-Rom.

Walker, R.G., Duke, W.L. & Leckie, D.A. 1983: Hummocky stratification: significance of its variable bedding sequences. Discussion. *Geological Society of America, Bulletin 94*, 1245–1251.

Watanabe, Y., Martini, J.E.J. & Ohmoto, H. 2000: Geochemical evidence for terrestrial ecosystems 2.6 billion years ago. *Nature 408*, 574–578.

Wellman, C.H., Osterloff, P.L. & Mohluddin, U. 2003: Fragments of the earliest land plants. *Nature 425*, 282–285.

Deep-sea trace fossils controlled by palaeo-oxygenation and deposition: an example from the Lower Cretaceous dark flysch deposits of the Silesian Unit, Carpathians, Poland

ALFRED UCHMAN

Uchman, A. 2004 10 25: Deep-sea trace fossils controlled by palaeo-oxygenation and de position: an example from the Lower Cretaceous dark flysch deposits of the Silesian Unit, Carpathians, Poland. *Fossils and Strata*, No. 51, pp. 39–57. Poland. ISSN 0300-9491.

Trace fossil associations and ichnofabrics have been studied in three lithostratigraphic units of the Lower Cretaceous dark flysch deposits, i.e. within the Upper Cieszyn Beds (Valanginian–Hauterivian), the Verovice Shale (Barremian–lowermost Albian) and the Lgota Beds (Albian–Cenomanian). The associations differ from those of the Upper Cretaceous and Tertiary flysch deposits. The trace fossil association of the Upper Cieszyn Shale belongs to the *Nereites* ichnofacies, but the associations of the other two units do not; therein the trace fossil diversity is distinctly lower, and graphoglyptids or horizontal meandering pascichnia are absent or very rare. These trace fossil associations supposedly changed as a result of general evolutionary processes that influenced deep-sea trace makers after the Early Cretaceous, and which intensified the aforementioned differences.

The composition of trace fossil associations and the vertical extent of the bioturbated zone in turbidite–hemipelagite beds were strongly affected by changing oxygenation on the deep-sea floor during the global Early Cretaceous anoxic events. The Verovice Shale is dominated by anoxic sediments interlayered with rare, thin, bioturbated horizons. In the Upper Cieszyn Shale and in the Lgota Beds, most tops of turbidite–hemipelagite rhythms are bioturbated. Non-bioturbated rhythms record anoxia, but their occurrence is influenced by the frequency of turbiditic deposition.

Protovirgularia obliterata and *Protovirgularia pennata* occur in the Verovice Shale in the deepest tier below *Chondrites*. These trace fossils were probably produced below the redox boundary by chemosymbiotic bivalves. The studied associations show that the deep-sea environment is influenced by many factors that change with time, which records the large-scale dynamics of deep-sea ecological processes.

Key words: Ichnology; bioturbation; tiering; anoxic events; turbidites.

Alfred Uchman [fred@ing.uj.edu.pl], Institute of Geological Sciences, Jagiellonian University, Oleandry 2a; 30-063 Kraków, Poland

Introduction

Since the 1950s it has become clear that flysch trace fossil communities are very diverse, even within basins and between stratigraphic units (e.g. Książkiewicz 1977; Crimes *et al.* 1981; Uchman 1999). Data for some periods are notably inadequate, especially for the Permian to Jurassic (Uchman 2003), but it should also be noted that data from the Lower Cretaceous flysch deposits are few in comparison with the Upper Cretaceous or Palaeogene. This gap can be partially filled by investigations of the

Lower Cretaceous flysch facies of the Silesian Unit in the Polish Carpathians, which has a continuous record of diverse facies throughout the Cretaceous and Palaeogene. The flysch facies contain numerous trace fossils whose taxonomy was partially investigated by Nowak (1957, 1959, 1961, 1962, 1970) and Książkiewicz (1970, 1977). Part of the Silesian flysch facies displays dark coloration that is suggestive of the influence of Cretaceous anoxic events. A lowered oxygenation during accumulation of the Lower Cretaceous sediments of the Silesian Unit was previously noticed by Książkiewicz (1977), who did not discuss the details. The dark flysch facies is important because literature on the relationship of flysch trace fossils to oxygenation changes is rather sparse (but see Leszczyński 1991; Uchman 1991, 1999; Wetzel & Uchman 1998) in comparison with pelagic and hemipelagic, fine-grained, post-Palaeozoic deposits (e.g. Bromley & Ekdale 1984; Savrda & Bottjer 1989).

In this paper, dark flysch deposits of the Upper Cieszyn Beds, Verovice Shale and Lgota Beds are considered in four representative sections at Poznachowice Dolne, Kaczyna, Zagórnik-Rzyki and Kozy (Fig. 1). Description of the trace fossils and the interpretation of the associations are the main aims of this paper. Particular attention is paid to the influence of oxygenation. Trace fossils

illustrated in this paper are housed at the Institute of Geological Sciences of the Jagiellonian University (prefix 167P).

Geological setting and stratigraphy

The Silesian Unit constitutes a large, complex nappe in the Ukrainian, Polish and Czech Flysch Carpathians. It contains thick, diverse deposits, mostly flysch, that accumulated in a deep-sea basin (Silesian Basin) from the late Kimmeridgian to the early Miocene. This basin, a part of the Western Tethys, was at least a few tens of kilometres wide and a few hundreds of kilometres long. Deposits of the Silesian Basin were folded and thrust northward during the Miocene.

The Lower Cretaceous deposits (Fig. 2) are represented by the:

- Cieszyn Limestone (upper Tithonian–Berriasian), 100–250 m thick, dominated by turbiditic, commonly

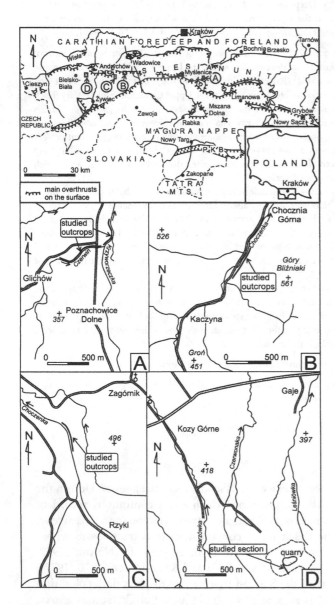

Fig. 1. Map of the western part of the Polish Carpathians (top, with inset location map) and detailed maps (below) showing the locations of the study areas. A: Poznachowice Dolne. B: Kaczyna. C: Zagórnik-Rzyki. D: Kozy. PKB, Pieniny Klippen Belt.

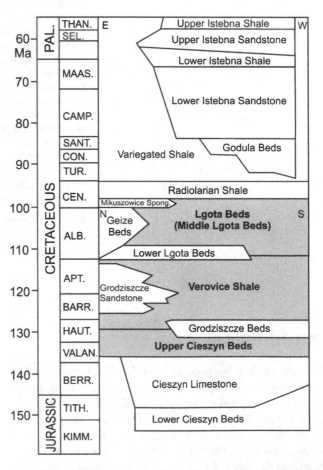

Fig. 2. Kimmeridgian–Palaeocene stratigraphy of the Silesian Unit (based on Ślączka, in Ślączka & Kaminski 1998). The investigated units are coloured grey.

sandy calcarenites and calcilutites interbedded with marly shales;

- Upper Cieszyn Beds (Valanginian–Hauterivian), about 300 m thick, dominated by dark grey marly mudstones alternating with regularly thin-bedded, calcareous sandstones;
- Grodziszcze Beds (Upper Hauterivian–Barremian), 95–140 m thick, represented by grey marly shales alternating with rare thin calcareous sandstone beds and marlstones. Locally, sandstone beds are more frequent. In some areas, these facies are replaced by thick-bedded calcareous sandstones and debris-flow deposits containing exotic pebbles and blocks;
- Verovice Shale (Barremian–lowermost Albian), about 200 m thick, composed of non-calcareous black mudstones interbedded with rare cross-laminated thin sandstone beds and sideritic concretions;
- Lgota Beds (Albian–Cenomanian), 300–350 m thick, dominated by thin- and medium-bedded, turbiditic sandstones and greenish grey spotty mudstones (dark spots are cross-sections of trace fossils visible on parting surfaces). Locally, the lower part of the unit exhibits thick-bedded sandstones, and the upper part of the unit is composed of spongiolithic cherts (Mikuszowice Cherts).

Upper Cieszyn Beds

The Upper Cieszyn Beds are about 300 m thick. They are dominated by dark grey to black marly mudstones interlayered with numerous very thin to thin layers (1–3 cm) of mostly cross-laminated turbiditic fine-grained sandstones, rare sandy limestones, and local sideritic claystones (e.g. Burtan 1978). The sandstones display sharp erosive bases and transitional gradation to overlying mudstones. These deposits are interpreted as turbiditic sandstone–mudstone couplets capped by hemipelagic mudstones. They are similar to the turbiditic facies C2.3 of Pickering *et al.* (1986), but the muddy part is thicker in the investigated deposits.

The Upper Cieszyn Beds are determined as having a Valanginian–Hauterivian age based on the benthic foraminiferids (Geroch & Nowak 1963; Nowak 1968). The sandstone turbidites were transported from the northwest and deposited at depths above the calcite compensation depth (CCD). The Upper Cieszyn Beds were investigated at Poznachowice Dolne (Fig. 1A), where a section about 200 m thick is exposed along the stream. This represents the longest section through the Upper Cieszyn Beds exposed in the Polish Carpathians.

Verovice Shale

The Verovice Shale is about 200 m thick (Książkiewicz 1951). It is composed of prevailing dark non-calcareous mudstones and siltstones, which are intercalated irregularly in thin- to medium-bedded fine-grained sandstones. Locally, horizontal lamination occurs in the shales. The sandstones display ripple cross-lamination. In places, ferruginous concretions are present. The mudstones represent the E2 (locally C2) facies of Pickering *et al.* (1986), which are typical of basin plains. Probably the sandstone beds were deposited by episodic bottom currents. The lack of calcium carbonate in the shales suggests deposition below the CCD. Lower bathyal depths are inferred on the basis of foraminiferids (Szydło 1997). Geochemical analyses indicate both a high input of plant detritus from adjacent lands and high phytoplankton production (Gucwa & Wieser 1980).

The Verovice Shale is assigned a Barremian–earliest Albian age (Szydło 1996, 1997). The Verovice Shale has been investigated at Zagórnik (Fig. 1), the best locality being where about 70 m of the top of the unit is exposed (Cieszkowski *et al.* 2001).

Lgota Beds

The Lgota Beds have been subdivided historically into the Lower Lgota Beds, the Middle Lgota Beds and the Upper Lgota Beds or Mikuszowice Cherts. The locally occurring Lower Lgota Beds, generally 80 m thick, are composed of thick-bedded, commonly amalgamated sandstones interbedded with packages of thin- to medium-bedded, commonly graded turbidites. This unit is characterised by thickening-up sequences, with typically thick-bedded, channelised sandstones in the upper part (facies C2.1, C2.2 of Pickering *et al.* 1986).

Of the three units, the Middle Lgota Beds form the dominant part. They are composed of thin- and medium-bedded, mostly fine-grained sandstones with well-developed Bouma (1962) intervals and interbedded dark to green mudstones. Isolated thick sandstone beds are locally present. The sandstone/mudstone ratio is approximately 1:1 (Unrug 1959). Facies C2.3 of Pickering *et al.* (1986) prevails. The Middle Lgota Beds are 220 m thick in the Kozy quarry.

The Upper Lgota Beds (Mikuszowice Cherts) are 50 m thick, and occur only locally. They consist of thin-bedded (rarely medium-bedded) sandstones that contain a considerable amount of biogenic silica, mainly as opal and chalcedony cement. In some beds, the silica predominates and forms spongiolite chert bands. The sandstones are regularly interbedded with mudstones. Facies C2.3 prevails. In the northern marginal part of the Silesian Unit, the Lgota Beds are replaced by the Geize Beds, which consist mainly of thick-bedded siliceous sandstones with spongiolites.

The Middle Lgota Beds are dominated by turbidites (Unrug 1959). Lenticular, mainly cross-laminated, beds of fine-grained, well-sorted sandstones beds intercalated in the shales have been interpreted as tractionites (Unrug 1959, 1977). Upper bathyal depths are proposed on the basis of foraminiferids (Książkiewicz 1975), but deposition below the CCD is suggested because of non-calcareous hemipelagites atop turbidite–hemipelagite rhythms. The Lgota Beds were probably deposited in distal, locally proximal, depositional lobes and fan fringe settings.

Until recently, the Lgota Beds were assigned to the Albian–?Cenomanian interval (Geroch & Nowak 1963). The middle Cenomanian age of the top of this unit was proved by Bąk *et al.* (2001) on the basis of foraminiferids and radiolarians, and by Gedl (2001) using dinocysts. For additional data based on dinocysts, see Jaminski (1995). The Lgota Beds have been investigated at Kaczyna, Rzyki, and the Kozy quarry (Fig. 1), where they are represented by the lithofacies typical of the Middle Lgota Beds.

Synopses of ichnotaxa

Belorhaphe zickzack (Heer 1877) (Fig. 3I): hypichnial, angular meanders with an apical angle of 45–85°, 7–9 mm wide and 4–6 mm high, in some cases with short appendages at the apices. The apices are slightly rounded and enlarged. Preserved as semi-reliefs. For a more complete discussion of this ichnospecies, see Uchman (1998).

Chondrites intricatus (Brongniart 1823) (Figs. 3A, 4B): occurs as a system of tree-like branching, downward-penetrating, markedly flattened tunnels, 0.4 mm in diameter. The tunnels form acute angles and show phobotaxis. In cross-section it occurs as patches of circular to elliptical spots and short bars. Commonly, the fill of the trace fossil is darker than the host rock. For a more extensive discussion of ichnogenus *Chondrites*, see Fu (1991) and Uchman (1999).

Chondrites targionii (Brongniart 1828) (Fig. 4A): endichnial, tubular, flattened tunnels branched in a dendroid manner. Branches are commonly slightly curved. The tunnels are 1.8–2.0 mm wide.

?Chondrites isp. (Fig. 3H): preserved as hypichnial, horizontal, short, straight to slightly curved ridges that are 0.6–1.0 mm wide and up to 10 mm long. They densely cover soles of sandstone beds. Most probably they are washed out and cast *Chondrites* burrow systems.

Gordia isp. (Fig. 3B): represented by simple hypichnial winding galleries; preserved in semi-relief, and shows crossings. The galleries are 0.5 mm wide. For a discussion of *Gordia*, see Fillion & Pickerill (1990) and Pickerill & Peel (1991).

Helminthopsis abeli Książkiewicz 1977 (Fig. 3C): simple, cylindrical, irregularly winding, smooth tunnels preserved in semi-relief; 5–10 mm wide.

Helminthopsis isp.: hypichnial winding, smooth gallery, 2.5 mm wide.

Helminthopsis hieroglyphica Wetzel & Bromley 1996 (Fig. 3F): hypichnial semicircular gallery that displays first- and second-order windings. The second-order windings are sharp but display low amplitude. The string is about 1.2–2.0 mm wide, and constant within individuals.

Helminthopsis tenuis Książkiewicz 1977 (Fig. 3E): hypichnial semicircular string, which displays alternating wide and narrow irregular meanders. The gallery is 2.5–5.0 mm wide. The width of the gallery is constant in a given specimen. For a discussion of *Helminthopsis*, see Han & Pickerill (1995) and Wetzel & Bromley (1996).

Lorenzinia isp.: hypichnial form consisting of six ridges radiating from a central flat area. The ridges are up to 4 mm long and about 1 mm wide. The trace fossil is about 13 mm wide.

Lorenzinia plana (Książkiewicz 1968) (Fig. 3G): hypichnial star-shaped trace fossil composed of straight or slightly curved ridges radiating from a central flat area. The ridges are semicircular, 5–30 mm long and 2–3 mm wide. There are more than 20 ridges per trace fossil. The central area is about 40 mm across, and the whole trace fossil is about 110 mm across. For a discussion of *Lorenzinia*, see Uchman (1998).

?Lorenzinia isp. (Fig. 4D): endichnial structure composed of a semicircular wreath of short flattened cylinders, 2.5–4.0 mm wide, up to 7 mm long, radiating from an indistinct area, and with dark infilling.

Megagrapton isp. (Fig. 5E): hypichnial irregular net, at least 65 mm across. Individual galleries are 1.2–2.0 mm thick. For a discussion of this ichnogenus, see Uchman (1998).

Paleodictyon strozzii Meneghini in Savi & Meneghini 1850 (Fig. 5F): hypichnial hexagonal net, whose maximum mesh size ranges from 3.0 to 5.5 mm; gallery diameter from 0.8 to 1.0 mm. For a recent review of *Paleodictyon*, see Uchman (1995).

Phycodes bilix (Książkiewicz 1977) (Fig. 5A): hypichnial trace fossil composed of a bunch of horizontal to sub-horizontal flattened cylinders branching from one horizontal to subhorizontal stem. The cylinders, 6–12 mm wide, display a knobby wall, covered with small

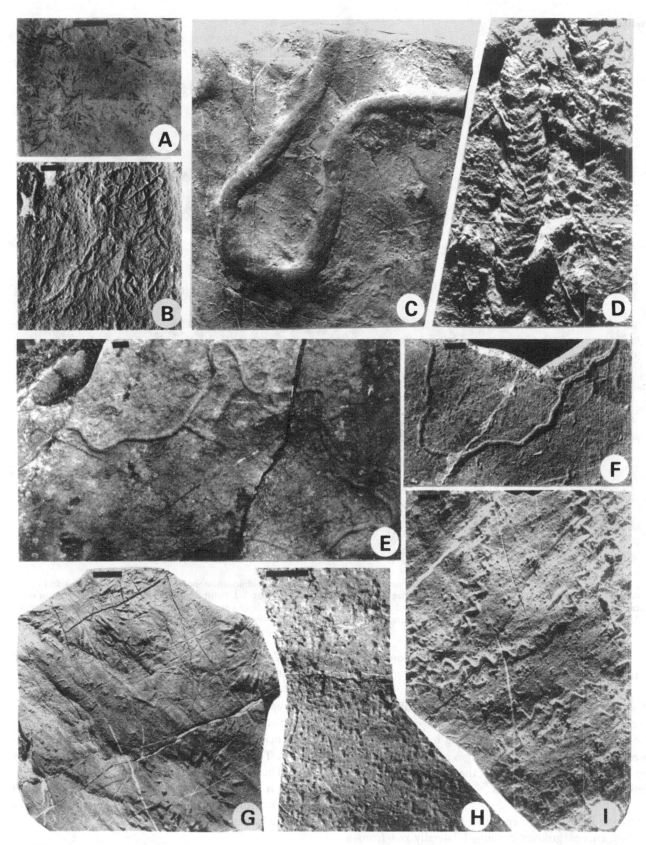

Fig. 3. Trace fossils from the Upper Cieszyn Beds at Poznachowice Dolne. Parting surface in shale in (A); soles of sandstone turbidites in (C)–(I). Scale bars 10 mm in (A)–(F) and (H)–(I); 40 mm in (G). A: *Chondrites intricatus.* 174P19. B: *Gordia* isp. 174P12. C: *Helminthopsis abeli.* 174P11. D: *Protovirgularia pennata.* 174P7. E: *Helminthopsis tenuis*; field photograph. F: *Helminthopsis hieroglyphica.* 174P12. G: *Lorenzinia plana.* 174P1. H: ?*Chondrites* isp.; field photograph. I: *Belorhaphe zickzack.* 174P2.

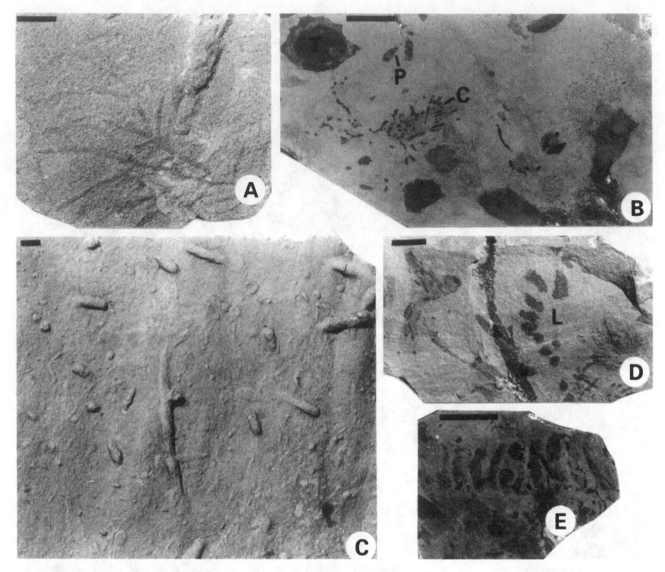

Fig. 4. Trace fossils from the Lgota Beds. A: *Chondrites targionii.* Parting surface in a turbiditic sandstone. Kozy. 174P9. B: *Thalassinoides* isp. (T), *Planolites* isp., form A (P) and *Chondrites intricatus* (C) against totally bioturbated background. Polished and oiled horizontal surface from shale. Kaczyna. 174P44. C: *Thalassinoides* isp. (thick ridges) and *Planolites* isp. (thin ridges). Sole of a turbiditic sandstone bed. Kozy; field photograph. D: ?*Lorenzinia* isp. (L) and *Planolites* isp., form B (P). Parting surface in shale. Kaczyna. 174P42. E: *Taenidium* isp. Parting surface in shale. Kaczyna.

(1–2 mm) irregular mounds. For a discussion of this ichnospecies, see Uchman (1998).

Phycodes isp. (Fig. 5B): hypichnial, semicircular straight to curved ridges diverging from one area, spreading out in the form of a fan, and then plunging into the bed, and preserved as a full-relief. The ridges are 9–13 mm wide, up to 90 mm long. Some of them are covered with delicate longitudinal wrinkles.

?*Phycodes* isp. (Fig. 5C): epichnial system of semicircular slightly curved grooves coming bilaterally out from a central straight stem. The grooves are inclined in the same direction in respect to the stem and ascend distally from the stem. They are up to 60 mm long, and 10–15 mm

wide. Very probably this trace fossil is a preservational variant of *Phycodes* isp.

Phycosiphon incertum Fischer-Ooster 1858 (Fig. 6A): preserved as small horizontal lobes, up to 5 mm wide, each encircled by a thin, less than 1 mm thick, marginal tunnel. They occur on the upper surface of sandstone beds. This trace fossil, produced by a deposit feeder, is common in fine-grained deep-sea and deeper shelf deposits. More information about *Phycosiphon* can be found in Wetzel & Bromley (1994).

Planolites isp. (Figs. 4B–D, 6B): represented by two morphotypes. Form A (Fig. 4B, C) is a variably oriented, but mostly horizontal, cylindrical burrow without a wall,

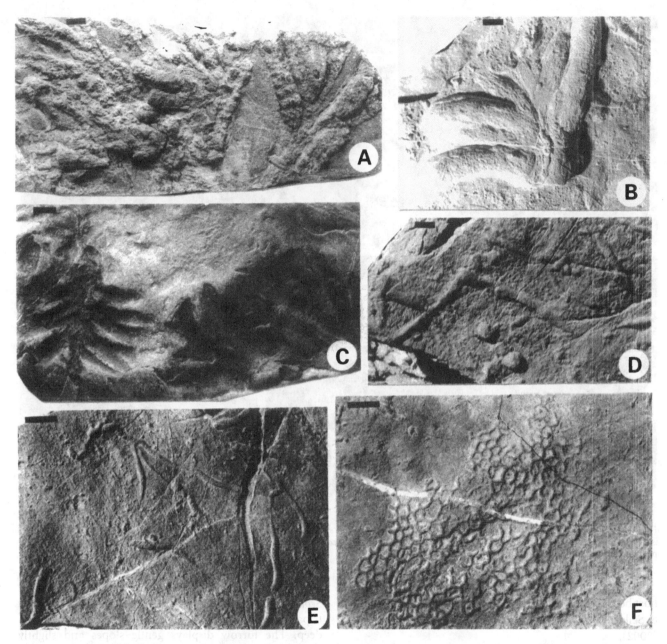

Fig. 5. Other trace fossils from the Upper Cieszyn Beds at Poznachowice Dolne. Soles of sandstone turbidites in (A), (B) and (D)–(F); top of sandstone turbidites in (C). Scale bars 10 mm in (A), (B) and (D)–(F); 40 mm in (C). A: *Phycodes bilix.* 174P20. B: *Phycodes* isp. 174P22. C: ?*Phycodes* isp. 174P26. D: *Thalassinoides* isp.; field photograph. E: *Megagrapton* isp. 174P10. F: *Paleodictyon strozzii.* 174P4.

1–3 mm in diameter. In cross-sections it appears as oval spots contrasting in colour against the surrounding rock. It is also preserved on soles of sandstone beds in full-relief as straight to slightly curved semicircular ridges. Form B (Figs. 4D, 6B) is represented by straight to slightly curved, strongly flattened endichnial cylinders with sharp margins, 6–9 mm wide, filled with lighter sediment than the surrounding rock in the Verovice Shale. In the Lgota Beds, the cylinders are 3–5 mm wide, filled with darker material that is locally burrowed preferentially by *Chondrites* (composite trace fossil). Short ridges or knobs, 5–8 mm in diameter, preserved on the lower surfaces of

sandstone beds as semi-relief, probably belong to *Planolites* (?*Planolites* isp.). For a discussion of *Planolites*, see Pemberton & Frey (1982) and Keighley & Pickerill (1995).

Protovirgularia obliterata (Książkiewicz 1977): hypichnial straight to slightly curved heart-shaped cylinder, 7–9 mm wide, preserved in full-relief. It projects from the sole as an angular ridge with indistinct median furrow and fine, dense chevron ribs on the slopes. Locally, the ridge displays primary successive branches *sensu* D'Alessandro & Bromley (1987) that are probes plunging up into the

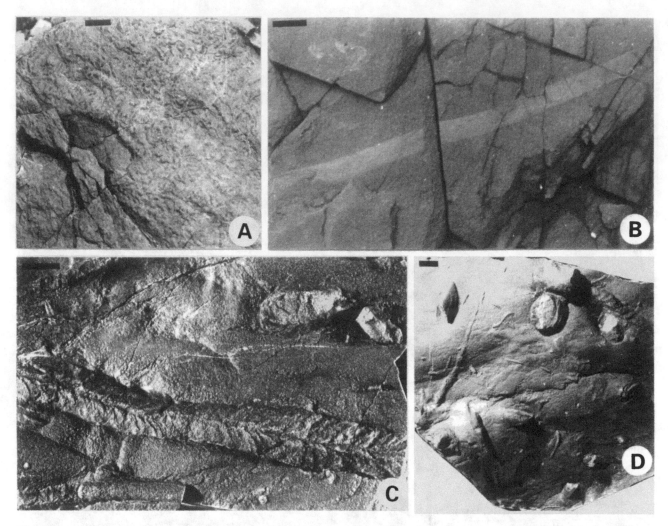

Fig. 6. Trace fossils from the Verovice Beds at Zagórnik. A: *Phycosiphon incertum.* Parting surface in siltstone, top view; field photograph. B: *Planolites* isp., form B. Parting surface in mudstone, top view; field photograph. C: *Protovirgularia pennata.* Sole of a sandstone bed. 174P27. D: *Protovirgularia pennata* crossing sole of a sandstone bed. 174P50.

host bed. For a discussion of *Protovirgularia*, see Seilacher & Seilacher (1994), Uchman (1998) and Mángano *et al.* (2002).

Protovirgularia pennata (Eichwald 1860) (Figs. 3D, 6C, D): hypichnial to endichnial straight to slightly curved, almond-shaped cylinder preserved in full-relief, oriented parallel or oblique to the bedding. It projects from the sole as a triangular ridge, 5–12 mm wide, with an indistinct, discontinuous, very narrow median furrow and fine, dense, chevron ribs on the slopes. The ridges display successive branches that are probes plunging up into the host bed.

Protovirgularia isp.: hypichnial straight or curved cylinder, almond-shaped in outline and preserved in full-relief, and sticks out on the sole as an angular ridge. The ridge is smooth, 3–4 mm wide.

Scolicia plana Książkiewicz 1970 (Fig. 7A): epichnial tripartite winding furrow, about 30 mm wide and 5 mm deep. The furrow displays gentle slopes and slightly elevated floor. The floor is covered with dense perpendicular thin ribs, and dissected by an indistinct median furrow. This trace fossil represents the lower part of an irregular echinoid burrow. For a discussion of *Scolicia*, see Uchman (1995).

Scolicia strozzii (Savi & Meneghini 1850) (Fig. 7B): a winding, bilobate smooth hypichnial ridge, which is about 25 mm wide, up to 8 mm high. The ridge is subdivided by a central, semicircular, furrow.

Taenidium isp. (Fig. 4E): a horizontal row of shallow menisci, about 9 mm wide. They occur as black arcuate strips, which are 1.0–1.5 mm wide and spaced 2–3 mm apart. For a discussion of this ichnogenus, see D'Alessandro & Bromley (1987).

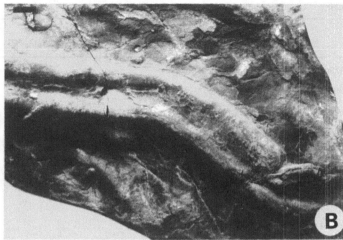

Fig. 7. Scolicia from the Lgota Beds. A: *Scolicia plana*. Upper surface of a turbiditic sandstone bed. Rzyki. UJ TF 621. B: *Scolicia strozzii*. Sole of a turbiditic sandstone bed. Kaczyna. UJ TF 898.

Thalassinoides isp. (Figs. 4B, C, 5D): hypichnial, horizontal, straight to slightly curved and branched cylinders without a wall, preserved in full-relief, 6–10 mm wide. In cross-section in the shale beds, it appears as oval spots. For a further discussion of this ichnogenus, see Ekdale (1992).

Trichichnus isp.: straight to winding, rarely branched, steeply vertical to oblique, thin cylinders filled with iron sulphides or oxides. The cylinder is about 1.0 mm in diameter. For a discussion of this ichnogenus, see Uchman (1999).

Trace fossil assemblages and ichnofabrics

Trace fossils of the Upper Cieszyn Beds are relatively abundant and diverse (Table 1). Most are preserved in semi- or full-relief on the lower surfaces of turbiditic sandstones. *Chondrites intricatus*, *Helminthopsis*, and *Planolites* are the most abundant ichnotaxa. Among the graphoglyptids, which are diverse, *Belorhaphe zickzack*, *Megagrapton* isp., and *Lorenzinia* ispp. are the most abundant. Common occurrence of *Protovirgularia obliterata* is characteristic. Other trace fossils are rare.

Totally bioturbated light layers (spotty layers *sensu* Uchman 1999) occur at the top of dark turbidite–hemipelagite couplets (Fig. 8). The couplets average 3 mm; most are less than 1 cm thick. This layer contains mostly *Planolites* cross-cut by very thin *Chondrites* overprinting a totally bioturbated background. In at least a few of the beds, the spotty layer does not occur at all.

In contrast, deposits of the Verovice Shale are dominated by laminated mudstones and siltstones nearly barren of distinctive trace fossils. Only locally less than 1 cm thick bioturbated horizons occur. The horizons are slightly lighter in colour and occur at the top of some depositional event beds. *Phycosiphon incertum*, *Chondrites intricatus*, and *Planolites* are the dominant, albeit usually poorly visible, ichnotaxa. Below some of the bioturbated horizons, *Protovirgularia pennata* and *Protovirgularia obliterata* occur in a zone a few centimetres thick, especially below sandstone beds, where they can be abundant (Fig. 9).

In the upper few metres of the section of the Verovice Shale at Zagórnik, a few centimetre thick layers of greenish, bioturbated, spotty shales occur. They contain *Planolites*, *Chondrites* and *Thalassinoides*. The appearance of the layers is typical of the overlying Lgota Beds. These occurrences would seem to indicate a general improvement in oxygenation.

The Lgota Beds contain a trace fossil assemblage of low diversity, dominated by *Planolites*, *Chondrites* and *Thalassinoides*. The oldest occurrence of *Scolicia* in the Flysch Carpathians was noted here by Książkiewicz (1977), but this and other trace fossils are rare.

Distinct spotty layers of totally bioturbated greenish-grey, non-calcareous mudstones occur at the top of turbidite–hemipelagite couplets (Fig. 10). Cross-sections of *Chondrites*, *Planolites*, and *Thalassinoides* are commonly visible against a totally bioturbated background. The last two ichnotaxa are commonly preserved in semi-reliefs on the lower surface of turbiditic sandstone beds. The spotty layer is absent in some couplets; instead dark mudstones occur, which are barren of trace fossils.

Discussion

The described trace fossil assemblages are markedly different from those of the younger deposits of the Alpine

Table 1.   Occurrence of trace fossils in the studied sections.

	Upper Cieszyn Shale Poznachowice Dolne	Verovice Shale Zagórnik	Lgota Beds Kaczyna	Rzyki	Kozy
[Arthrophycus strictus Książkiewicz]*				X	
Arthrophycus tenuis (Książkiewicz)*				X	
Belorhaphe zickzack (Heer)	R				
Chondrites intricatus (Brongniart) [Chondrites aequalis Sternberg]	C	R	C	C	C
Chondrites targionii (Brongniart) [Chondrites furcatus (Brongniart)]	R				R
?Chondrites isp.	R		C	C	C
Gordia isp.	R				
Helminthopsis abeli Książkiewicz [Helminthopsis hieroglyphica Książkiewicz]	C				
Helminthopsis hieroglyphica Wetzel & Bromley	R				
Helminthopsis isp.	R				
Helminthopsis tenuis Książkiewicz	R				
Lorenzinia isp.	R				
Lorenzinia plana (Książkiewicz) [Sublorenzinia plana Książkiewicz]	R				
?Lorenzinia isp.			R		
Megagrapton isp.	R				
Paleodictyon strozzii Meneghini	R				
Phycodes bilix (Książkiewicz)	R				
Phycodes isp.	R				
?Phycodes isp.	R				
Phycosiphon incertum Fischer-Ooster		C			
Planolites isp.	C	R	C	C	F
Protovirgularia isp.	R				
Protovirgularia obliterata (Książkiewicz)	C	C			
Protovirgularia pennata (Eichwald)	R				
Scolicia plana Książkiewicz				X	
Scolicia strozzii (Savi & Meneghini) [Taphrhelminthopsis vagans Książkiewicz]			X		
Taenidium isp.			R		
Thalassinoides isp. [Sabularia rudis Książkiewicz; Buthotrephis aff. succulens Hall]	C		C	C	F
Trichichnus isp.					R

X, data by Książkiewicz (1977) (his original ichnotaxonomic determinations are revised here, and included in square brackets); C, common; F, frequent; R, rare (present author data).

*According to Seilacher (2000) and Rindsberg & Martin (2003), the flysch Arthrophycus is indeterminate because of a lack of evidence of an internal structure.

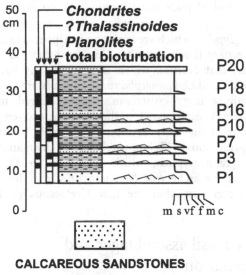

Fig. 8. A short sedimentological log with the vertical extent of trace fossils from the Upper Cieszyn Beds at Poznachowice Dolne. Grain-size scale: m, mud; s, silt; vf, very fine sand; f, fine sand; m, medium sand; c, coarse sand. Samples (P1–P20) are also shown.

region (e.g. Crimes *et al.* 1981; Leszczyński & Seilacher 1991; Tunis & Uchman 1996a, b; Uchman 1995, 1999, 2001). Their diversity is lower and they do not contain some trace fossils, based on the detailed documentation of other formations, as might be expected, for instance *Scolicia*, *Nereites irregularis* or *Ophiomorpha rudis*. Most probably these differences resulted from lowered oxygenation and evolutionary processes (Uchman 2004).

Ichnofacies problem: some evolutionary aspects and specificity of the trace fossil assemblages

The classical ichnofacies model (Seilacher 1967; Frey & Seilacher 1980) can only be applied to the investigated deposits to a certain extent. The Upper Cieszyn Beds contain the typical *Nereites* ichnofacies with a significant

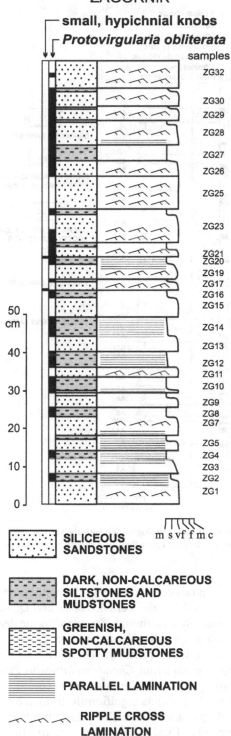

Fig. 9. A short sedimentological log with the vertical extent of trace fossils from the Verovice Shale at Zagórnik, where *Protovirgularia obliterata* is exceptionally abundant. Grain-size scale as in Fig. 8.

contribution of graphoglyptids (about 20% of ichnotaxa), typical of the *Paleodictyon* ichnosubfacies Seilacher 1978).

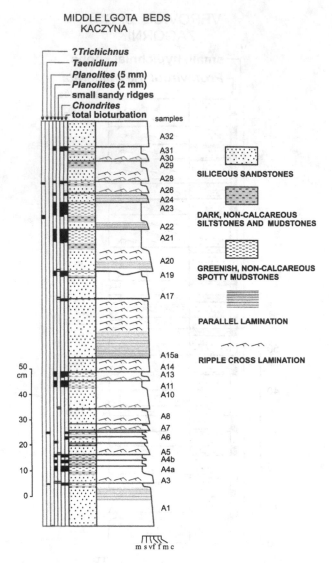

Fig. 10. A short sedimentological log with the vertical extent of trace fossils from the Middle Lgota Beds at Kaczyna. Grain-size scale as in Fig. 8.

There is an obvious difficulty in the application of the ichnofacies model to the Verovice Shale and Lgota Beds. The Verovice Shale, although deposited in the deep sea, contains no graphoglyptid or meandering trace fossils, which are the index forms of the *Nereites* ichnofacies. *Protovirgularia obliterata*, *Chondrites intricatus*, *Planolites*, and *Phycosiphon incertum* are most common. This ichnoassemblage differs significantly from most of the modern deep-sea ichnofacies of pelagic muds at depths exceeding the CCD. The latter, apart from rare *Chondrites*, contain common *Teichichnus* and *Zoophycos* (Ekdale *et al.* 1984; Wetzel 1991), but these do not occur in the Verovice Shale.

The trace fossil association of the Lgota Beds is dominated by *Chondrites*, *Planolites* and *Thalassinoides*. The occurrence of meandering forms (*Scolicia*, *Helminthopsis*) and graphoglyptids (?*Lorenzinia*) is poorly

documented. This ichnoassociation can only provisionally be ascribed to a very impoverished version of the *Nereites* ichnofacies.

The studied trace fossil assemblages do not contain *Scolicia* ispp., *Ophiomorpha rudis*, or numerous graphoglyptids, which are common in Upper Cretaceous and Tertiary flysch deposits. This can be explained by the fact that *Scolicia*, which is produced by irregular echinoids, and *Ophiomorpha rudis*, which is produced by a relatively large crustacean, only appeared in the Tithonian (latest Jurassic) for the first time, and remained very rare in the Early Cretaceous, probably because of widespread anoxia (Tchoumatchenco & Uchman 2001). Many graphoglyptids, which are characteristic trace fossils of the archetypal *Nereites* ichnofacies, had their first occurrences after the Late Cretaceous (Uchman 2003).

Trace fossil assemblages and oxygenation

Oxygenation on the deep-sea floor fluctuates. In an oxygen-restricted setting, it increases after turbiditic flows, which introduce oxygenated water (Dżułyński & Ślączka 1958), and decreases between turbiditic events. This is considered as the crucial phenomenon that, together with changes in benthic food supply, governs the sequential colonisation of thin turbiditic beds by bioturbating organisms (Wetzel & Uchman 2001).

The general oxygenation level of the deep-sea floor can be considered on the scale of lithostratigraphic units. The oxygenation level is reflected by the diversity and composition of the trace fossil assemblages, and by the vertical extent of the bioturbated zone in turbidite–hemipelagite couplets. However, each of these parameters is partially problematic for flysch deposits. The diversity of trace fossil assemblages corresponds to the diversity of infaunal organisms, but it is also influenced by preservation potential. In thin- and medium-bedded flysch, where delicate scouring and casting are common during deposition, the preservational potential is higher than in shaly or thick-bedded flysch.

In this study, the highest preservational potential is inferred for the Upper Cieszyn Beds, which have the greatest total number and frequency of deposition of thin-bedded turbidites. The lowest preservational potential is inferred for the Verovice Beds, where the number of sandstone beds is comparatively small. In the Lgota Beds, the total number and frequency of turbidites are lower than in the Upper Cieszyn Beds. However, a few thousand thin turbidites in the former unit should enable preservation of all burrow systems, which are normally scoured and cast.

The composition of trace fossil assemblages can also be taken into account. Assemblages dominated by ichnotaxa related to opportunistic r-selected organisms are typical of stressed environments, commonly related to lowered oxygenation (Ekdale 1985a, 1988). Indeed, the contribution of graphoglyptids, which are related to the K-selected organisms (Ekdale 1985a), is relatively low in the investigated sediments. It is highest in the Upper Cieszyn Beds. In the Lgota Beds, the presence of graphoglyptids is problematic and in the Verovice Shale they are absent.

According to the model of sequential colonisation of turbidites (Wetzel & Uchman 2001), a newly deposited turbiditic bed contains the maximum content of oxygen and benthic food, which decreases during subsequent colonisation. In the middle and shallow tiers, the bed is colonised at first by less efficient mobile deposit feeders producing *Phycosiphon*, which exploit small areas encircled by marginal tunnels. They are followed by larger, more efficient, trace makers of *Nereites irregularis*, which produce tight meanders. When benthic food is exploited, the bed is colonised by stationary, possibly chemotrophic forms producing *Chondrites* (Wetzel & Uchman 2001). Such a succession is expressed by the cross-cutting relationships between trace fossils, which enables reconstruction of the tiering pattern (Fig. 11). The mobile deposit feeders do not have a permanent connection to the sea floor, and therefore they need oxygenated sediments.

It is intriguing that in the Upper Cieszyn Beds and Lgota Beds, *Nereites* and *Phycosiphon* do not occur at all. *Phycosiphon* only occurs in some horizons of the Verovice Shale. The absence of *Nereites irregularis* may be an evolutionary effect. It occurs more abundantly after the Lower Cretaceous, and is rare or problematic in older sediments (Uchman 2004). Alternatively, another explanation might be that *Nereites* follows the redox boundary (Wetzel 2002) and the redox boundary in the studied sediments was too shallow or too unstable for the *Nereites* producers to become active. The absence or rarity of *Phycosiphon*, which is abundant in older and younger sediments, may be caused by a low oxygen content within the middle and shallow tiers of the turbiditic beds. Probably, the sediment was only bioturbated in a very thin near-surface layer, where no distinct trace fossils could be preserved owing to the low shear strength caused by the high water content. In the Lgota Beds, the middle tier is commonly occupied by small *Thalassinoides*, which was an open-burrow system, into which the crustacean trace maker may have pumped oxygenated water (cf. Ekdale 1988).

Scolicia, a trace fossil produced by irregular echinoids, occurs in a few beds in the Lgota Beds for the first time in the Carpathian Flysch (Książkiewicz 1977). Echinoids usually burrow close to the redox boundary (Bromley *et al.* 1995). The absence of their burrows could indicate a

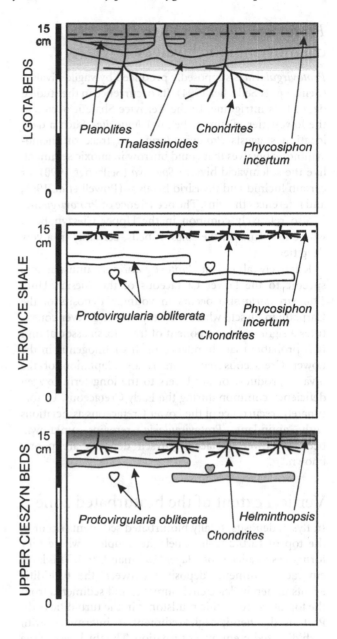

Fig. 11. Tiering pattern of the commonest trace fossils in the investigated units. The shaded area corresponds to the totally bioturbated sediments. Note that the dashed horizontal line depicts the occurrence of *Phycosiphon incertum.*

very shallow redox boundary that does not allow echinoids to completely bury themselves.

Thus, the absence of mainly horizontal trace fossils, which are not interpreted as open burrows, may have been caused by the poor oxygenation of the shallow and middle tiers. This is in basic agreement with the model by Ekdale & Mason (1988), who considered sediments with pascichnial *Phycosiphon* and *Scalarituba* (=*Nereites*) as better oxygenated than sediments with fodinichnial *Chondrites* and *Zoophycos*. This model, however, does not explain the absence of *Zoophycos* in the investigated sections.

Protovirgularia as a burrow of chemosymbiotic bivalves

Protovirgularia is supposedly produced by vagile bivalves (Seilacher & Seilacher 1994). Its occurrence in the studied deposits is intriguing. In the Verovice Shale it occurs in the lowest tier, distinctly below *Chondrites*. Such a deep location suggests *Protovirgularia* as a trace of chemosymbiotic bivalves that could burrow in anoxic sediment, like the solemyacid bivalve *Solemya* (Seilacher 1990), or certain lucinid and thyasirid bivalves (Powell *et al.* 1998, and references therein). The occurrence of *Protovirgularia* is also relatively common in the Upper Cieszyn Beds, where it occurs on the soles of beds, certainly in a very deep tier.

Relatively abundant *Protovirgularia* is unusual and specific to the Lower Cretaceous of the Silesian Unit. *Protovirgularia* also occurs in younger deposits of the Carpathian Flysch, where it is very rare and never constitutes a significant component of trace fossil associations. It is probable that abundance of this ichnogenus in the Lower Cretaceous sediments is an adaptation of the bivalve producer or producers to the long-term oxygen deficiency common during the Early Cretaceous. Unfortunately, recurrence of the Lower Cretaceous associations with abundant *Protovirgularia* remains unknown, because Lower Cretaceous flysch deposits are poorly known.

Vertical extent of the bioturbated zone

In flysch deposits, totally bioturbated sediments occur at the top of turbidite–hemipelagite couplets, where they form the so-called spotty layer (Uchman 1999). This layer embraces sediments deposited between the turbiditic events in hemipelagic environments and sediments from the top of the turbiditic mudstones in the turbiditic beds. Part of the hemipelagic sediment is intermixed with turbiditic sediment by bioturbation. The thickness of the spotty layer was related to oxygenation, i.e. a thicker spotty layer could indicate better oxygenation (Uchman 1999). This view is reconsidered and corrected here.

Owing to their low depositional rate, hemipelagic sediments are usually completely bioturbated even where bioturbating action is neither strong nor deep (Berger *et al.* 1979). All background sediments have a relatively long residence time on the sea floor and can be biologically disturbed several times. Therefore, the thickness of hemipelagic sediments cannot be related to the thickness of the oxygenated zone because the thickness depends mostly on the time of deposition (see the discussion by Wetzel 1991). In contrast, rapidly deposited siliciclastic turbiditic mud is colonised from the top down, and the thickness of a totally bioturbated layer in such a mudstone bed can be related to oxygenation. Often it is

difficult to delineate macroscopically between pelagic and turbiditic mud because of negligible contrast between the grain sizes. However, changes in colour and calcium carbonate content can help to distinguish them.

When the bioturbated spotty layer is absent, several possibilities can be invoked. One is that there was not enough time for colonisation, as when turbidites were deposited frequently one after another. However, deposition of two or more turbiditic beds one after another is generally quite rare and cannot explain the common absence of the spotty layer in many depositional rhythms. Another possibility is that erosion of the turbiditic flow removed the spotty layer. This should be indicated by common erosional structures. Moreover, some remnants of the spotty layer would be expected, because erosion usually acts unevenly. Where such features are lacking, the absence of the spotty layer can be explained by anoxic conditions on the sea floor, especially in dark sediments (cf. Bromley & Ekdale 1984). Similar interpretations were proposed previously for the Albian–Eocene flysch of northern Spain (Leszczyński 1991, 1993) and for the Upper Cretaceous Inoceramian Beds of the Polish Carpathians (Uchman 1992).

The spotty layer is common in the Upper Cieszyn Beds, absent from the Verovice Shale, and relatively uncommon in the Lgota Beds. Erosional structures, such as small flute casts or groove marks, occur in the Upper Cieszyn Beds and in the Lgota Beds, but are the same on the soles of sandstone beds above the couplets regardless of the presence or absence of a spotty layer. Therefore, the absence of couplets with a spotty layer in these units is referred to anoxic conditions. At first glance, the frequency of the couplets without a spotty layer could be related theoretically to the frequency or the duration of anoxic events, but the frequency of the turbidites seems to be the foremost influence. For the Upper Cieszyn Beds, the turbidites are almost 10 times more frequent than in the Lgota Beds (Table 2). In other words, it appears that 10 turbidites were deposited in the Upper Cieszyn Beds while only one accumulated in the Lgota Beds during an anoxic episode of the same duration. If this is true, then the spotty layers of the Lgota Beds should be on average 10 times thicker than those of the Upper Cieszyn Beds because the duration between turbiditic events was on average 10 times longer. In fact, the spotty layers are only five times thicker (on average 15 and 3 mm, respectively). This might be caused by differences either in the rate of accumulation of hemipelagic mud or in the depth of total bioturbation in the turbiditic mud. Because distinguishing between turbiditic and hemipelagic mudstones in these units is problematic, the issue remains unresolved. However, the overall evaluation of the trace fossil association in these units favours the second possibility, i.e. that the smaller thickness of totally bioturbated turbiditic muds in the Lgota Beds is recording lower oxygenation than in the Upper Cieszyn Beds.

Table 2. Some sedimentological parameters for the lithostratigraphic units studied plus the Godula Beds (Turonian). The expected thickness of the hemipelagite layer is calculated for a sedimentation rate of 1 mm/1000 years.

Lithostratigraphic units	Thickness and duration of accumulation	Average number of turbidites/m and approximate total number of turbidites	Frequency of turbiditic events (years/turbidite)	Expected thickness of the hemipelagite layer (mm)	Average thickness of the spotty layer (mm)
Godula Beds	2000 m 5.5 Myr	10 20,000	275	0.3	30
Lgota Beds	250 (locally 450) m 18 Myr	15 3750 (locally 6750)	4800 (locally 2666)	2.5–5	15
Verovice Shale	300 m 15 Myr	5 1500	10,000	10	0.5
Upper Cieszyn Beds	300 m 9.5 Myr	65 19,500	487	0.5	3

The spotty layers from the younger Godula Beds (Turonian), which accumulated in the same basin, are much thicker than those from the Upper Cieszyn Beds. However, the frequency of turbidites in these units is almost the same. The expected thickness of the bioturbated hemipelagic sediments deposited at the same rate would be about 20 times greater in the Verovice Shale than in the Upper Cieszyn Beds, but in reality it is less in the Verovice Shale than in the Upper Cieszyn Beds (Table 2). Most probably, the original vertical extent of bioturbation in the Verovice Shale was much less than for the Upper Cieszyn Beds.

In summary, there was a gradation of generalised oxygenation of the deep-sea floor during deposition of the investigated units (Fig. 12). Sediments of the Verovice Shale were deposited mostly in anoxic conditions with short oxic events that enabled colonisation of the deep-sea floor by burrowers. The trace fossil assemblage of this unit shows the lowest diversity. The thicker layers of greenish, bioturbated, spotty shales at the top of this unit indicate a general improvement in oxygenation. The Lgota Beds were deposited in relatively more oxic conditions than in the Verovice Shale. Most couplets contain the spotty layer, but their trace fossil diversity is

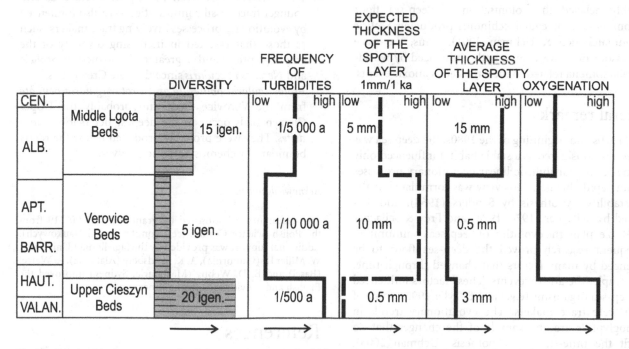

Fig. 12. Relationships between trace fossil diversity, frequency of turbidites and oxygenation in the investigated units. The frequency of turbidites is expressed by an average number of years per each turbidite flow (e.g. one turbidite flow/5000 years). The expected thickness is a theoretical thickness of the hemipelagic layer based on a sedimentation rate of 1 mm/1000 years (= 1 mm/1 ka). The average thickness refers to the hemipelagic layer measured in the field. The arrows show the direction of increasing values of the various parameters employed.

relatively low. The Upper Cieszyn Beds were deposited in even more oxic conditions than the Lgota Beds, apart from the higher number of rhythms without the spotty layer. The latter pattern resulted from about 10 times higher frequency of turbiditic deposition. The trace fossil assemblage of this unit shows the highest diversity.

Cretaceous anoxic events

The decrease in deep-sea floor oxygenation during sedimentation of the discussed deposits can be related to the widely known Early Cretaceous anoxic events (e.g. Arthur & Schlanger 1979; Jenkyns 1980). Periods of true anoxia are mostly short episodes (Ekdale 1985b), less than one million years in duration (Bralower *et al.* 1994). This is confirmed by the investigations presented here. Only the Verovice Shale contains evidence of longer anoxic periods, as indicated by the presence of thicker non-bioturbated units separated by thin bioturbated horizons. Verovice and Lgota deposition roughly corresponds to the Oceanic Anoxic Event 1 (OAE 1) of Arthur & Schlanger (1979). Longer oxygen deficiency intervals in the deep ocean during the late Barremian–early Aptian, i.e. during the deposition of the Verovice Shale, are a world-wide phenomenon (Bralower *et al.* 1994). Improvement in the oxygenation after the deposition of the Verovice Shale may be related to the late Aptian drop in global temperature (Scott 1995; Price *et al.* 1998), which may have resulted in acceleration of oceanic circulation. The Early Cretaceous anoxic events probably delayed the colonisation of deep-sea floor environments by irregular echinoids producing *Scolicia* (Tchoumatchenco & Uchman 2001). Thus, the trace fossil associations were apparently influenced by world-wide phenomena related to the Cretaceous anoxic events.

General remarks

In the 1970s and beginning of the 1980s, the deep sea was commonly considered as a stable habitat influenced only by long-term probabilistic linear evolutionary processes without rapid changes. This view was formulated as the time-stability hypothesis by Sanders (1968), and was invoked by Seilacher (1976, 1978) and Frey & Seilacher (1980) to explain the evolution of deep-sea communities. Subsequent research proved the deep-sea floor to be influenced by many factors that changed through time, for example, the anoxic events. These factors influenced the deep-sea organisms (e.g. Gage & Tyler 1992), some of which were trace makers. The evolutionary trends in graphoglyptids are an example of the changes that do not fit the time-stability hypothesis (Uchman 2003). The trace fossil associations discussed in this paper are influenced by many factors, such as oxygenation changes, preservation potential, frequency of turbiditic deposition,

and evolutionary changes in trace makers, which cause significant disturbances through time, and influence changes in the trace fossil record from unit to unit.

Conclusions

1. Trace fossil associations from the Lower Cretaceous dark flysch deposits of the Upper Cieszyn Beds, Verovice Shale, and Lgota Beds differ from most trace fossil communities known from Upper Cretaceous and Tertiary flysch successions. Only the trace fossil association of the Upper Cieszyn Shale (Valanginian–Hauterivian) can be ascribed to the *Nereites* ichnofacies. In the other two studied units, graphoglyptids and horizontal meandering pascichnia, the typical components of this ichnofacies, are absent or rare.
2. The composition of the trace fossil communities and the vertical extent of bioturbation in the turbidite–hemipelagite couplets were strongly affected by lowered oxygenation on the deep-sea floor. The lowered oxygenation appears to correspond to global Early Cretaceous anoxic events.
3. The Verovice Shale is dominated by anoxic strata interlayered with rare, thin bioturbated horizons. In the Upper Cieszyn Shale and Lgota Beds, most tops of turbidite–hemipelagite rhythms are bioturbated. Unbioturbated couplets record anoxia. Their frequency is correlated with that of turbiditic deposition.
4. The differences between the Upper Cretaceous and younger trace fossil communities were also influenced by evolutionary processes involving trace makers, such as those that resulted in increasing diversity of the graphoglyptids, and a greater abundance of *Scolicia* and *Nereites irregularis* since the Late Cretaceous.
5. *Protovirgularia obliterata* and *Protovirgularia pennata* from the Verovice Shale and probably the Upper Cieszyn Beds occur in the deepest tier below *Chondrites*. They were probably produced below the redox boundary by chemosymbiotic bivalves.

Acknowledgements

This research was sponsored by grant 6 PO4D 002 19 from the Polish Science Committee (Komitet Badań Naukowych). Additional support was provided by the Jagiellonian University. W. Miller III (California), A. K. Rindsberg (Alabama), A. Wetzel (Basel) and B. D. Webby (Macquarie, Sydney) improved the English and provided helpful comments.

References

Arthur, M.A. & Schlanger, S.O. 1979: Cretaceous 'oceanic anoxic events' as casual factors in development of reef-reservoired giant oil fields. *American Association of Petroleum Geologists, Bulletin 63*, 870–885.

Bąk, K., Bąk, M. & Paul, Z. 2001: Barnasiówka Radiolarian Shale Formation – a new lithostratigraphic unit in the Upper Cenomanian–lowermost Turonian of the Polish Outer Carpathians (Silesian Series). *Annales Societatis Geologorum Poloniae 71*, 75–103.

Berger, W.H., Ekdale, A.A. & Bryant, P.F. 1979: Selective preservation of burrows in deep-sea carbonates. *Marine Geology 32*, 205–230.

Bouma, A.H. 1962: *Sedimentology of Some Flysch Deposits*. Elsevier, Amsterdam.

Bralower, T.J., Arthur, M.A., Leckie, R.M., Sliter, W.V., Allard, D.J. & Schlanger, S.O. 1994: Timing and paleoceanography of oceanic dysoxia/anoxia in the late Barremian to early Aptian (Early Cretaceous). *Palaios 9*, 335–369.

Bromley, R.G. & Ekdale, A.A. 1984: *Chondrites*: a trace fossil indicator of anoxia in sediments. *Science 224*, 872–874.

Bromley, R.G., Jensen, M. & Asgaard, U. 1995: Spatangoid echinoids: deep-tier trace fossils and chemosymbiosis. *Neues Jahrbch für Geologie und Paläontologie, Abhandlungen, 195*, 25–35.

Brongniart, A.T. 1823: Observations sur les Fucoïdes. *Société d'Historie Naturelle de Paris, Mémoire 1*, 301–320.

Brongniart, A.T. 1828: *Histoire des végétaux fossiles ou recherches botaniques et géologiques sur les végétaux renfermés dans les diverses couches du globe, Vol. 1*. G. Dufour & E. d'Ocagne, Paris.

Burtan, J. 1978: *Objaśnieniu do Szczegółowej Mapy Geologicznej Polski. Arkusz Mszana Dolna (1016), 1: 50 000*. Wydawnictwa Geologiczne, Warszawa.

Cieszkowski, M., Gedl, E., Ślączka, A. & Uchman, A. 2001: Stop C2 – Rzyki village. *In* Cieszkowski, M. & Ślączka, A. (eds): *Silesian & Subsilesian Units. 12th Meeting of the Association of European Geological Societies & LXXII Zjazd Polskiego Towarzystwa Geologicznego, Field Trip Guide*, 115–118. Panstwowy Instytut Geologiczny, Kraków.

Crimes, T.P., Goldring, R., Homewood, P., Stuijvenberg, J. & Winkler, W. 1981: Trace fossil assemblages of deep-sea fan deposits, Gurnigel and Schlieren flysch (Cretaceous–Eocene). *Eclogae Geologicae Helvetiae 74*, 953–995.

D'Alessandro, A. & Bromley, R.G. 1987: Meniscate trace fossils and the *Muensteria–Taenidium* problem. *Palaeontology 30*, 743–763.

Dżułyński, S. & Ślączka, A. 1958: Directional structures and sedimentation of the Krosno Beds (Carpathian flysch). *Rocznik Polskiego Towarzystwa Geologiczengo 28*, 205–260.

Eichwald, E. 1860: *Lethaea Rossica ou Paléontologie de la Russie. Décrite et Figurée, Vol. 1*. E. Schweizerbart, Stuttgart. [Plates published in 1868.]

Ekdale, A.A. 1985a: Paleoecology of the marine endobenthos. *Palaeogeography, Palaeoclimatology, Palaeoecology 50*, 63–81.

Ekdale, A.A. 1985b: Trace fossils and mid-Cretaceous anoxic events in the Atlantic Ocean. *Society of Economic Paleontologists and Mineralogists, Special Publication 35*, 333–342.

Ekdale, A.A. 1988: Pitfalls of paleobathymetric interpretation based on trace fossil assemblages. *Palaios 3*, 464–472.

Ekdale, A.A. 1992: Muckraking and mudslinging: the joys of deposit-feeding. *Short Courses in Paleontology 5*, 145–171.

Ekdale, A.A., Bromley, R.G. & Pemberton, G.S. 1984: Ichnology: the use of trace fossils in sedimentology and stratigraphy. *Society of Economic Geologists and Paleontologists, Short Course 15*, 1–317.

Ekdale, A.A. & Mason, T.R. 1988: Characteristic trace-fossil association in oxygen-poor sedimentary environments. *Geology 16*, 720–723.

Fillion, D. & Pickerill, R.K. 1990: Ichnology of the Upper Cambrian? to Lower Ordovician Bell Island and Wabana groups of eastern Newfoundland, Canada. *Palaeontographica Canadiana 7*, 1–119.

Fischer-Ooster, C. 1858: *Die fossilen Fucoiden der Schweizer-Alpen, nebst Erörterungen über deren geologisches Alter*. Huber, Bern.

Frey, R.W. & Seilacher, A. 1980: Uniformity in marine invertebrate ichnology. *Lethaia 13*, 183–207.

Fu, S. 1991: Funktion, Verhalten und Einteilung fucoider und lophoctenoider Lebensspuren. *Courier Forschung. Institut Senckenberg 135*, 1–79.

Gage, J.D. & Tyler, P.A. 1992: *Deep-sea Biology: a Natural History of Organisms at the Deep-sea Floor*. Cambridge University Press, Cambridge.

Gedl, E. 2001: Late Cretaceous dinocysts of the siliceous limestones from Rzyki (Silesian Nappe, western Outer Carpathians, Poland). *Biuletyn Państwowego Instytutu Geologicznego 396*, 48–49.

Geroch, S. & Nowak, W. 1963: Profil dolnej kredy śląskiej w Lipniku kolo Bielska. *Rocznik Polskiego Towarzystwa Geologicznego 32*, 241–264.

Gucwa, I. & Wieser, T. 1980: Geochemia i mineralogia skal osadowych fliszu karpackiego zasobnych w materię organiczną. *Polska Akademia Nauk – Oddział w Karkowie, Prace Mineralogiczne 69*, 3–43.

Han, Y. & Pickerill, R.K. 1995: Taxonomic review of the ichnogenus *Helminthopsis* Heer 1877 with a statistical analysis of selected ichnospecies. *Ichnos 4*, 83–118.

Heer, O. 1877: *Flora Fossilis Helvetiae. Vorweltliche Flora der Schweiz*. J. Wurster & Co., Zürich.

Jaminski, J. 1995: The mid-Cretaceous palaeoenvironmental conditions in the Polish Carpathians – a palynological approach. *Review of Palaeobotany and Palynology 87*, 43–50.

Jenkyns, H.C. 1980: Cretaceous anoxic events: from continents to oceans. *Journal of the Geological Society of London 137*, 171–188.

Keighley, D.G. & Pickerill, R.K. 1995: The ichnotaxa *Palaeophycus* and *Planolites*: historical perspectives and recommendations. *Ichnos 3*, 301–309.

Książkiewicz, M. 1951: *Objaśnienia do Arkusza Wadowice*. Państwowy Instytut Geologiczny, Warszwa.

Książkiewicz, M. 1968: O niektórych probleatykach z fliszu Karpat polskich, Część III [On the problematic organic traces from the flysch of the Polish Carpathians, Part 3]. *Rocznik Polskiego Towarzystwa Geologicznego 38*, 3–17.

Książkiewicz, M. 1970: Observations on the ichnofauna of the Polish Carpathians. *Geological Journal Special Issue 3*, 288–322.

Książkiewicz, M. 1975: Bathymetry of the Carpathian Flysch basin. *Acta Geologica Polonica 25*, 309–367.

Książkiewicz, M. 1977: Trace fossils in the Flysch of the Polish Carpathians. *Palaeontologia Polonica 36*, 1–208.

Leszczyński, S. 1991: Oxygen-related controls on predepositional ichnofacies in turbidites, Guipúzcoan flysch (Albian–lower Eocene), northern Spain. *Palaios 6*, 271–280.

Leszczyński, S. 1993: Ichnocoenosis versus sediment colour in upper Albian to lower Eocene turbidites, Guipúzcoan province, northern Spain. *Palaeogeography, Palaeoclimatology, Palaeoecology 100*, 251–265.

Leszczyński, S. & Seilacher, A. 1991: Ichnocoenoses of a turbidite sole. *Ichnos 1*, 293–303.

Mángano, G.M., Buatois, L., West, R.R. & Maples, C.G. 2002: Ichnology of a Pennsylvanian equatorial tidal flat – the Stull Shale Member at Waverly, eastern Kansas. *Kansas Geological Survey, Bulletin 245*, 1–133.

Nowak, W. 1957: Kilka hieroglifów gwiaździstych z zewnętrznych Karpat fliszowych [Quelques hiéroglyphes étoilés des Karpates de

Flysch extérieures]. *Rocznik Polskiego Towarzystwa Geologicznego 26*, 187–224.

Nowak, W. 1959: *Palaeodictyum* w Karpatach fliszowych [*Palaeodictyum* in the Flysch Carpathians]. *Kwartalnik Geologiczny 3*, 103–125.

Nowak, W. 1961: Z badań nad hieroglifami fliszu karpackiego. I. Niektóre hieroglify z warstw cieszyńskich i grodziskich. *Sprawozdanie z Posiedzeń Komisji Nauk, Oddz. Polskiej Akademii Nauk w Krakowie*, 226–228.

Nowak, W. 1962: Z badań nad hieroglifami fliszu karpackiego. Niektóre hieroglify warstw godulskich, lgockich i wierzowskich. *Sprawozdanie z Posiedzeń Komisji Nauk, Oddzialu Polskiej Akademii Nauk w Krakowie, Styczeń-Czerwiec 1962*, 263–264.

Nowak, W. 1968: Stomiosphaerids of the Cieszyn Beds (Kimmeridgian–Hauterivian) in the Polish Silesia and their stratigraphical value. Summary. *Annales Societatis Geologorum Poloniae 38*, 275–328.

Nowak, W. 1970: Spostrzeżenia nad problematykami *Belorhaphe* i *Sinusites* z dolnokredowego i paleogeńskiego fliszu Karpat Polskich [Problematical organic traces of *Belorhaphe* and *Sinusites* in the Carpathian Lower Cretaceous and Paleogene flysch deposits] *Kwartalnik Geologiczny 14*, 149–162.

Pemberton, G.S. & Frey, R.W. 1982: Trace fossil nomenclature and the *Planolites–Palaeophycus* dilemma. *Journal of Paleontology 56*, 843–881.

Pickerill, R.K. & Peel, J.S. 1991: *Gordia nodosa* isp. n. and other trace fossils from the Cass Fjord Formation (Cambrian) of north Greenland. *Grønlands Geologiske Undersøgelse, Rapport 150*, 15–28.

Pickering, K., Stow, D., Watson, M. & Hiscott, R. 1986: Deep-water facies, processes and models: a review and classification scheme for modern and ancient sediments. *Earth-Science Review 23*, 75–174.

Powell, E.N., Callendar, W.R. & Stanton, R.J., Jr. 1998: Can shallow- and deep-water chemoautotrophic and heterotrophic communities be discriminated in fossil record? *Palaeogeography, Palaeoclimatology, Palaeoecology 144*, 85–114.

Price, G.D., Valdes, P.J. & Sellwood, B.W. 1998: A comparison of GCM simulated Cretaceous 'greenhouse' and 'icehouse' climates: implications for the sedimentary record. *Palaeogeography, Palaeoclimatology, Palaeoecology 142*, 123–138.

Ringsberg, A.K. & Martin, A.J. 2003: *Arthophycus* in the Silurian of Alabama (USA) and the problem of compound trace fossils. *Palaeogeography, Palaeoclimatology, Palaeoecology 192*, 187–219.

Sanders, H.L. 1968: Marine benthic diversity: a comparative study. *American Naturalist 100*, 243–282.

Savi, P. & Meneghini, G.G. 1850: Osservazioni stratigrafiche e paleontologische concernati la geologia della Toscana e dei paesi limitrofi. – Appendix. *In* Murchison, R.I. (ed.): *Memoria sulla struttura geologica delle Alpi, degli Apennini e dei Carpazi*, 246–528. Stamperia granducale, Firenze.

Savrda, E.C. & Bottjer, J.D. 1989: Trace-fossil model for reconstructing oxygenation histories of ancient marine bottom waters: application to Upper Cretaceous Niobrara Formation, Colorado. *Palaeogeography, Paleoclimatology, Palaeoecology 74*, 49–74.

Scott, R.W. 1995: Global environmental controls on Cretaceous reefal ecosystems. *Palaeogeography, Palaeoclimatology, Palaeoecology 119*, 187–199.

Seilacher, A. 1967: Bathymetry of trace fossils. *Marine Geology 5*, 413–428.

Seilacher, A. 1976: Evolution von Spuren-Vergesellschaftungen. *Zentralblatt für Geologie und Paläontologie, Teil II 1976*, 396–402.

Seilacher, A. 1978: Evolution of trace fossil communities in the deep sea. *Neues Jahrbuch für Geologie und Paläontologie, Abhandlungen 157*, 251–255.

Seilacher, A. 1990: Aberrations in bivalve evolution related to photo- and chemosymbiosis. *Historical Biology 3*, 289–311.

Seilacher, A. 2000: Ordovician and Silurian arthrophycid ichnostratigraphy. *In* Sola, M.AS. & Worsley, D. (eds.), *Geological Exploration in Murzuq Basin*, 237–258. Elsevier, Amsterdam.

Seilacher, A. & Seilacher, E. 1994: Bivalvian trace fossils: a lesson from actuopaleontology. *Courier Forschungsinstitut Senckenberg 169*, 5–15.

Ślączka, A. & Kaminski, M.A. 1998: A guidebook to excursions in the Polish Flysch Carpathians. *Grzybowski Foundation Special Publication 6*, 1–171.

Szydło, A. 1996: Charakterystyka mikropaleontologiczna utworów dolnej kredy z rejonu Bielska-Białej i Żywca. *Posiedzenia Naukowe Państwowego Instytutu Geologicznego 52*, 67–68.

Szydło, A. 1997: Biostratigraphical and palaeoecological significance of small foraminiferal assemblages of the Silesian (Cieszyn) Unit, Western Carpathians. *Annales Societatis Geologorum Poloniae 64*, 325–337.

Tchoumatchenco, P. & Uchman, A. 2001: The oldest deep-sea *Ophiomorpha* and *Scolicia* and associated trace fossils from the Upper Jurassic–Lower Cretaceous deep-water turbidite deposits of SW Bulgaria. *Palaeogeography, Palaeoclimatology, Palaeoecology 169*, 85–99.

Tunis, G. & Uchman, A. 1996a: Trace fossil and facies changes in the Upper Cretaceous–Middle Eocene flysch deposits of the Julian Prealps (Italy and Slovenia): consequences of regional and world-wide changes. *Ichnos 4*, 169–190.

Tunis, G. & Uchman, A. 1996b: Ichnology of the Eocene flysch deposits in the Istria peninsula, Croatia and Slovenia. *Ichnos 5*, 1–22.

Uchman, A. 1991: Trace fossils from stress environments in Cretaceous–Paleogene flysch of the Polish Outer Carpathians. *Annales Societatis Geologorum Poloniae 6*, 207–220.

Uchman, A. 1992: An opportunistic trace fossil assemblage from the flysch of the Inoceramian beds (Campanian–Palaeocene), Bystrica Zone of the Magura Nappe, Carpathians, Poland. *Cretaceous Research 13*, 539–547.

Uchman, A. 1995: Taxonomy and palaeoecology of flysch trace fossils: the Marnoso-arenacea Formation and associated facies (Miocene, Northern Apennines, Italy). *Beringeria 15*, 3–115.

Uchman, A. 1998: Taxonomy and ethology of flysch trace fossils: a revision of the Marian Ksiazkiewicz collection and studies of complementary material. *Annales Societatis Geologorum Poloniae 68*, 105–218.

Uchman, A. 1999: Ichnology of the Rhenodanubian Flysch (Lower Cretaceous–Eocene) in Austria and Germany. *Beringeria 25*, 65–171.

Uchman, A. 2001: Eocene flysch trace fossils from the Hecho Group of the Pyrenees, northern Spain. *Beringeria 28*, 3–41.

Uchman, A. 2003: Trends in diversity, frequency and complexity of graphoglyptid trace fossils: evolutionary and palaeoenvironmental aspects. *Palaeogeography, Palaeoclimatology, Palaeoecology 192*, 123–142.

Uchman, A. 2004: Phanerozoic history of deep-sea trace fossils. *Geological Society, London, Special Publication 228*, 125–139.

Unrug, R. 1959: Spostrzezenia nad sedymentacją warstw lgockich [On the sedimentation of the Lgota beds (Bielsko area, Western Carpathians)]. *Rocznik Polskiego Towarzystwa Geologicznego 29*, 197–225.

Unrug, R. 1977: Ancient deep-sea traction current deposits in the Lgota beds (Albian) of the Carpathian Flysch. *Rocznik Polskiego Towarzystwa Geologicznego 47*, 355–370.

Wetzel, A. 1991: Ecologic interpretation of deep-sea trace fossil communities. *Palaeogeography, Palaeoclimatology, Palaeoecology 85*, 47–69.

Wetzel, A. 2002: Modern *Nereites* in the South China Sea – ecological association with redox conditions in the sediment. *Palaios 17*, 507–515.

Wetzel, A. & Bromley, R.G. 1994: *Phycosiphon incertum* revisited: *Anconichnus horizontalis* is its junior subjective synonym. *Journal of Paleontology 68*, 1396–1402.

Wetzel, A. & Bromley, R.G. 1996: A re-evaluation of ichnogenus *Helminthopsis* Heer 1877 – new look at the type material. *Palaeontology 39*, 1–19.

Wetzel, A. & Uchman, A. 1998: Trophic level in the deep-sea recorded by ichnofabrics: an example from Palaeogene flysch in the Carpathians. *Palaios 13*, 533–546.

Wetzel, A. & Uchman, A. 2001: Sequential colonization of muddy turbidites: examples from Eocene Belove•a Formation, Carpathians, Poland. *Palaeogeography, Palaeoclimatology, Palaeoecology 168*, 171–186.

Trace fossils from the Triassic–Jurassic deep water, oceanic radiolarian chert successions of Japan

YOSHITAKA KAKUWA

Kakuwa, Y. **2004 10 25**: Trace fossils from the Triassic–Jurassic deep water, oceanic radiolarian chert successions of Japan. *Fossils and Strata*, No. 51, pp. 58–67. Japan. ISSN 0300-9491.

Trace fossils are reported from the Triassic to Jurassic radiolarian chert successions exposed in central Japan, deposited in a deep oceanic setting. Spathian and Anisian siliceous claystones and cherts with carbonaceous claystone interbeds yield numerous burrows such as zigzag-shaped, irregular-shaped forms other than ?*Chondrites* and *Planolites*, while Griesbachian to Smithian rocks are rarely bioturbated due to poor oxygenation of the oceanic bottom. Bioturbated structures are rarely encountered in radiolarian cherts of Middle Triassic to Middle Jurassic age, despite a well-oxygenated environment. The types of burrow and the degree of bioturbation in the radiolarian mudstones increase from the Middle to Upper Jurassic owing to the increase in food supply for the burrowers. Upper Jurassic black sandy mudstones, on the other hand, contain few burrows, probably due to the high sedimentation rate. The high frequency of bioturbation and the emergence of peculiar burrows such as the zigzag-shaped forms in the Spathian claystones may reflect the anomalous environment after the Permian–Triassic mass extinction event.

Key words: Burrows; radiolarian chert; Triassic; Jurassic; Japan; mass extinction.

Yoshitaka Kakuwa [kakuwa@chianti.c.u-tokyo.ac.jp], Graduate School of Arts and Sciences, University of Tokyo, Komaba, Tokyo 153-0041, Japan

Introduction

Studies of trace fossils in deep-sea deposits are concentrated mainly in turbidite successions that accumulated near continents (e.g. Bottjer *et al.* 1988; Wetzel & Uchman 1997), with the exception of cores from the Pacific, Atlantic and other oceans (e.g. Chamberlain 1975; Ekdale 1978; Wetzel 1983). The record from deep-sea cores extends back to the Middle Jurassic. Older oceanic rocks have a less complete record because they have been either subducted into the mantle or partly accreted to continents making the mountain belts.

Radiolarian bedded cherts exposed in orogenic belts are considered to be one of the typical accreted sedimentary rock types of the deep oceanic environment. Previous reports of trace fossils from radiolarian chert successions are, however, limited mainly to occurrences in the Upper Jurassic radiolarites of the circum-Alpine orogenic belts (Folk & McBride 1978; McBride & Folk 1979; Barret 1982; Baltuck 1983; Vecsei *et al.* 1989). Also, McBride & Thomson (1970), Webb (1986), Dec *et al.* (1992) and Bruchert *et al.* (1994) reported the presence of bioturbation in Palaeozoic cherts. Thus, trace fossils from established deep-sea deposits of pre-Middle Jurassic age remain mostly unknown. The main part of the Japanese Islands is comprised of accretionary complexes of different ages, and radiolarian cherts are widely exposed. This paper is a preliminary report on the Triassic and Jurassic trace fossils found in the radiolarian chert sequences of the Jurassic accretionary complexes in central Japan.

Geological setting

Jurassic accretionary complexes of central Japan

The Jurassic accretionary complexes are distributed in narrow belts extending east-northeast to west-southwest across central Japan. Two main belts are recognised: the Ashio–Mino–Tamba Belt and the Chichibu Belt (Fig. 1). The Chichibu Belt is further subdivided into

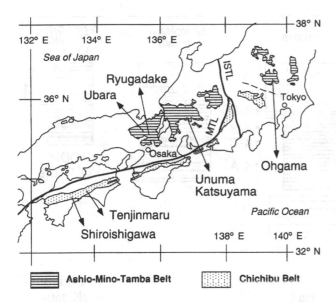

Fig. 1. Index map showing the locations of the key Carboniferous to Jurassic sections studied in the two main belts of central Japan. ISTL = Itoigawa–Shizuoka Tectonic Line; MTL = Median Tectonic Line.

three: Northern, Middle and Southern. The Upper Carboniferous to Middle Jurassic radiolarian bedded chert successions are underlain by basalt and overlain by thick terrigenous clastic rocks. The bedded chert successions occur as segmented blocks in a muddy matrix and/or in thrusted sheets within both belts (Isozaki 1996) (Fig. 2A).

The radiolarian bedded cherts are commonly composed of 1–15 cm thick chert beds, comprised mainly of radiolarians, with less than 1 cm thick shale partings. The radiolarian bedded cherts in Jurassic accretionary complexes were continuously deposited about 160 million years ago, without influx of coarse terrigenous materials, below the carbonate compensation depth (CCD), and are interpreted as pelagic deep ocean deposits of Panthalassa (Matsuda & Isozaki 1991). The continuity of deposition of the radiolarian bedded cherts was interrupted by black carbonaceous claystone and grey siliceous claystone, with rare radiolarian skeletons between the end-Permian mass extinction event and the full recovery of radiolarians in the Anisian (Kakuwa 1996) (Fig. 2B). The gradation from bedded radiolarian cherts to radiolarian mudstones associated with acidic tuffs, siltstones, and finally sandstones, with the increase in terrigenous muds and coarse clastic materials reflects the increased proximity to a continental source (Fig. 2A). Similar sequences of the chert–clastic association occur repeatedly in the same belt, but the age of the lithological change from chert to mudstone varies from Middle to Late Jurassic. This difference in the timing of the lithological change reflects differences in the timing of accretion events (Matsuoka 1984, 1992; Isozaki 1996, 1997).

Description of key sections

Upper Permian to Lower Triassic chert and argillaceous rocks were examined in the Ubara section (35°11'07"N, 135°15'14"E) of the Tamba Belt and in the Tenjinmaru section (33°52'10"N, 134°10'49"E) of the Northern Chichibu Belt. The 1 m thick grey radiolarian chert grades upwards into a 1 m thick grey siliceous claystone, and then to a more than 0.8 m thick unit of black carbonaceous claystone in the Ubara section. The grey chert and grey siliceous claystone is Changhsingian in age, and the black carbonaceous claystone is Griesbachian in age (Fig. 2B) (Kuwahara *et al.* 1991; Yamakita *et al.* 1999). The Tenjinmaru section is comprised of Upper Permian grey chert, then about 2 m of black carbonaceous claystone and followed by a 1 m thick grey siliceous claystone. The black carbonaceous claystone is Griesbachian to lower Smithian age, and the grey siliceous claystone is Smithian in age (Fig. 2B) (Yamakita 1993; Yamakita & Kadota 2001).

Characteristic grey siliceous claystone of Smithian to Spathian age is widely distributed in Japan (Koike 1979; Kakuwa 1994). The siliceous claystone is mainly composed of fine-grained illite without coarse terrigenous clastics and tuffaceous material (Kakuwa 1991). Grey siliceous claystone with black carbonaceous claystone and radiolarian chert interbeds was examined in the 11 m thick Ryugadake section of the Tamba Belt (35°4'45"N, 135°37'53"E) and in the 4 m thick Ohgama section of the Ashio Belt (Fig. 2B; 36°28'31"N, 139°34'10"E). In the latter section, the deposits are Spathian in age (Maeda *et al.* 2000).

The Lower Triassic to Upper Jurassic chert–clastic succession is continuously exposed along the Kiso River near Inuyama City, representing a classical locality for Triassic and Jurassic radiolarian biostratigraphy in Japan. The samples examined are from the 5 m thick Unuma section (Fig. 2B), which corresponds to the bottom of the CH2 section of Yao *et al.* (1980) and outcrop number Un1 of Yao & Kuwahara (1997; 35°23'45"N, 136°56'49"E). The grey bedded chert changes colour up-section to red, and the age is late Anisian to early Ladinian based on the presence of *Triassocampe cornata* (Yao & Kuwahara 1997). Also examined were grey to red Anisian cherts from the 13 m thick Katsuyama section (Fig. 2B), which corresponds to Section M of Sugiyama (1997) along the Kiso River, northeast of the Unuma section (35°25'9"N, 136°58'16"E).

The Unuma CH3 section in the Inuyama area consists of about a 100 m thickness of Triassic to Early Jurassic variegated chert, 20 m of Bajocian to Bathonian dark red radiolarian mudstone, 10 m of Bathonian to Callovian dark green grey silty radiolarian mudstone, and over 50 m of black sandy mudstone and sandstone (Fig. 2C; 35°24'47"N, 136°57'56"E) (Matsuoka *et al.* 1994). The

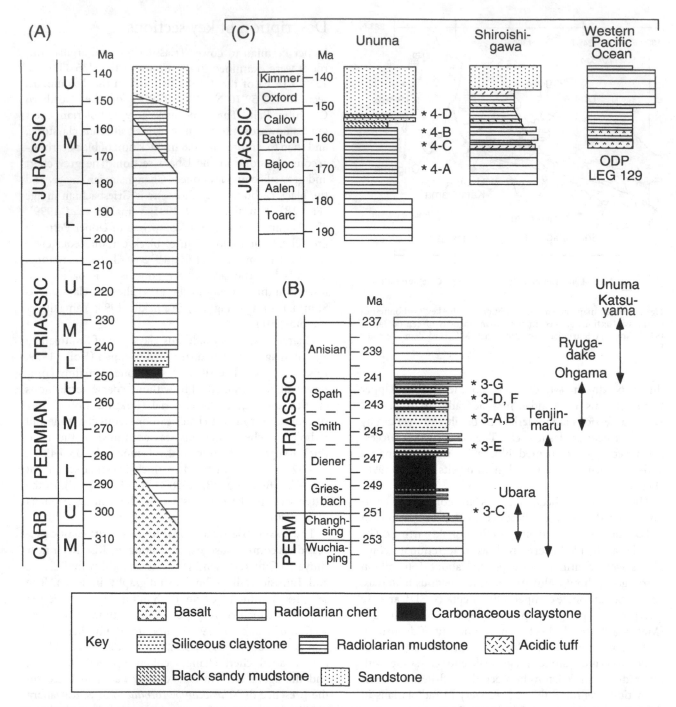

Fig. 2. A: Schematic lithological column and calibrated time scale for the chert succession in the Jurassic accretionary complex of central Japan. B: Composite lithological column and calibrated time scale for the chert succession in the Late Permian to Early Triassic of central Japan, and showing the span of the main sampled sections (individual stratigraphic ranges indicated by the length of double arrows). C: Lithological columns and calibrated time scale for the Unuma and Shiroishigawa sections in central Japan, and the core samples for the ODP LEG 129 in the western Pacific. The asterisks accompanying 3-A to 3-G and 4-A to 4-D on the right side of the columns in (B) and (C) identify the sampled horizons in Figs. 3 and 4.

dark red radiolarian mudstone is associated with rare coarse terrigenous clastics and some tuffaceous material. The content of terrigenous clastic material increases upwards, and thin beds of graded sandstone with parallel and cross-lamination commonly occur in the black sandy mudstone interval. These are then overlain by massive

or cross-laminated, thick-bedded sandstone with water escape structures.

A similar chert–clastic rock succession is recorded in the Southern Chichibu Belt (Fig. 1). The section examined at Shiroishigawa is composed of an 18 m thick Bajocian to Bathonian red chert, a 35 m thick Bathonian to Oxfordian greenish grey radiolarian mudstone and siltstone unit with frequent acidic tuff beds, and a massive sandstone unit at the top (Fig. 2C; 33°29'29"N, 133°7'51"E) (Matsuoka 1983).

Methods

Trace fossils in the cherts and the siliceous rocks are usually invisible in the field, either to the naked eye or through a hand lens. An etching method using hydrofluoric acid was applied, like that commonly used by radiolarian biostratigraphers to extract radiolarian fossils. Samples were cut, roughly polished and immersed in diluted hydrofluoric acid (about 5%) for less than 1 day. They were then rinsed with water, dried and observed under a stereoscopic microscope.

Images were taken with a digital camera, and processed by image analysis software to make it easier to recognise small or compacted trace fossils by enlargement and decompaction. The compaction rate of cherts is estimated to be around 25% (Iijima *et al.* 1989; Tada 1991) (Fig. 3A, B).

Description of trace fossils

Lower Triassic claystone and chert

Bioturbation structures are found in a grey siliceous claystone bed at the Permian–Triassic boundary in the Ubara section (Fig. 3C). This grey siliceous claystone occurs in the basal part of the laminated black carbonaceous claystone. The bioturbation at the Permian–Triassic boundary has also been observed in rocks of the Tenjinmaru section (Kakuwa 1996) and in another section of the Ashio–Mino–Tamba Belt.

Rare, small, sub-millimetre-sized, trace fossils are observed in the Griesbachian and Dienerian black carbonaceous claystone of the Tenjinmaru section. The trace fossils are unlined, elliptical in shape, elongated parallel or subparallel to the bedding plane, and filled with organic-rich matter without internal texture. They appear like cross-sections of *Chondrites*, but the branching system that characterises the *Chondrites* ichnogenus has not been found. Larger and better defined bioturbation structures are represented in the uppermost Dienerian and Smithian grey siliceous claystones and cherts with an interbedded black carbonaceous claystone bed in the Tenjinmaru section. They comprise two obliquely aligned, crosscutting, tubular, unlined, burrows in a dark grey chert bed (Fig. 3E). The burrows are filled with black, organic-rich matter with no internal texture. Radiolarians are sparse in the surrounding matrix. The black organic matter in the burrows was probably reworked from the overlying carbonaceous claystone bed.

Trace fossils are distinct in the Smithian to Spathian grey siliceous claystone-dominated Ryugadake and Ohgama sections, and in the Anisian chert-dominated Unuma and Katsuyama sections. The irregular concentration of radiolarians in the discontinuous lamination of Smithian siliceous claystone in the Ryugadake section is evidently the result of bioturbation (Fig. 3A, B). Zigzag-shaped (Fig. 3D), S-shaped (Fig. 3F), straight, curved and irregularly shaped trace fossils (Fig. 3G) in addition to *Chondrites*, *Planolites* and *Teichichnus* are recognised in the siliceous claystone and chert associated with the carbonaceous claystone interbeds. The zigzag-shaped burrow is filled with black organic-rich matter, unlined, with no internal texture and unbranched. The burrow can be traced about 7 cm laterally and 3 cm in a vertical direction. The burrow tunnel meanders obliquely with respect to the bedding plane. The burrowing depth is estimated to have been more than 12 cm prior to compaction. The diameter of the burrows is most commonly less than 2 mm. The size and variety of burrows and the degree of bioturbation increases upwards into the Spathian (Kakuwa 2000).

Bioturbation is much more clearly defined where the grey siliceous claystone is overlain by the black carbonaceous claystone than where the grey siliceous claystone is succeeded by grey shale. This is due partly to the colour contrast between the grey host rock and the trace fossils defined by the black carbonaceous material they derived from the overlying bed in the course of their burrowing activities. The large burrowers tend to develop through the interval associated with the sedimentation of organic-rich, black carbonaceous claystones.

Hemipelagic rocks of the Middle to Upper Jurassic

Four types of trace fossil are recognised in the radiolarian mudstones from the Unuma section of Bajocian to Bathonian age (Fig. 4). The first type consists of sharply delineated and segmented traces of about 0.1 mm in diameter and more than 0.3–0.4 mm in length on the slab surface (type 1, Fig. 4A, C). They are composed of less siliceous, finer, darker material, and lack radiolarian skeletons. They also tend to have a patchy distribution. These trace fossils possibly belong to *Phycosiphon* or *Chondrites*. The second type is similar, but larger, and shows no clear tendency to form concentrations (type 2, Fig. 4A). This second type resembles a variety of discontinuous lamination, and may be correlated to the "woody texture" reported from the Middle and Upper Jurassic radiolarites of the Pacific Ocean (Behl *et al.* 1992; Ogg *et al.* 1992). The third type of trace also resembles the first and second types, but is larger and aligned more obliquely to the bedding plane (type 3, Fig. 4A, C). This third type

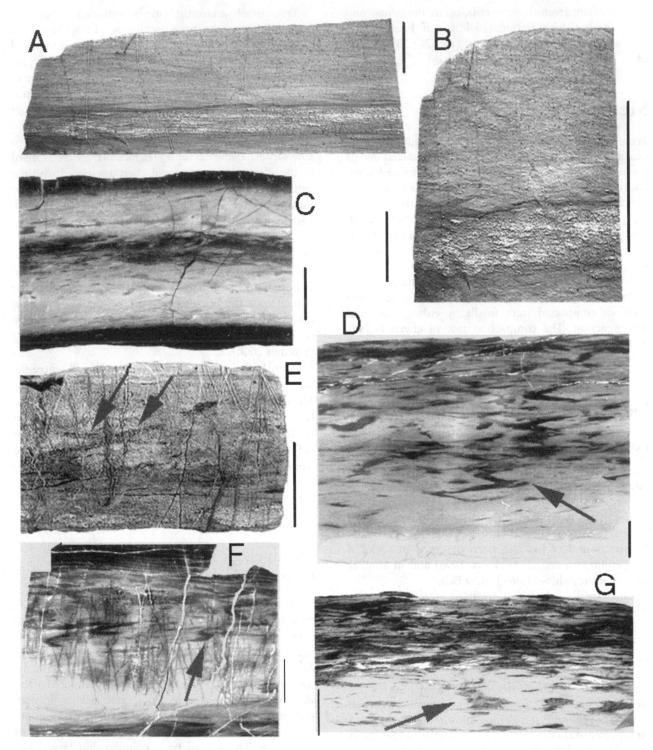

Fig. 3.   Trace fossils and bioturbation structures from the Lower Triassic siliceous claystone and chert in Ubara, Tenjinmaru, Ohgama and Ryugadake sections of central Japan. A: Smithian siliceous claystone of the Ryugadake section showing discontinuous lamination. B: Decompacted (and enlarged) image of part of (A) showing that the discontinuous lamination of radiolarians is a bioturbated structure. C: Bioturbated siliceous claystone interbedded in carbonaceous claystone of the basal Griesbachian, Ubara section. D: Zigzag-shaped burrow (arrow) in the Spathian siliceous claystone of the Ryugadake section. E: Two linear burrow tunnels (arrows) running obliquely to the bedding plane and cross-cutting each other in dark grey chert (Dienerian age) of the Tenjinmaru section. F: S-shaped burrow (arrow) in the Spathian radiolarian chert of the Ohgama section. G: Irregular-shaped burrow (arrow) in the Spathian siliceous claystone of the Ryugadake section. Scale bars = 1 cm.

Fig. 4. Trace fossils and bioturbation structures from the Middle to Upper Jurassic radiolarian mudstones of the Unuma section, central Japan. A: Small *Chondrites*-type burrows in red radiolarian mudstone of Bajocian age. B: Large burrows in green grey radiolarian mudstone of Bathonian age. C: *Chondrites*- and *Planolites*-type burrows in green grey radiolarian mudstone of Bathonian age. D: *Phycosiphon* isp. is the diagnostic trace fossil in black sandy mudstone of Callovian age. Note the numerical labels for arrows in (A) and (C) that identify the four types of trace fossil described in the text. Scale bars = 1 cm.

is 0.3 mm in diameter and 2 mm in length on the slab surface. The fourth type is ellipsoidal in shape and 0.7 mm in diameter, 2–4 mm in length (type 4, Fig. 4C). The trace fossil is slightly darker than the host rock and lacks a clearly defined wall; it possibly represents a *Planolites*.

The four types described contain few, if any, radiolarian skeletons. Type 2 traces are the most common and have been reported from grey green radiolarian mudstones in another section of the same age (Kakuwa 1991). Microphotographs of the trace fossils show that they are composed of finer clayey materials and framboidal pyrite that lack associated radiolarian skeletons (Kakuwa 1991).

The trace fossils in the upper Bajocian to Bathonian green grey radiolarian mudstones (Fig. 4C) are larger than those in the Bajocian dark red radiolarian mudstones (Fig. 4A). Still larger are the traces in the upper Bathonian to Callovian radiolarian mudstones (Fig. 4B). These are 2–3 mm in diameter and 18 mm in length on the slab surface. This establishes a pattern of size and abundance increase in trace fossils upwards through the Bajocian to lower Callovian succession, as represented in both the Unuma and Shiroishigawa sections.

The black sandy mudstones overlying the radiolarian mudstones contain few radiolarians and trace fossils. The only trace fossil found is filled with black fine-grained organic-rich material with a lighter mantle (Fig. 4D). It is 0.8 mm in diameter and has a maximum length of 9 mm. A lighter mantle is diagnostic of *Phycosiphon* isp. (e.g. Wetzel & Bromley 1994). This occurs concentrated in a few horizons and is limited to the rocks with fine sandy lamination. The decline to one ichnofossil type in these sandy mudstones indicates a lowering of trace fossil diversity as compared with lower levels of the hemipelagic Middle Jurassic succession.

Trace fossils in cherts

Bioturbation structures occur in the Spathian cherts and in some of the Smithian cherts, but they are barely visible in the dark grey cherts of the Upper Permian or in the red cherts of the Middle Triassic to Upper Jurassic. Rare, small *Chondrites*-like trace fossils are found in red radiolarian cherts of the Shiroishigawa section that correlates with the radiolarian mudstones of the Unuma section. The latter have the most varied trace fossil associations.

Discussion

Anomalous trace fossils in the Lower Triassic

The zigzag-shaped trace fossil in the Lower Triassic claystones and cherts is not found in the Jurassic

radiolarian mudstones. Sedimentation rate is believed to have had an influence on trace fossils (e.g. Wetzel 1991), but it is difficult to calculate the rates in accreted sedimentary rocks because of the many associated faults. Sedimentation rates of the Lower Triassic and Jurassic argillaceous rocks are calculated as follows, based on estimates of thickness, and absolute ages with ties to biostratigraphic zonations. The sedimentation rate for the Smithian to Spathian siliceous claystones of the Ryugadake section is estimated to be a little higher than 2.2 mm per thousand years given that the thickness is 11 m and the duration is slightly less than 5 million years (Smithian–Spathian inclusive; Ross *et al.* 1994). The sedimentation rate of the Middle Jurassic radiolarian mudstone is estimated to be somewhat higher than 2.8 mm per thousand years based on a thickness of 42 m and a duration of a little less than 15 million years (Matsuoka 1995). Thus, the estimated sedimentation rates of both successions of argillaceous rocks are approximately the same. Consequently, the differences in the trace fossil composition are not likely to be influenced markedly by the sedimentation rates.

Organic-rich claystones prevail in the Lower Triassic chert succession. The large zigzag-shaped trace fossil occurs only in the siliceous claystones associated with carbonaceous claystones. Deposition of carbonaceous material, on the other hand, is much reduced in the Middle and Upper Jurassic radiolarian mudstone. The abundance of organic matter as a source of food in the Lower Triassic claystones probably influenced the development of the zigzag-shaped burrowers. Indeed, it probably stimulated the appearance of new benthic animals, and the increase in their size and levels of bioturbation in the deep-sea environment, as well as oxygenation (e.g. Ekdale 1978; Wetzel 1991).

Bioturbation is most distinctive in the Lower Triassic cherts, whereas it is more obscure in Middle Triassic to Upper Jurassic cherts. Bioturbation of Permian radiolarian cherts is rarely found based on a rough examination of the rocks. Recrystallisation, pervasive diagenesis including intensive cementation and formation of numerous quartz veins, and tectonic deformation obscure the trace fossils of radiolarian cherts. However, diagenesis and tectonic deformation also affected the Lower Triassic cherts. The burrow-bearing Lower Triassic cherts are grey and black in colour and associated with organic-rich shales, while the Middle Triassic to Upper Jurassic cherts are red or green and generally lack organic-rich shales. The total organic carbon percentages of the two Lower Triassic grey cherts are 0.3 and 0.7%, while the three Middle to Upper Triassic red chert samples have much smaller amounts of 0.05, 0.07 and 0.12%. Consequently, the abundant food supply may have been a most important factor in allowing the trace makers to develop large and conspicuous burrows in the Lower Triassic cherts.

The environmental perturbation of the Panthalassan Ocean following the mass extinction event of the Permian–Triassic boundary continued throughout the Early Triassic with anomalous organic-rich sediments being deposited on the oceanic bottom. Oxygen levels decreased to almost anoxic during the Griesbachian to Dienerian time and most of the benthic animals disappeared from the oceanic bottom, but then the oxygen levels recovered gradually by the Smithian to Spathian times (Kakuwa 2000), with anomalous benthic animals flourishing temporarily due to the supply of high organic matter during the recovery interval. Similar observations of secular changes in oxygenation with trace fossils are reported from the shallow-marine environments after the Permian–Triassic boundary event (Twichett & Wignall 1996; Twichett 1999).

Stratigraphic variations in bioturbation during the Jurassic

The diversity and size of trace fossils increased upwards through the Jurassic sections studied, from occurrences in the Lower Jurassic cherts to the Middle Jurassic radiolarian mudstones, and to the Upper Jurassic radiolarian mudstones. However, small ?*Chondrites* are rare in the Middle Jurassic red cherts, whereas larger and much more abundant trace fossils are common in the contemporary radiolarian mudstones. This suggests that occurrences of trace fossils depended more on lithological changes than on age relationships. The abundance of radiolarian skeletons, for example, decreased upwards through sections in accordance with an increase in content of terrigenous siliciclastics supplied from the continents, as evidenced by the increase in Al_2O_3 concentration (Kakuwa 1994). However, the increase in terrigenous clayey materials may not be the controlling factor for trace-making benthic animals to flourish. Probably, the influx of organic matter along with fine siliciclastics from land areas increased as the oceanic plate approached a land area. Then the activity of benthic animals was enhanced, so the diversity of trace fossils also increased.

The degree of bioturbation and the ichnodiversity decreased in the black sandy mudstones that overlie the radiolarian mudstones. The black colour reflects the higher content of organic matter compared with the red, green and grey radiolarian mudstones. Consequently, the availability of food was probably not the controlling factor in this particular case. The high sedimentation rate, which is implied by the cross-lamination and graded bedding in the sandy mudstones, may have limited the extent of colonisation by the various trace makers.

Comparison with deep-sea cores

The oldest sedimentary rocks recovered from the deep-sea drilling cores are Middle Jurassic radiolarites and cherts from the western Pacific Ocean (Shipboard Scientific Party 1990). The radiolarites include radiolarian mudstones, cherts and porcelanites, but no radiolarian bedded cherts such as those exposed in central Japan (Shipboard Scientific Party 1990).

Mottling is reported from the silicified metalliferous clayey radiolarites of the upper Bathonian interflow sediments, the interbedded red radiolarites and claystones of Callovian age and the radiolarites of Oxfordian to lower Hauterivian age (Shipboard Scientific Party 1990). In fig. 20 of Shipboard Scientific Party (1990) and fig. 4 of Ogg *et al.* (1992), *Planolites*-like lenticular trace fossils of up to 6 mm in diameter are shown. The "woody" texture, which is interpreted as compressed subhorizontal burrows, is common in the Callovian to Lower Cretaceous radiolarites (Behl *et al.* 1992; Ogg *et al.* 1992). A similar texture is observed in the accreted Bajocian radiolarian mudstones.

Chamberlain (1975) summarised the trace fossils observed in the photographs of the Deep Sea Drilling Project (DSDP) cores from 109 sites and reported the presence of nine ichnofossil types: *Chondrites*, composite burrows, *Planolites*, *Helminthoida*, halo burrows, rind burrows, large pyritised burrows, *Teichichnus* and *Zoophycos*. Burrows referred to *Chondrites*, *Planolites*, *Teichichnus*, and a composite pattern of activity have been found in siliceous rocks of accreted terranes. However, three differences are noted. First, unquestionable *Zoophycos* has not been found in Triassic to Jurassic chert successions. Second, the size of *Chondrites* and *Planolites* is generally larger in the deep-sea cores than those in accreted radiolarian chert successions. Third, the zigzag-shaped burrows of the Lower Triassic siliceous claystones have not been reported from the deep-sea cores, nor from the Middle Triassic to Upper Jurassic accreted rocks.

Zoophycos is found in the Upper Cretaceous siliceous rocks of the deep-sea cores in the Pacific Ocean (Chamberlain 1975), but not in the Middle Jurassic radiolarian mudstones of either the deep-sea cores or the accreted radiolarian chert sequences. Vecsei *et al.* (1989) reported the occurrence of *Zoophycos* from 8 m thick radiolarites of Oxfordian/Kimmeridgian age interbedded in limestones and marls deposited in Austroalpine rifted continental margin of the Northern Calcareous Alps (Austria). This is the oldest record of *Zoophycos* from radiolarites or radiolarian cherts. *Zoophycos* is known to be present in slope and deep basin by the Early Silurian (Bottjer *et al.* 1988), but this "deep basin" does not include associated chert-type pelagic deposits like those found in deeper habitats on the ocean floor. Based on these data, it seems likely that the invasion of *Zoophycos* into such pelagic deposits of the deep-sea realm did not occur until Late Jurassic time. Alternatively, a gap exists in the distribution of *Zoophycos* in the deep-sea pelagic environments from Early Triassic to Middle Jurassic.

The size of burrows in deep-sea cores was estimated from the photographs shown in the *Initial Reports* (Warme *et al.* 1973; van der Lingen 1973; Nielsen & Kerr 1976; Ekdale 1978; Wetzel 1987). The associated Cretaceous to Tertiary lithologies are mostly calcareous, excepting the turbidites. The diameters of *Chondrites* and *Planolites* tunnels in deep-sea cores range from 0.5 to 5 mm and from 0.5 to 30 mm, respectively. They are significantly larger than those in the rocks of the Jurassic accretionary complex. The size of specific trace fossils is influenced by oxygen and benthic food availability (e.g. Ekdale 1978; Wetzel 1991). Palaeo-oxygenation of Early Triassic oceanic bottom is considered to be low (Kakuwa 2000), and the small size may be ascribed to poor oxygenation. The occurrence of red cherts and red radiolarian mudstones in the Middle to Late Jurassic successions, however, implies that a well-oxygenated environment was widespread during this time. If the size of burrows depends on the amount of available food available for trace-making organisms to live in the pelagic sediments, then the calcareous and/or the post-Jurassic sediments could probably afford to sustain much larger benthic animals than the siliceous and/or pre-Cretaceous pelagic sediments. *Zoophycos* is absent from sediments that contain less benthic food, while it occurs predominantly in sediments enriched in a benthic food supply (Wetzel 1991). If this observation from off northwest Africa is adopted, the gap in distribution of *Zoophycos* is consistent with a limited food supply on the deep-sea floor.

The third factor raised above relating to the presence of zigzag-shaped burrows is ascribed to an anomalous environment in the Early Triassic ocean after the mass extinction event. Twichett & Wignall (1996) and Twichett (1999) described anomalous absences of trace fossils in shallow-marine carbonate rocks such as *Chondrites* and *Zoophycos* after the end-Permian mass extinction. The effect of the mass extinction and the duration of post-extinction stress were the causes for the anomalous patterns (Twichett 1999). The long-term history of occurrences of trace fossils in pelagic environments remains poorly understood. The radiolarian chert successions following the other mass extinction events in the Phanerozoic, such as those across the Ordovician–Silurian, Frasnian–Famennian and Triassic–Jurassic boundaries, need to be studied to establish whether similar trace fossil records exist or that the Permian–Triassic event is unique.

Acknowledgements

I express sincere gratitude to Dr Maria Gabriela Mángano and Dr Barry D. Webby who gave me the opportunity to make this presentation at the First International Palaeontological Convention in Sydney during 2002, and provided useful comments at various stages of the review process. I also thank Dr Alfred Uchman and Dr Richard J. Twichett for their critical reviews of the paper and other helpful suggestions.

References

Baltuk, M. 1983: Some sedimentary and diagenetic signatures in the formation of bedded radiolarite. *In* Iijima, A., Hein, J.R. & Siever, R. (eds): *Siliceous Deposits in the Pacific Region*, 299–316. Elsevier, Amsterdam.

Barret, T.J. 1982: Stratigraphy and sedimentology of Jurassic bedded chert overlying ophiolites in the North Apennines, Italy. *Sedimentology 29*, 353–373.

Behl, R.J. & Smith, B.M. 1992: Silicification of deep-sea sediments and the oxygen isotope composition of diagenetic siliceous rocks from the western Pacific Pigafetta and East Mariana Basins, LEG 129. *In* Larson, R.L. *et al.* (eds): *Proceedings of ODP, Scientific Results, 129*, 81–116. Ocean Drilling Program, College Station, TX.

Bottjer, D.J., Droser, M.L. & Jablonski, D. 1988: Paleoenvironmental trends in the history of trace fossils. *Nature 333*, 252–255.

Bruchert, V., Delano, J.W. & Kidd, W.S.F. 1994: Fe- and Mn-enrichment in Middle Ordovician hematitic argillites preceding black shale and flysch deposition: the Shoal Arm Formation, north-central Newfoundland. *Journal of Geology 102*, 197–214.

Chamberlain, C.K. 1975: Trace fossils in DSDP cores of the Pacific. *Journal of Paleontology 49*, 1074–1096.

Dec, T., Swinden, H.S. & Floyd, J.D. 1992: Sedimentological, geochemical and sediment-provenance constraints on stratigraphy and depositional setting of the Strong Island Chert (Exploit subzone, Notre Dame Bay). *Current Research, Newfoundland Department of Mines, Geological Survey Branch Report 92-1*, 85–96.

Ekdale, A.A. 1978: Trace fossils in Leg 42A cores. *In* Hsu, K. *et al.* (eds): *Initial Report of DSDP 42, Part 1*, 821–827. US Government Printing Office, Washington.

Folk, R.F. & McBride, E.F. 1978: Radiolarites and their relation to subjacent "oceanic crust" in Liguria, Italy. *Journal of Sedimentary Petrology 48*, 1069–1102.

Iijima, A., Kakuwa, Y. & Matsuda, H. 1989: Silicified wood from the Adoyama Chert, Kuzuh, central Honshu, and its bearing on compaction and depositional environment of radiolarian bedded chert. *In* Hein, J.R. & Obradovic, J. (eds): *Siliceous Deposits of the Tethys and Pacific Regions*, 151–168. Springer, New York.

Isozaki, Y. 1996: Anatomy and genesis of subduction-related orogen: a new view of geotectonic subdivision and evolution of the Japanese Islands. *The Island Arc 5*, 289–320.

Isozaki, Y. 1997: Jurassic accretion tectonics of Japan. *The Island Arc 6*, 25–51.

Kakuwa, Y. 1991: Lithology and petrology of Triasso-Jurassic bedded cherts of the Ashio, Mino and Tamba belts in southwest Japan. *Scientific Papers of College of Arts and Sciences, University of Tokyo 41*, 7–57.

Kakuwa, Y. 1994: Sedimentary petrographical, geochemical and sedimentological aspects of Triassic–Jurassic bedded cherts in southwest Japan. *In* Iijima, A., Abed, A.M. & Garrison, R.E. (eds): *Proceedings of the 29th International Geological Congress Part C*, 233–248.

Kakuwa, Y. 1996: Permian–Triassic mass extinction event recorded in bedded chert sequence in southwest Japan. *Palaeogeography, Palaeoclimatology, Palaeoecology 121*, 35–51.

Kakuwa, Y. 2000: Evaluation of paleo-oxygenation of oceanic bottom across the Permo-Triassic boundary by ichnofabric analysis of radiolarian chert sequence. *In* Carter, E.S., Whalen, P., Noble, P.J. & Crafford, A.E.J. (eds): *Abstract of the 9th Meeting of the International Association of Radiolarian Paleontologists*, 39.

Koike, T. 1979: Biostratigraphy of Triassic conodonts. *Prof. Kanuma Memorial Volume*, 21–27 [in Japanese].

Kuwahara, K., Nakae, S. & Yao, A. 1991: Late Permian 'Toishi-type' siliceous mudstone in the Mino–Tamba Belt. *Journal of Geological Society of Japan 97*, 1005–1008 [in Japanese].

Maeda, H., Kamata, Y., Karasawasan & Project Research Group. 2000: Lithostratigraphy and conodont fossils of Lower Triassic siliceous sediments in the Ashio Belt, central Japan. *Abstract of the 7th Meeting of Radiolarian Symposium*, 9 [in Japanese].

Matsuda, T. & Isozaki, Y. 1991: Well-documented travel history of Mesozoic pelagic chert in Japan: from remote ocean to subduction zone. *Tectonics 10*, 475–499.

Matsuoka, A. 1983: Middle and Late Jurassic radiolarian biostratigraphy in the Sakawa and adjacent areas, Shikoku, southwest Japan. *Journal of Geosciences, Osaka City University 26*, 1–48.

Matsuoka, A. 1984: Togano Group of the Southern Chichibu Terrane in the western part of Kochi Prefecture, southwest Japan. *Journal of Geological Society of Japan 90*, 455–477 [in Japanese].

Matsuoka, A. 1992: Jurassic–Early Cretaceous tectonic evolution of the Southern Chichibu terrane, southwest Japan. *Palaeogeography, Palaeoclimatology, Palaeoecology 96*, 71–88.

Matsuoka, A. 1995: Jurassic and lower Cretaceous radiolarian zonation in Japan and in the western Pacific. *The Island Arc 4*, 140–158.

Matsuoka, A., Hori, R., Kuwahara, K., Hiraishi, M., Yao, A. & Ezaki, Y. 1994: Triassic–Jurassic radiolarian-bearing sequences in the Mino Terrane, central Japan. *In* Organizing Committee of INTERRAD 7 (eds): *Guide Book for the 7th Meeting of the International Association of Radiolarian Paleontologists Field Excursion*, 19–61.

McBride, E.F. & Folk, R.F. 1979: Features and origin of Italian Jurassic radiolarites deposited on continental crust. *Journal of Sedimentary Petrology 49*, 837–868.

McBride, E.F. & Thomson, A. 1970: The Caballos Novaculite, Marathon Region, Texas. *Geological Society of America, Special Paper 22*, 1–129.

Nielsen, T.H. & Kerr, D.R. 1976: Turbidites, redbeds, sedimentary structures, and trace fossils observed in DSDP LEG 38 cores and the sedimentary history of the Norwegian–Greenland Sea. *In* Talwani, M. *et al.* (eds): *Initial Reports of DSDP, Supplements to Volumes 38, 39, 40 and 41*, 259–288. US Government Printing Office, Washington.

Ogg, J.G., Karl, S.M. & Behl, R.J. 1992: Jurassic through Early Cretaceous sedimentation history of the central equatorial Pacific and of SITES 800 and 801. *In* Larson, R.L. *et al.* (eds) *Proceedings of ODP, Scientific Results, 129*, 571–613. Ocean Drilling Program, College Station, TX.

Ross, C.A., Baud, A. & Menning, M. 1994: A time scale for project Pangea. *Canadian Society of Petroleum Geologists, Memoir 17*, 81–83.

Shipboard Scientific Party 1990: Site 801. *In* Lancelot, Y. & Larson, R.L. (eds): *Proceedings of ODP, Initial Report 129*, 91–170. Ocean Drilling Program, College Station, TX.

Sugiyama, K. 1997: Triassic and Lower Jurassic radiolarian biostratigraphy in the siliceous claystone and bedded chert units of the southeastern Mino Terrane, central Japan. *Bulletin of Mizunami Fossil Museum No. 24*, 79–193.

Tada, R. 1991: Compaction and cementation in siliceous rocks and their possible effect on bedding enhancement. *In* Einsele, G., Ricken, W. & Seilacher, A. (eds): *Cycles and Events in Stratigraphy*, 480–491. Springer, Heidelberg.

Twichett, R.J. 1999: Palaeoenvironments and faunal recovery after the end-Permian mass extinction. *Palaeogeography, Palaeoclimatology, Palaeoecology 154*, 27–37.

Twichett, R.J. & Wignall, P.B. 1996: Trace fossils and the aftermath of the Permo-Triassic mass extinction: evidence from northern Italy. *Palaeogeography, Palaeoclimatology, Palaeoecology 124*, 137–151.

van der Lingen, G. 1973: Ichnofossils in deep-sea cores from the Southwest Pacific. *In* Burns, R.E. *et al.* (eds): *Initial Report of DSDP LEG 21*, 693–700. US Government Printing Office, Washington.

Vecsei, A., Frisch, W., Pirzer, M. & Wetzel, A. 1989: Origin and tectonic significance of radiolarian chert in the Austroalpine rifted tectonic margin. *In* Hein, J.R. & Obradovic, H. (eds): *Siliceous Deposits of the Tethys and Pacific Region*, 64–80. Springer, New York.

Warme, J.E., Kennedy, W.J. & Schneidermann, N. 1973: Biogenic sedimentary structures (trace fossils) in LEG 15 cores. *In* Edger, N.T. *et al.* (eds): *Initial Reports of DSDP LEG 4*, 813–831. US Government Printing Office, Washington.

Webb, J.A. 1986: Ordovician bedded cherts of eastern Victoria, Australia. *Abstract of 12th International Sedimentological Congress*, A8.

Wetzel, A. 1983: Biogenic structures in modern slope to deep-sea sediments in the Sulu Sea basin (Philippines). *Palaeogeography, Palaeoclimatology, Palaeoecology 42*, 285–304.

Wetzel, A. 1987: Ichnofacies in· Eocene to Maestrichtian sediments from Deep Sea Drilling Project Site 605, off the New Jersey coast. *In* van Hinte, J.E. *et al.* (eds): *Initial Reports of DSDP 93*, 825–835. US Government Printing Office, Washington.

Wetzel, A. 1991: Ecologic interpretation of deep-sea trace fossil communities. *Palaeogeography, Palaeoclimatology, Palaeoecology 85*, 47–69.

Wetzel, A. & Bromley, R.G. 1994: *Phycosiphon incertum* revisited: *Anconichnus horizontalis* is its junior subjective synonym. *Journal of Paleontology 68*, 1396–1402.

Wetzel, A. & Uchman, A. 1997: Ichnology of deep-sea fan overbank deposits of the Ganei Slates (Eocene, Switzerland) – a classical flysch trace fossil locality studied first by Oswald Heer. *Ichnos 5*, 139–162.

Yamakita, S. 1993: Conodonts from P/T boundary sections of pelagic sediments in Japan. *Abstract of 100th Annual Meeting of the Geological Society of Japan*, 64–65 [in Japanese].

Yamakita, S. & Kadota, N. 2001: Early Triassic conodonts from the upper part of the black carbonaceous claystone layer of the Tenjinmaru P/T boundary section in Northern Chichibu Belt, Southwest Japan. *Abstract of 150th Regular Meeting of the Palaeontological Society of Japan*, 61 [in Japanese].

Yamakita, S., Kadota, N., Kato, T., Tada, R., Ogihara, S., Tajika, E. & Hamada, Y. 1999: Confirmation of the Permian/Triassic boundary in deep-sea sedimentary rocks; earliest Triassic conodonts from black carbonaceous claystone of the Ubara section in the Tamba Belt, Southwest Japan. *Journal of the Geological Society of Japan 105*, 895–898.

Yao, A. & Kuwahara, K. 1997: Radiolarian faunal change from Late Permian to Middle Triassic times. *News of Osaka Micropaleontologists Special Volume No. 10*, 87–96 [in Japanese].

Yao, A., Matsuda, T. & Isozaki, Y. 1980: Triassic and Jurassic radiolarians from the Inuyama Area, central Japan. *Journal of Geosciences, Osaka City University 23*, 135–154.

Palaeoenvironmental implications of trace fossils in estuary deposits of the Cretaceous Bluesky Formation, Cadotte region, Alberta, Canada

STEPHEN M. HUBBARD, MURRAY K. GINGRAS & S. GEORGE PEMBERTON

Hubbard, S.M., Gingras, M.K. & Pemberton, S.G. **2004 10 25**: Palaeoenvironmental implications of trace fossils in estuary deposits of the Cretaceous Bluesky Formation, Cadotte region, Alberta, Canada. *Fossils and Strata*, No. 51, pp. 68–87. USA. ISSN 0300-9491.

Estuarine settings are characterised by numerous physical and chemical stresses that can strongly influence the behaviour of burrowing organisms. Although lowered salinity and fluctuating salinity levels normally represent the chief stresses recognised in bays and estuaries, high sedimentation rates, high current energy, turbidity, and low levels of oxygen in bottom and interstitial waters are known to be significant factors that strongly influence the resultant ichnofossil assemblages. This study builds on earlier research and suggests that the effects of each of these parameters can be observed in the rock record through trace fossil analysis.

The subsurface lower Albian Bluesky Formation in the Cadotte region of Alberta has been interpreted to represent an estuarine deposit. Examination of the trace fossil assemblages from various facies therein suggests that physicochemical stresses were variable across the ancient estuary and mainly constituted the following: (1) low salinity and fluctuating salinity levels, which are interpreted to have contributed to patterns of low ichnofossil diversity and burrow dimunition proximal to the fluvial point source(s) in the upper and central parts of the depositional system; (2) high sedimentation rates and current energy, evidenced ichnologically by sporadic, penetrative bioturbation and the rare preservation of opportunistic sediment stirring, are most significant in the vicinity of the tidal inlet in the lower estuary and in the bayhead delta of the upper estuary; (3) high turbidity associated with the turbidity maximum in tidal channels of the central reaches of the estuary probably inhibited suspension-feeding behaviours locally; and, (4) low levels of dissolved oxygen in quiescent-water embayments and lagoons, often represented in brackish-water deposits as a *Trichichnus*, *Palaeophycus*, and *Diplocraterion* assemblage.

Key words: Estuary deposits; trace fossils; Alberta; Lower Cretaceous; Bluesky Formation.

Stephen M. Hubbard [stevehub@pangea.stanford.edu], Department of Geological and Environmental Sciences, Stanford University, Stanford, CA 94305, USA

Murray K. Gingras [mgingras@ualberta.ca] & S. George Pemberton [george.pemberton@ualberta.ca], Department of Earth and Atmospheric Sciences, University of Alberta, Edmonton, AB T6G 2E3, Canada

Introduction

Ichnology is useful for enhancing sedimentological interpretations because it provides information that cannot be derived from analysis of the physical sedimentary structures alone. Ichnological signatures are known to represent composite animal responses to several physical and chemical parameters, including salinity and its fluctuations, strong currents, water turbidity, sedimentation rate, oxygenation, depositional system episodicity, the nature of the available food supply, and temperature locally. Animals also show behavioural responses to changes in substrate cohesiveness and sediment grain size [thorough reviews are provided in Pemberton *et al.* (1992a) and Taylor *et al.* (2003)]. Thus, trace fossil assemblages potentially contain a great deal of information. However, the various animal behaviours indicated by trace fossils represent a complex adaptive landscape; some behaviours are used serially, whereas other ichnofossils might represent a specific faunal response to a particular environmental parameter. Thus, deciphering the information latent in a trace fossil assemblage can present a daunting task.

Due to intense hydrocarbon exploration throughout the last five decades, many sedimentary deposits in the Western Canada Sedimentary Basin (WCSB) have been scrutinised with respect to their ichnological significance. Among the most commonly applied ichnological paradigms in the WCSB is the brackish-water model (Pemberton *et al.* 1982), which strives to characterise brackish-water ichnofossil assemblages. However, the utility of that model has, at times, undermined the general appreciation that trace fossil assemblages are the result of *multiple* physical and/or chemical stresses, and not just those associated with salinity. This paper uses a case study to present an interpretation of the various stresses exerted on the trace-making fauna of a Cretaceous estuary. It is not intended to challenge the brackish-water model, but rather to add to it, suggesting the potential utility of considering various other environmental factors to hone interpretations of ancient estuarine deposits.

Physicochemical stresses in marginal-marine settings

Salinity stress is a prominent environmental parameter in brackish-water settings. Nevertheless, numerous other factors potentially influence associated trace fossil assemblages. The effects of what are considered herein to be the most important in terms of their significance in imparting a signature into the (marginal-marine) rock record are discussed. These include salinity stresses, high sedimentation rates and current energy, excessively turbid conditions, and low oxygenation of bottom and interstitial waters.

Salinity stresses

In the 1970s, researchers working on east coast (Georgia) estuaries of the USA began to recognise the distinguishing characteristics of burrow assemblages made by infaunal organisms in brackish-water settings (e.g. Howard & Frey 1973, 1975; Howard *et al.* 1975; Dörjes & Howard 1975). Howard & Frey (1973) identified specific burrowing activities that were associated with different sediment textures and salinity gradients across marginal-marine depositional systems.

Brackish water is considered to represent water chemistries that are characterised by intermediate salinities, thus representing a continuum between fresh (<0.5‰) and normal marine (~34‰) water. Trace fossil assemblages associated with sediments deposited in brackish water are typically characterised by one or more

diagnostic attribute. A summary of these characteristics was presented by Pemberton *et al.* (1982): (a) a low diversity of trace fossil forms is common, (b) forms present typically represent an impoverished marine assemblage, (c) a dominance of morphologically simple, vertical and horizontal structures, (d) the presence of assemblages dominated by a single ichnofossil species, (e) traces are diminutive relative to fully marine counterparts, and (f) some forms may be found in high densities. Numerous authors have observed a mixed *Skolithos–Cruziana* ichnofacies in brackish-water deposits (e.g. Howard & Frey 1975, 1985; Ekdale *et al.* 1984; Frey & Pemberton 1985; MacEachern & Pemberton 1994; Buatois *et al.* 2002). Trace fossils commonly associated with brackish-water deposits include *Gyrolithes, Palaeophycus, Cylindrichnus, Skolithos, Planolites, Thalassinoides,* and *Teichichnus.*

Work on the McMurray Formation of northern Alberta by Pemberton *et al.* (1982) represents the first successful attempt to recognise and interpret the physical manifestation of salinity stress in ancient ichnofossil assemblages. Throughout the last two decades, several researchers have demonstrated the utility of this palaeoenvironmental tool in numerous study areas (e.g. Frey & Pemberton 1985; Wightman *et al.* 1987; Beynon *et al.* 1988; Ranger & Pemberton 1992; Pemberton *et al.* 1994; MacEachern & Pemberton 1994; Wightman & Pemberton 1997; Hubbard *et al.* 1999; MacEachern *et al.* 1999a; Buatois *et al.* 2002).

Since the initial work in the 1970s there has been a general paucity of ichnological studies in modern brackish-water sedimentary environments (neoichnology). An exception is Gingras *et al.* (1999), who revisited the effects of brackish water on benthic organisms and the concurrent traces they leave in siliciclastic sediments at Willapa Bay, Washington State, USA. The authors complemented the analysis of modern sediments with a Pleistocene data set in their study area, enabling many of the physicological and physical processes responsible for trace fossil suites generated in brackish-water settings to be determined.

In addition to the effect of lowered water salinity, fluctuating salinity levels related to the tidal exchange of marine and brackish (or fresh) water also impart a significant stress on burrowing organisms. Deep burrows (e.g. *Skolithos, Thalassinoides* and *Cylindrichnus*) are commonly constructed by organisms to buffer the effects of salinity fluctuation (which can range from fresh to hypersaline) resulting from tidal exchange of water across the tidal flat, precipitation, and discharge of groundwater (Rhoads 1975; Pemberton & Wightman 1992).

Sedimentation stresses

High current energy, water turbidity, and high sedimentation rates are not necessarily directly related. In sedimentary systems that contain large quantities of mobile sand, however, current energy is directly proportional to the degree of shifting a substrate might suffer. Thus, for the purpose of this study (the Bluesky Formation is characteristically sandy in the study area), these parameters are assessed together. Important to palaeoecological interpretations of trace fossil assemblages, the preservation potential of burrows excavated in shallow tiers and by deeper-seated slow-burrowing infauna is significantly reduced by the rapid migration of bedforms in regions dominated by strong currents. Conditions related to the stresses that high sedimentation rates and high current energy impart on burrowing infauna include: (1) high current velocities that limit filter feeding for many soft-bodied organisms. In settings such as the tidal inlet, organic detritus is swept from the sediment–water interface and grazing and other activities on the sea floor are essentially eliminated. (2) Episodic sedimentation associated with storms may bury organisms so quickly that they are unable to adjust to the new sediment–water interface. Such challenging sedimentary conditions are deleterious to the presence of burrowing organisms and normally result in lower burrowing intensities (Howard 1975; Frey & Pemberton 1985). For example, at Willapa Bay, Washington, Gingras *et al.* (1999) noted that (i) low biogenic reworking is observed in intertidal point bars characterised by high sedimentation rates, and (ii) burrowing is strictly limited to organisms able to adapt to rapid sedimentation in these localities, such as amphipods, threadworms and mobile polychaetes (such as *Nereis*).

In estuarine settings, turbidity is a prevalent environmental stress. The area of highest turbidity stress occurs within the turbidity maximum, which is associated with the saltwater wedge where the most intense salinity gradients occur (Meade 1969; Howard *et al.* 1975). The turbidity maximum results from converging fluvially derived freshwater and tidally derived saltwater. This zone of convergence, which often occurs in the vicinity of the middle estuary, is characterised by high rates of clay/organic material flocculation, and the overall accumulation of fine-grained sediments from suspension (Kranck 1981; Rahmani 1988; Allen 1991). The boundary of the saltwater intrusion can fluctuate, significantly impacting burrowing infauna unable to cope with regular environmental changes. Turbid conditions probably exert a significant environmental stress on burrowing organisms, especially filter feeders (Rhoads *et al.* 1972; Rhoads 1973; Ranger & Pemberton 1992; Buatois & López Angriman 1992). However, the effect of the turbidity maximum on burrowing organisms is probably variable. As the turbidity maximum is associated with the landward extreme of saltwater intrusion, it has been suggested that it defines

the boundary between marine-dominated assemblages and persistent brackish- to freshwater assemblages (Allen 1991). Therefore, in the central to upper reaches of the estuary, the saline water that permits organisms to live is inherently associated with turbid conditions.

The exclusion of suspension feeding under extremely turbid conditions suggests that burrowing behaviour is biased towards deposit-feeding strategies. In fact, this relationship has been observed in the rock record by numerous authors (Rhoads *et al.* 1972; Buatois & López Angriman 1992; Gingras *et al.* 1998; Coates & MacEachern 1999). In those examples, sedimentary facies that accumulated in turbid waters are dominated by low-diversity trace fossil assemblages, which only record deposit-feeding strategies. Many benthic organisms are thought to adjust their feeding strategies to survive under turbid conditions, and in some cases organisms are suspected to thrive during times of high turbidity (Wilber 1983).

Oxygen stresses

Oxygen deficiency is most characteristic in deep- and/or quiet-water environments, dominated by the slow accumulation of fine-grained sediments. However, marginal-marine sedimentary settings, such as the wave-dominated estuary in the Bluesky Formation, can contain large quantities of terrigenous organic detritus, which, in the process of decomposition, may lower oxygen levels near the sediment–water interface (Leithold 1989; Saunders *et al.* 1994; Coates & MacEachern 1999). Low oxygen concentrations in bottom and interstitial waters influence trace fossil size and diversity; as oxygen content decreases there is a noticeable reduction in the size of the burrows present and the diversity of organisms also decreases (Rhoads & Morse 1971; Savrda & Bottjer 1989; Wignall 1991). A dominance of deposit-feeding burrows that maintain an open conduit to the sediment–water interface is considered characteristic of oxygen-deficient sediments (Ekdale & Mason 1988). Common trace fossil forms in such settings include *Chondrites*, *Zoophycos*, *Teichichnus* and *Trichichnus* (Ekdale & Mason 1988; Savrda & Bottjer 1989; Wignall 1991; Savrda 1992).

Working on the shallow-marine Carmel Formation (Middle Jurassic) in central Utah, de Gibert & Ekdale (1999) interpreted a low-diversity trace fossil assemblage, which included *Arenicolites*, *Chondrites*, *Gyrochorte*, *Lockeia*, *Planolites*, *Protovirgularia*, *Rosselia*, *Scalarituba*, *Skolithos*, *Taenidium*, and *Teichichnus*, to have resulted from greater than normal marine salinities coupled with depletion of oxygen in pore waters. In general, the assemblage consisted mostly of non-specialised ichnotaxa with small burrow sizes dominating the assemblages. The amount of bioturbation was reported to be low.

Brackish-water deposits in the WCSB

Brackish-water deposits are common in Lower to Middle Cretaceous strata of the WCSB. This has been demonstrated by numerous authors using various techniques (e.g. micro-palaeontology: Finger 1983; Mattison & Wall 1993; MacEachern *et al.* 1999a; isotopic analysis: Longstaffe *et al.* 1992; Holmden *et al.* 1997; McKay & Longstaffe 1997; palynology: Wightman *et al.* 1987; MacEachern *et al.* 1999a; organic matter characterisation: Reidiger *et al.* 1997; and ichnology: Pemberton *et al.* 1982; Wightman *et al.* 1987; Beynon *et al.* 1988; MacEachern & Pemberton 1994; MacEachern *et al.* 1999a).

Extensive brackish-water deposits in the Cretaceous record of Alberta are related to the preservation of numerous, marginal-marine sedimentary systems in which salinity gradients (from fresh to marine) were characteristic. The depositional environments were varied and included estuaries, incised valleys, embayments and deltas (Zaitlin *et al.* 1995). Basin physiography is also considered to have been a contributing factor, because the shallow waters of the Western Intracontinental Seaway were probably sensitive to heavy rainfall and runoff from the rising Cordillera, as well as from the topographically emergent Canadian Shield. Tidal forces are known to have been active in the basin (Klein & Ryer 1978; Smith 1988), and would have been responsible for the exchange of saline water into the area.

From an economic perspective, brackish-water deposits in Cretaceous strata of the WCSB are extremely important.

Heavy oil deposits in the region are largely encased within associated strata (e.g. Beynon *et al.* 1988; Zaitlin & Shultz 1990; Wightman & Pemberton 1997; Hubbard *et al.* 1999), as are numerous conventional oil- and gas-bearing sandstones (e.g. Brownridge & Moslow 1991; MacEachern & Pemberton 1994; Zaitlin *et al.* 1995; Gingras *et al.* 1998).

Database and previous work

The study area is located in the Cadotte region of the central Alberta plains, between townships 83 and 86 and ranges 15 and 20 west of the fifth meridian (Fig. 1). The Bluesky Formation in the area occurs in the subsurface between 500 and 600 m depth (from the surface). The sedimentological database consists of 81 cored intervals (Figs. 1, 2).

Siliciclastic sediments of the Bluesky Formation are interpreted to represent an ancient wave-dominated estuary (Fig. 3A) (Hubbard 1999; Hubbard *et al.* 1999, 2002). The previously published work provides a well-documented sedimentological and stratigraphical framework into which this ichnological study is integrated. For the purposes of this paper, the Bluesky deposits are separated into three discrete facies divisions reflecting the upper, middle and lower estuary (Fig. 3B).

Upper estuary deposits consist primarily of facies deposited on the bayhead delta, in quiescent bays, and on supratidal and intertidal flats (Fig. 3). The upper estuary represents the head of the estuary most closely related

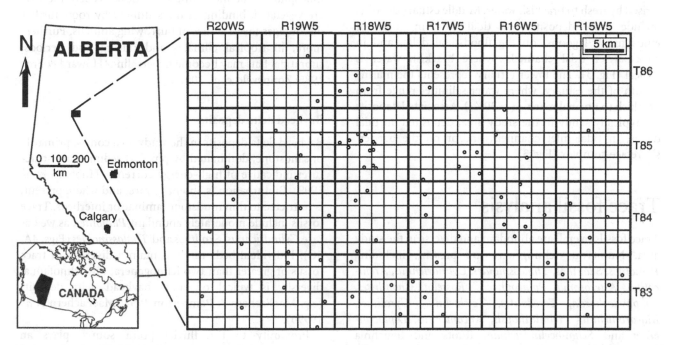

Fig. 1. Location map of the study area and distribution of examined cores (black circles).

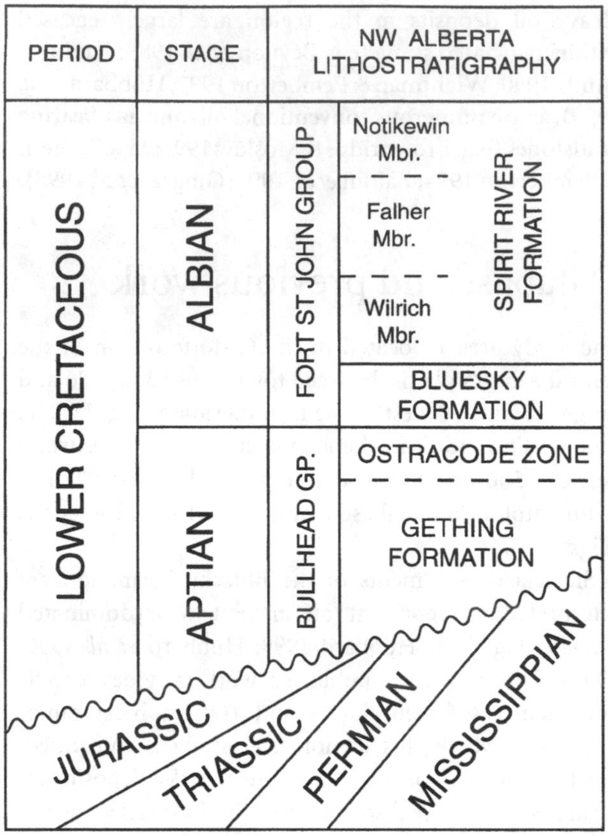

PERIOD	STAGE	NW ALBERTA LITHOSTRATIGRAPHY	
LOWER CRETACEOUS	ALBIAN	FORT ST JOHN GROUP	Notikewin Mbr. / Falher Mbr. / Wilrich Mbr. (SPIRIT RIVER FORMATION)
			BLUESKY FORMATION
	APTIAN	BULLHEAD GP.	OSTRACODE ZONE
			GETHING FORMATION

JURASSIC TRIASSIC PERMIAN MISSISSIPPIAN

Fig. 2. Subsurface Lower Cretaceous stratigraphic nomenclature for the Cadotte region of Alberta. Not to scale.

Trace fossil analysis

Trace fossils overall are abundant and diverse in the study area. Numerous ichnogenera are present, including *Planolites, Skolithos, Thalassinoides, Cylindrichnus, Arenicolites, Asterosoma, Palaeophycus, Rosselia, Chondrites, Subphyllochorda, Teredolites, Teichichnus, Gyrolithes, Macaronichnus, Helminthopsis, Anconichnus, Diplocraterion* and *Schaubcylindrichnus*. Roots and fugichnia (escape traces) are also present. Trace fossil diversity

to (and including) the fluvial point source. In this region, water salinities are considered to have been low, characterised by fresh to brackish water. Middle estuary deposits include those deposited on tidal flats, in quiescent embayments/lagoons, on the flood-tidal delta, and in back-barrier tidal channels, where brackish water is typical (Fig. 3). The lower estuary represents the marine mouth of the estuary, where water salinities range from brackish to normal marine (> 30‰). Associated deposits are typically sandy in the study area, deposited on tidal deltas, in tidal inlet channels, and/or on the barrier/shoreface complex (Fig. 3).

within a particular facies is variable, generally not exceeding six diagnostic ichnogenera. The distribution of trace fossils across the study area provides a rationale for the palaeoenvironmental reconstruction of the estuary. Figure 4 provides a summary of the trace fossils present in the Bluesky Formation in the Cadotte region of Alberta.

Upper estuary – observations and interpretations

Herein, the upper estuary is considered to be the part of the ancient deposit that was most strongly influenced by fluvial processes but within the realm of observable tidal reworking. Sediments in the upper estuary accumulated in numerous subenvironments, thereby under various physicochemical conditions. The four subenvironments are: supratidal flat, bayhead delta, intertidal/subtidal flat, and quiescent bay.

Supratidal flat deposits

Supratidal flat deposits are moderately to pervasively bioturbated, with a trace fossil assemblage consisting solely of roots (Figs. 4A, 5E, F). Pedogenic slickensides suggest the incipient development of soils, or at least pedogenic alteration in associated lithofacies.

This low-diversity assemblage of traces is typical of supratidal areas, and is well documented from studies of Georgia marshes (e.g. Howard & Frey 1985). Freshwater input from rain, sporadic storm surges of brackish water into the area, and episodes of desiccation induce inhospitable infaunal living conditions. Where the flats are vegetated, binding of the sediment by roots further discourages the activities of burrowing animals. Furthermore, in tropical climates and during arid periods, supratidal flats may become hypersaline (Howard & Frey 1985; Zonneveld *et al.* 2001).

Bayhead delta deposits

Bayhead delta deposits in the study area consist primarily of sandstone, dominated by physical sedimentary structures indicative of high-energy currents (Hubbard *et al.* 1999). Bioturbation is absent to rare, and where present, is restricted to thin mudstone laminae or interbeds. Trace fossils include moderately abundant *Planolites*, as well as rare *Cylindrichnus, Skolithos* and *Thalassinoides* (Figs. 4A, 5C). Individual beds rarely contain a diversity of trace fossils of greater than two ichnogenera; low ichnofaunal diversity is typical of sandy bayhead delta deposits (e.g. MacEachern & Pemberton 1994; MacEachern *et al.* 1999b).

Proximity to the fluvial point source plays an important role in lowering overall salinities in the delta,

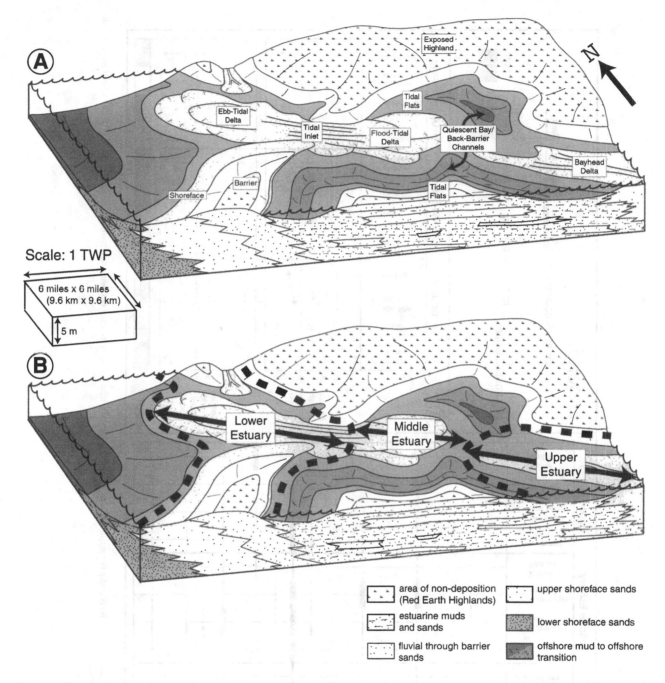

Fig. 3. A: The wave-dominated estuary facies model interpreted for the Bluesky Formation in the Cadotte study area (modified from Hubbard *et al.* 2002). B: Estuarine divisions (upper, middle and lower) as defined in this study.

thereby having the potential to exert a significant environmental stress on organisms (Gingras *et al.* 1998). The observed trace fossil assemblage is, in fact, consistent with salinity-stressed ichnofacies documented from other deposits of similar age, most notably the McMurray Formation of the WCSB (e.g. Ranger & Pemberton 1992). In these deposits, the trace fossil assemblage is dominated by ethologies associated with organisms able to adapt to changing environmental conditions. In particular,

Skolithos, Planolites and *Thalassinoides* are all burrow architectures that, in the modern, can be constructed relatively quickly and are commonly occupied by robust, efficient burrowers. The trace fossil assemblage is thereby interpreted to reflect brackish-water conditions accompanied by energetic currents and high sedimentation rates during deposition; conditions that eliminate the potential for many trace-making organisms to inhabit sediments of the bayhead delta (Nanson *et al.* 2000).

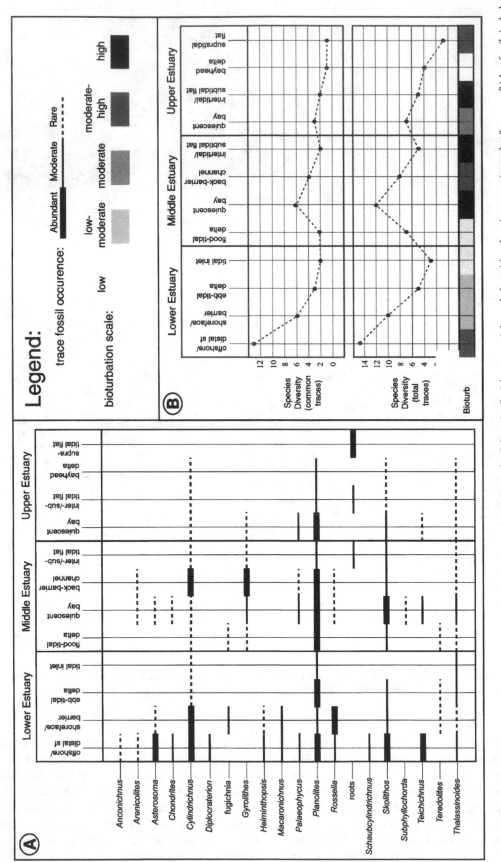

Fig. 4. A: Palaeoenvironmental distribution of trace fossils in the Bluesky Formation. B: Trace fossil diversity (by ichnogenera) for each depositional environment (note that "common" ichnofossils include those that are either abundant or moderately abundant).

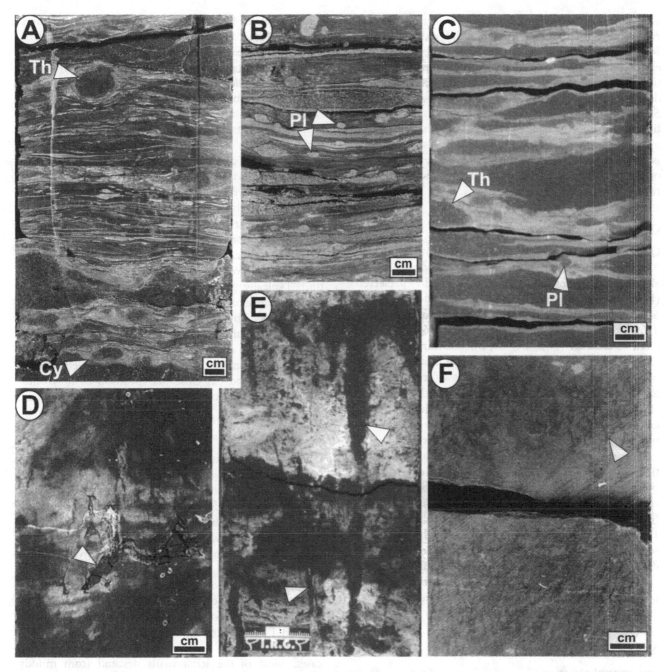

Fig. 5. Trace fossils of the upper estuary in the Bluesky Formation, including *Thalassinoides* (Th), *Cylindrichnus* (Cy), *Planolites* (Pl) and roots (indicated by arrows in D, E, and F). Note that in each photograph, bitumen-stained sandstone is the dark lithology and muddy siltstone is the light lithology. Black mudstone laminae are present in (A) and (B). A, B: Low-abundance and diversity trace fossil assemblages in quiescent bay deposits (core locations – A: 07-09-84-16W5, 654.45 m; B: 06-36-83-16W5, 627.2 m). C: Bioturbated mudstone beds associated with bayhead delta distributary channel point bar deposits (07-30-83-15W5, 645.0 m). D: Intertidal flat, burrow-mottled sandstone and mudstone with coalified roots (06-25-83-15W5, 628.5 m). E, F: Rooted supratidal flat deposits (E: 04-10-85-18W5, 597.5 m; F: 06-33-83-17W5, 625.9 m).

Tidal flat deposits

The facies associated with subtidal and intertidal flat deposition in the upper estuary is characterised by a mottled texture related to extensive biogenic reworking coupled with slow sedimentation rates (Frey & Pemberton 1985; Gingras *et al.* 1999). Individual trace fossils are rarely discernible. Trace fossils recognised include common *Planolites* and rhizoliths, with subordinate *Cylindrichnus*, *Skolithos* and *Thalassinoides* (Figs. 4A, 5D).

This low-diversity trace fossil suite results from reduced salinities, as well as fluctuating currents and sedimentation rates (Weimer *et al.* 1982). Such suites are similar to *Planolites/Arenicolites/Skolithos* assemblages reported from upper estuarine tidal flats at Willapa Bay, western USA (Gingras *et al.* 1999) and the Bay of Fundy, eastern Canada (Gingras 2002). At those locales, fluctuation in water salinity can also result in the construction of deeper burrow refuges for many organisms, such as thin *Skolithos* produced by capitellid polychaetes, an observation that echoes Rhoads (1975). Notably, sediments on upper estuarine tidal flats are commonly dysaerobic. This is the result of a relatively high proportion of terrestrial organic detritus accumulating in the sediment and local rooting in higher (?supratidal) zones.

Quiescent bay deposits

Quiescent bay deposits of the upper estuary in the Bluesky Formation are typically moderately bioturbated, although vestigial lamination is generally preserved (Figs. 4B, 5A, B). Common *Planolites*, *Palaeophycus* and *Skolithos*, as well as rare *Cylindrichnus*, *Gyrolithes*, *Teichichnus* and *Thalassinoides* characterise the trace fossil assemblage (Figs. 4A, 5A, B). The number of trace fossil forms observed within individual units is typically no more than four. The simple burrow forms common to these deposits, as well as the low-diversity/low-abundance assemblage of trace fossils indicate a stressed environment, capable of only supporting organisms utilising trophic generalist feeding behaviours.

Current energies and sedimentation rates are low in this comparatively sheltered setting. Consequently, the predominant stresses on burrowing organisms probably reflect reduced salinity and possibly escalated turbidity (e.g. MacEachern & Pemberton 1994; MacEachern *et al.* 1999a). Heightened turbidity is inferred, in part, from sedimentological observations, including common mud drapes and intercalated mud beds. Mud linings associated with *Cylindrichnus* and *Palaeophycus*, and the persistence of a deposit-feeding component of the ichnofossil assemblage also support this interpretation.

Middle estuary – observations and interpretations

In the study area, the middle estuary represents all sediments that accumulated within the bay, between the bayhead and mouth. Facies of the middle estuary in the Bluesky Formation include those deposited on intertidal/subtidal flats, in low-energy back-barrier channels, in quiescent bays, and on the flood-tidal delta. Examples of typical trace fossils from facies of the middle estuary are presented in Figs. 6 and 7.

Tidal flat deposits

Subtidal and intertidal flat sediments in the middle estuary are typically completely reworked due to biogenic activity, with few physical sedimentary structures preserved (Fig. 6E, F). Trace fossils are generally unidentifiable. However, *Planolites*, *Skolithos* and roots are moderately abundant (Fig. 6E). *Cylindrichnus*, *Gyrolithes* and *Thalassinoides* are observed rarely (Fig. 4A).

Sediment homogenisation due to extensive burrowing is common to stressed, brackish-water settings, particularly on tidal flats where sedimentation rates are typically very slow (Fig. 4B) (Edwards & Frey 1977; Frey & Pemberton 1985; Gingras *et al.* 1999). Stresses on organisms in the intertidal realm are principally considered to be related to lowered salinity, and short-term fluctuations in salinity levels related to precipitation, discharging groundwater, and tidal influence; suppressed oxygen levels, desiccation, temperature, and predation may also be important locally (Martini 1991; Cadée 1998; Gingras *et al.* 1999). However, food resources (algae) may be plentiful. Typically, the ichnological assemblage associated with this range of stresses and an abundant food resource is characterised by: (1) a sporadic distribution of intensely churned biogenic textures that range between fabrics dominated by a single diminutive ichnogenus and those that result from four or five comparatively robust ichnogenera; and (2) close juxtaposition of horizontal and vertical biogenic structures in high burrow densities.

Tidal channel deposits

Low-energy, back-barrier tidal channel deposits comprise point bar deposits characterised by a prevalence of inclined heterolithic stratification. Bioturbation varies from moderate to high, with finer-grained units typically more thoroughly reworked. The trace fossil suite consists of common *Cylindrichnus*, *Gyrolithes*, *Planolites* and *Skolithos* (Figs. 4A, 6). *Arenicolites*, *Palaeophycus*, *Rosselia* and *Thalassinoides* are also present locally. In many examples, strata from ancient point bar deposits are characterised by monospecific trace fossil assemblages, which are often associated with brackish-water deposits (Fig. 6B, D) (Pemberton *et al.* 1992a). Trace fossil diversities range between two and four ichnogenera in most cases. Most of the ichnofossils descend from muddy horizons, and are defined by muddy fills (Fig. 6A–D).

During periods of significant sand deposition and reworking, it is inferred that conditions were unfavourable for infaunal colonisation. Seasonal pulses of increased sedimentation, high water turbidity and potentially lowered salinity may all be related to increased fluvial discharge into the estuary via the distributary network of the bayhead delta. Workers in modern estuaries have explained seasonal variations in levels of bioturbation in this way (Dalrymple *et al.* 1991; Gingras *et al.* 1999, 2002). The observations of the Bluesky Formation, however, may also reflect a taphonomic bias, in that mud-lined burrows are more easily seen, especially in oil–sand intervals. The trace fossil assemblage is dominated by

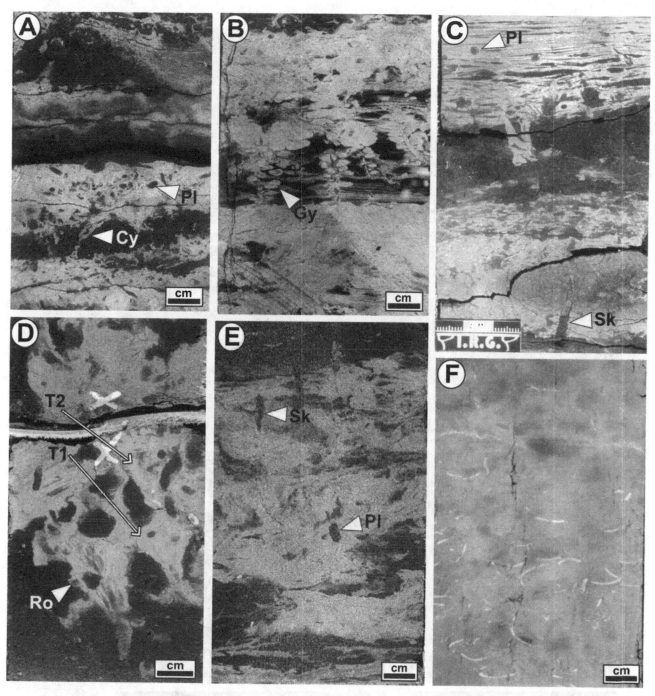

Fig. 6. Trace fossils of the middle estuary (low-energy, back-barrier tidal channels and tidal flats) in the Bluesky Formation, including *Cylindrichnus* (Cy), *Planolites* (Pl), *Gyrolithes* (Gy), *Skolithos* (Sk), and *Rosselia* (Ro). Note that in each photograph, bitumen-stained sandstone is the dark lithology and muddy siltstone is the light lithology. A–C: Heterolithic stratification (mud beds are pervasively reworked) associated with tidal channel (point bar) deposits (core locations – A: 06-33-83-17W5, 633.3 m; B: 05-33-83-19W5, 606.5 m; C: 11-17-83-17W5, 689.0 m). D: Readjustment of a *Rosselia* in tidal channel deposits, probably constructed in response to high sedimentation rates (T1 = time 1; T2 = time 2, after burrow readjustment) (11-17-83-18W5, 625.8 m). E, F: Tidal flat deposits are commonly highly bioturbated. However, individual trace fossil genera are commonly not discernible (E: 12-15-85-18W5, 574 m; F: 04-24-84-16W5, 632.6 m). Note the abundance of pelecypod shells in (F).

simple filter-feeding and interface-feeding structures (e.g. *Cylindrichnus, Gyrolithes, Skolithos, and Arenicolites*). Deposit-feeding ethologies are simple (e.g. *Planolites* and *Thalassinoides*) and reflect mobile trace makers (e.g. polychaetes and decapods).

Quiescent bay deposits

Facies associated with quiescent bay sedimentation in the middle estuary are characterised by moderate to high levels of bioturbation (Fig. 4B). The trace fossil suite is

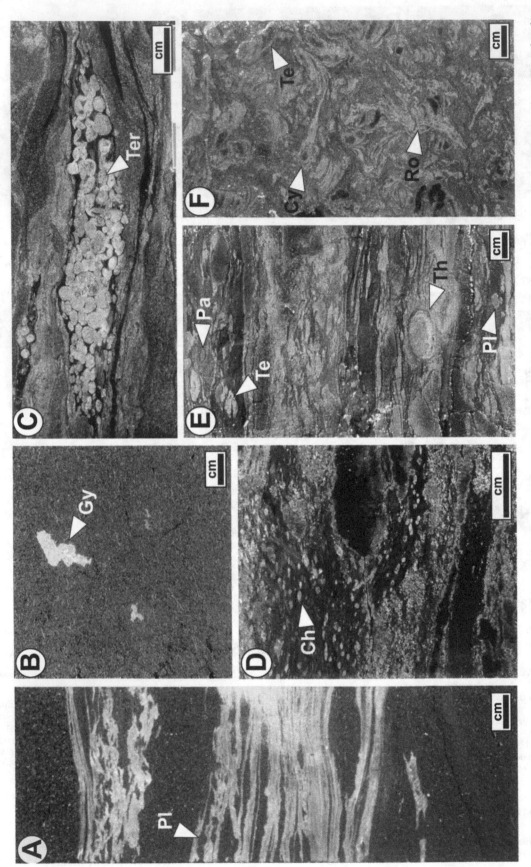

Fig. 7. Trace fossils of the middle estuary (flood-tidal delta and quiescent bay deposits) in the Bluesky Formation, including *Planolites* (Pl), *Gyrolithes* (Gy), *Teredolites* (Ter), *Chondrites* (Ch), *Palaeophycus* (Pa), *Teichichnus* (Te), *Thalassinoides* (Th), *Cylindrichnus* (Cy) and *Rosselia* (Ro). Note that in each photograph, bitumen-stained sandstone is the light lithology. Black mudstone-stained sandstone is the dark lithology and muddy siltstone is the light lithology. Black mudstone laminae are present in (D) and (E). A, B: Flood-tidal delta distributary channel deposits are characterised by rare, diminutive trace fossils (core locations – A: 04-05-85-17W5, 604.9 m; B: 14-21-85-18W5, 582.44 m). C: Allochthonous coal with pyritised *Teredolites* (10-23-83-18W5, 611.0 m). D: *Chondrites* in deposits interpreted to have accumulated in an oxygen-stressed, quiescent embayment setting (13-22-84-19W5, 566.5 m). E, F: Moderately through to pervasively bioturbated quiescent bay deposits (E: 04-24-84-16W5, 635.5 m; F: 11-17-83-18W5, 627.0 m).

diverse, consisting of common *Planolites*, *Skolithos*, *Gyrolithes*, *Palaeophycus*, *Teichichnus* and *Thalassinoides*, as well as rare *Arenicolites*, *Asterosoma*, *Chondrites*, *Cylindrichnus*, *Rosselia* and *Subphyllochorda* (Figs. 4A, 7D–F). Individual beds rarely contain more than five or six trace fossil forms.

The trace fossil suite in the Bluesky Formation is similar to that recognised by MacEachern *et al.* (1999a) in central bay deposits of Viking Formation incised valley fills, Alberta. Supported by micro-palaeontological and palynological data in their study, those suites were interpreted to be highly indicative of deposition in brackish water. In this study, the presence of *Subphyllochorda*, *Asterosoma*, and *Rosselia* suggests that the water chemistries were somewhat more marine.

Flood-tidal delta deposits

Flood-tidal delta deposits in the Bluesky Formation are characterised by sparse bioturbation (Fig. 4B). A low-diversity trace fossil assemblage is present, with diminutive forms common (Fig. 7B). *Planolites* and *Skolithos* are most commonly observed, restricted to thin mudstone beds (Fig. 7A). Other trace fossils include *Cylindrichnus*, fugichnia, *Gyrolithes*, *Thalassinoides* and *Teredolites* (Fig. 4A). The occurrence of *Teredolites* is restricted to wood fragments and coaly debris, and is probably not *in situ* within tidal delta sandstones (Fig. 7C). Trace fossil diversity within a single unit is low, ranging from one to two ichnogenera.

Along with high sedimentation and low salinity, turbid water may also contribute to reduced burrowing activity in settings such as tidal deltas. Trace fossil diversity is directly affected by turbidity as filter-feeding strategies may be hindered due to excessive sediment in the water column (Wilber 1983; Moslow & Pemberton 1988; Gingras *et al.* 1998; Coates & MacEachern 1999).

Lower estuary/marine – observations and interpretations

Subenvironments of the lower estuary, as defined in this study, specifically include the tidal inlet, ebb-tidal delta, shoreface/barrier and offshore/distal shoreface. There is a paucity of core data in the westernmost part of the study area (Fig. 1); consequently, offshore/distal shoreface deposits are poorly constrained. These deposits preserve complex facies architectures, and features from different cores suggest deposition in lower shoreface, offshore transition or offshore settings, based on the generalised shoreline profile model defined by Reading & Collinson (1996). The other lower estuarine deposits are more extensively cored, and are sandier overall (Hubbard *et al.* 2002).

Tidal inlet deposits

Upper flow regime conditions are interpreted to have persisted in the tidal inlet, based on the presence of pervasive horizontal planar parallel to low-angle cross-beds in these facies (Hubbard *et al.* 2002). Consequently, bioturbation is low to absent in associated units (Fig. 4B). The only trace fossils observed in tidal inlet deposits are *Planolites*, *Thalassinoides* and *Cylindrichnus* (Figs. 4A, 8F).

Trace fossil diversities are low, in part due to high currents and sedimentation rates in tidal inlet settings (e.g. MacEachern & Pemberton 1994; Savrda *et al.* 1996). It is likely that physical processes overshadowed the effects of other potential stresses on the community of burrowers. The paucity of trace fossils observed in tidal inlet sandstones of the Bluesky Formation is consistent with research in the Georgia estuaries of the eastern USA, which concluded (Howard & Frey 1973) that burrows are rare to absent in coarse sands deposited by energetic currents in the lower estuary.

Ebb-tidal delta deposits

Bioturbation in ebb-tidal delta deposits is low to moderate overall, and predominantly associated with mudstone intervals (Fig. 4B). The low-diversity trace fossil suite consists of common *Planolites*, *Skolithos* and *Thalassinoides*, as well as rare *Cylindrichnus* and *Teredolites* (Figs. 4A, 8E).

Ebb-tidal delta deposits are rare in the rock record, and their inaccessibility in the modern has resulted in a general paucity of information about their facies characteristics, especially with regard to burrowing organisms. No more than three trace fossil forms were observed in individual units interpreted to have been deposited in ebb-tidal deltas. Such a reduced diversity suggests stressed environmental conditions. Variable sedimentation rates, salinity fluctuations, brackish water, and heightened water turbidity are all likely to have contributed to the hostile environment indicated by the trace fossil assemblage.

Shoreface/barrier deposits

Sandy shoreface/barrier deposits are predominantly characterised by low to moderate bioturbation intensities (Fig. 4B), although biogenic reworking is pervasive in some facies (Fig. 8G). Robust, mud-lined *Cylindrichnus* and *Rosselia* are commonly present in bitumen-stained barrier sandstones of the Bluesky Formation (Fig. 8H, I). These vertical burrows predominantly represent domiciles of filter-feeding organisms, and the trace fossil assemblage is assigned to the *Skolithos* ichnofacies (*cf.* Pemberton *et al.* 1992a). Other common traces include fugichnia, *Macaronichnus*, *Planolites* and *Skolithos* (Fig. 4A).

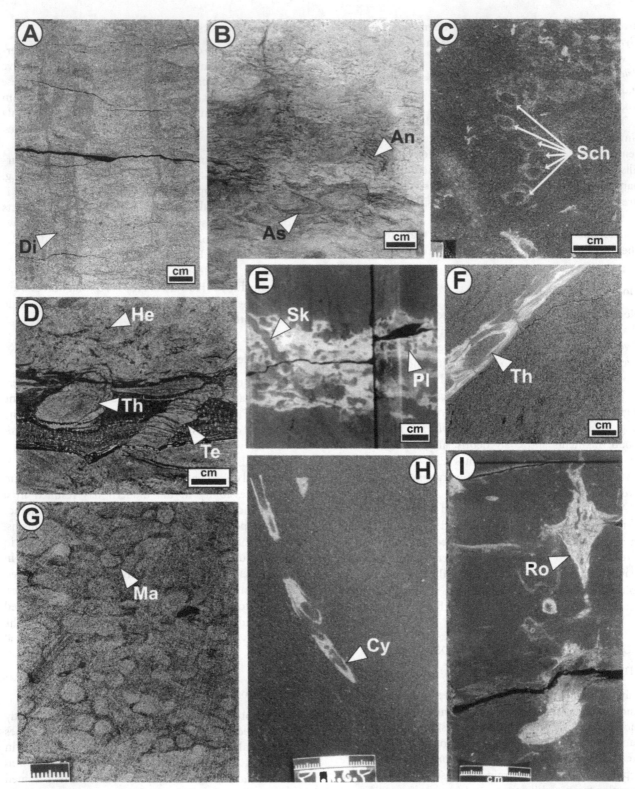

Fig. 8. Trace fossils of the lower estuary in the Bluesky Formation, including *Diplocraterion* (Di), *Anconichnus* (An), *Asterosoma* (As), *Schaubcylindrichnus* (Sch), *Macaronichnus* (Ma), *Helminthopsis* (He), *Teichichnus* (Te), *Thalassinoides* (Th), *Skolithos* (Sk), *Planolites* (Pl), *Cylindrichnus* (Cy) and *Rosselia* (Ro). Note that in each photograph, bitumen-stained sandstone is the dark lithology and muddy siltstone is the light lithology. Black mudstone laminae are present in (D). A–D: Offshore/distal shoreface deposits are characterised by diverse trace fossil assemblages, commonly associated with complex feeding and dwelling structures (core locations – A: 16-08-84-20W5, 525.1 m; B: 16-08-84-20W5, 529.8 m; C: 04-10-85-20W5, 520.2 m; D: 13-17-83-20W5, 572.8 m). E: Complete reworking of a mudstone bed in an ebb-tidal delta deposit (04-10-85-20W5, 554.9 m). F: Bioturbation is rare in tidal inlet deposits, restricted primarily to mudstone interbeds (note that the angle of bedding is exaggerated as the core is from a deviated well) (09-16-85-18W5, 627.75 m). G: Upper shoreface sandstone completely reworked by *Macaronichnus* (14-02-84-20W5, 553.1 m). H, I: Robust, mud-lined trace fossils in deposits of the fossil barrier–bar complex (H: 11-17-83-18W5, 627.2 m; I: 14-11-83-20W5, 600.5 m).

Macaronichnus is known to be a useful indicator of highly oxygenated surface waters and sediments, common to a narrow range of environments, including the upper shoreface, foreshore and estuary channel bars (e.g. Clifton & Thompson 1978; Curran 1985; Saunders *et al.* 1994). Rarer trace fossil elements present in shoreface/ barrier deposits of the Bluesky Formation include *Asterosoma*, *Helminthopsis*, and *Thalassinoides*, as well as *Teredolites* in transported wood clasts (Fig. 4A).

Distal shoreface/offshore transition deposits

Offshore/distal shoreface deposits in the study area are divided into two dominant lithofacies: (1) fine-grained siltstone/mudstone and (2) sandstone. Moderate to high bioturbation intensities are characteristic of these facies, with high diversities of trace fossil forms common (up to eight ichnogenera in some units). The trace fossil assemblage in fine-grained units includes *Asterosoma*, *Cylindrichnus*, *Chondrites*, *Helminthopsis*, *Palaeophycus*, *Planolites*, *Skolithos*, *Teichichnus* and *Thalassinoides*. This assemblage is assigned to the *Cruziana* ichnofacies, consisting mainly of deposit-feeding structures.

Common trace fossils present in the sand-rich facies include *Asterosoma*, *Cylindrichnus*, *Diplocraterion*, *Helminthopsis*, *Palaeophycus*, *Planolites*, *Rosselia*, *Schaubcylindrichnus*, *Skolithos*, and *Teichichnus* (Figs. 4A, 8A–C). Locally present are *Anconichnus*, *Arenicolites* and *Macaronichnus*. The high diversity of burrows and the predominance of vertical, cylindrical and U-shaped burrows lead to the assignment of the assemblage into the *Skolithos*, or mixed *Skolithos–Cruziana* ichnofacies,

deposited in the lower shoreface to offshore environment (cf. MacEachern & Pemberton 1992; Pemberton *et al.* 1992b). Unbioturbated beds associated with this facies are characterised by oscillation ripples and hummocky cross-stratification, suggesting the working of sediments by fair-weather and storm waves.

Discussion

Estuaries comprise a complex association of sediments, deposited in various subenvironments (Howard & Frey 1973). The different physicochemical stresses on burrowing organisms throughout an estuary are numerous, including lowered salinity, fluctuating salinity, high current energy, increased sedimentation rates, high water turbidity, and reduced oxygenation of bottom and interstitial waters. These palaeoenvironmental factors, as a whole, shape the trace fossil assemblage that is ultimately preserved in the rock record (Rhoads *et al.* 1972; Pemberton *et al.* 1992a). Discerning the effects of these stresses on the assemblage is difficult, however, especially if it is conceded that the burrowing behaviour of an organism is also strongly influenced by sediment texture (e.g. Howard & Frey 1973; Taylor *et al.* 2003). Nevertheless, ichnological patterns are evident across the Bluesky Formation estuarine complex in the study area (Fig. 4). Ichnofabrics were influenced by the relative intensities of the various primary palaeoenvironmental stresses that were active across the depositional system (Fig. 9). A summary of the effects of various stresses on infaunal organisms and their reflections in the rock record is

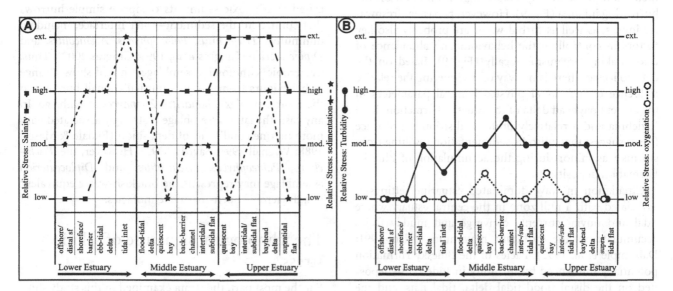

Fig. 9. A: Interpreted relative stresses on burrowing organisms due to lowered salinity and high sedimentation rates in the various subenvironments of the wave-dominated estuary studied. These are based on the observations made and interpretations discussed in this paper. Note that stresses related to lowered salinity increase towards the fluvial point source. B: Interpreted relative stresses on burrowing organisms due to high water turbidity and low oxygenation of bottom and interstitial waters.

presented below, based on observations made in this study.

The ichnological signature of rapid sedimentation and strong currents

Widespread coarse-grained deposits and the prevalence of sedimentary structures indicative of upper flow regime conditions lead to the interpretation that numerous facies of the estuary studied were deposited under energetic conditions (Hubbard *et al.* 2002). Accordingly, high current energy and sedimentation rates are interpreted to have limited burrowing activity in tidal inlet channels, on proximal tidal or bayhead deltas, and at the bases of back-barrier tidal channels.

In lower estuarine subenvironments, such as the tidal inlet where strong currents and rapid sedimentation are characteristic, burrows are absent despite the fact that most other (inferred) environmental parameters would be ideal for organisms to flourish (e.g. upper brackish to normal marine salinities, and oxygenated water and sediment). It is therefore interpreted that in the tidal inlet, high current energies and sedimentation (as well as migrating bedforms locally) comprise the primary stresses on burrowers. The significance of this environmental parameter is underscored by the fact that tidal inlet deposits are characterised by the least diverse trace fossil assemblage in the entire lower and middle estuary (Fig. 4B).

It is more difficult to assess the relative importance of the different environmental parameters in the bayhead delta area. There, strong currents and heightened sedimentation probably imparted a significant effect on the burrowing infauna (Fig. 9). However, persistent freshwater inputs, as well as turbid water are probably also key factors in controlling the behaviour and abundance of trace-making organisms locally (Fig. 9). Based on the data collected, there is no way of deciphering the relative influence of individual physicochemical parameters in the ancient bayhead delta of the Bluesky Formation. Rare fugichnia and a relatively uniform distribution of trace fossils suggest that episodic sedimentation resulting from storms was minor during the accumulation of Bluesky Formation deposits.

Circulation in wave-dominated estuarine settings is associated with a null zone in the central estuary where tidal and fluvial currents converge and muddy facies accumulate in the turbidity maximum (Allen 1991; Dalrymple *et al.* 1992). Facies of the Bluesky Formation potentially influenced by turbidity include those deposited on the distal flood-tidal delta, tidal flats, and the bayhead delta, as well as in quiescent bays and low-energy back-barrier tidal channels (Fig. 9). As discussed, burrows of deposit feeders normally predominate in sediments

accumulating in turbid environments, as filter-feeding organisms may not be able to cope with excessive material in the water column (Moslow & Pemberton 1988; Buatois & López Angriman 1992; Gingras *et al.* 1998). Trace fossils interpreted as the burrows of filter-feeding organisms such as *Skolithos*, *Gyrolithes* and *Cylindrichnus* persist across most of these middle to upper estuary facies, however, suggesting that, overall, turbidity may not have been a major stress in the fossil estuary (Fig. 4).

The ichnological signature of oxygenation stress

In the estuarine deposits studied, stresses on trace-making organisms due to lowered oxygen are considered to have been low overall (Fig. 9). Most subenvironments are characterised by sandy substrates, interpreted to have been associated with effective water circulation resulting from tidal currents, wave energy and fluvial input. Facies in the Bluesky Formation, associated with potentially low levels of oxygen in bottom or interstitial water, include rare mudstones deposited in quiescent embayments/lagoons (Fig. 7D). Ekdale & Mason (1988) suggested that deposit-feeding burrows with open connections to the sediment–water interface (e.g. *Chondrites* and *Zoophycos*) are typical of facies associated with extremely low oxygen levels in interstitial and bottom waters. Other than in central bay facies, burrows recording this behaviour are rare to absent (Fig. 4A).

It is likely, however, that burrow assemblages associated with dysaerobic sediment differ from those of poorly oxygenated sedimentary environments (as in de Gibert & Ekdale 1999). Modern shallow-water assemblages associated with anoxic sediments comprise simple burrows that connect to the sediment–water interface, including diminutive *Trichichnus*, *Palaeophycus*, *Arenicolites*, and *Diplocraterion* (Gingras *et al.* 1999; Gingras 2002). Thus, dysaerobic ichnofossil assemblages should show a range of behaviours from simple, unusually small burrows to the more complex *Chondrites–Zoophycos* assemblage. In any case, the latter assemblage is strongly associated with more marine conditions of deposition (Ekdale & Mason 1988; Wignall 1991; Savrda 1992). However, the *Trichichnus*, *Palaeophycus*, *Arenicolites*, and *Diplocraterion* assemblage may represent the brackish-water equivalent of a dysaerobic open-marine ichnofacies.

The ichnological signature of salinity stress

For the most part, the strata examined in this study show that salinity stress is the dominant biologically limiting factor in the estuary deposits. The diversity and size of trace fossils systematically decrease along the length of the fossil estuary, from facies associated with freshwater input in the region of the bayhead delta to facies deposited in the region of the tidal inlet. Also, previous tenets of the brackish-water model (e.g. Howard & Frey 1973, 1975; Dörjes & Howard 1975; Pemberton *et al.* 1982; Wightman *et al.* 1987; Beynon *et al.* 1988; Pemberton & Wightman 1992; MacEachern & Pemberton 1994; Gingras *et al.* 1999; Buatois *et al.* 2002) apply well to this marginal-marine Bluesky Formation deposit. These include: impoverished marine assemblages; vertical and horizontal structures present; monospecific horizons; somewhat diminutive trace fossils; locally prolific population densities; and the presence of the mixed *Skolithos–Cruziana* ichnofacies. The ichnological patterns documented across the ancient deposit and their potential importance are discussed below.

Ichnological variation across the estuary

Through the comparison of deposits from estuarine subenvironments with similar physical conditions, the effects of chemical variability across the depositional system can be assessed. In other words, if it can be established that physical processes are similar on tidal flats in the middle and inner estuary, it can be inferred that physical differences in the sedimentary facies are due primarily to hydrochemical changes in the depositional waters. In this section, ichnological comparisons of: quiescent bay deposits of the upper and middle estuary, and of the offshore; tidal flat deposits of the upper and middle estuary; and mudstone laminae associated with bayhead, flood-tidal and ebb-tidal deltas are attempted, in order to assess the relative effect of salinity stress.

Quiescent deposits: embayment and offshore

Facies that accumulated predominantly through quiet-water sedimentation, such as embayment and offshore deposits, potentially preserve the best evidence for recognising the effects of ancient salinity levels (MacEachern *et al.* 1999a). This is true mostly due to the reduced impact of some of the other stresses in these environments (Fig. 9). In the upper estuary, where water salinity is lowest proximal to the fluvial point source, *Planolites*, *Palaeophycus* and *Skolithos* represent the only commonly recurring trace fossils in the quiescent bay (Fig. 4A). The secondary components of this assemblage consist of four other ichnogenera. In the same depositional setting in the middle estuary, the recurring suite comprises six forms: *Planolites*, *Skolithos*, *Gyrolithes*, *Palaeophycus*, *Teichichnus* and *Thalassinoides* (Fig. 4A). Secondary components of this assemblage include six other ichnogenera. At the marine end of the system in the offshore, the diversity of trace fossils is much higher, with recurring elements

consisting of *Asterosoma*, *Cylindrichnus*, *Planolites*, *Skolithos*, *Teichichnus*, *Chondrites*, *Helminthopsis*, *Palaeophycus*, and *Thalassinoides* (Fig. 4A). In this setting, the complete assemblage also consists of up to six other, rarer components.

Some key observations can be summarised from the above information regarding salinity levels and the palaeoenvironment. The first is related to the diversity of trace fossils in the Bluesky Formation. At Willapa Bay, Washington, Gingras *et al.* (1999) documented the effects of lowering salinity on organism diversity and size from modern point bar sediments. They observed that species diversity and the size of the largest burrows decreased as salinity decreased up estuary (Fig. 10). The same trend is evident in similar curves plotted for the estuarine system present in the study area (Fig. 4B). Perturbations in these curves are associated with subenvironments where salinity stress was probably not the primary stress on burrowing organisms, such as in the tidal inlet (Fig. 9). Furthermore, the observation made by Gingras *et al.*

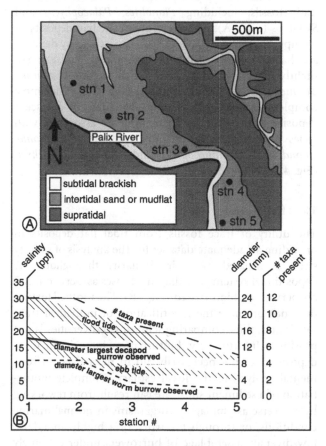

Fig. 10. A: Depositional environment map of the Palix River, Willapa Bay, Washington (western USA). At this location, observations of water salinity, burrowing organisms, and burrows were made at each of the five observation stations present. B: The diversity and size of burrowers/burrows decreases upriver where salinity is lowest. The flood- and ebb-tidal cycles are shaded, as they represent a range of values at each station (modified after Gingras *et al.* 1999).

(1999) regarding the largest burrow sizes is consistent with the observations in the Bluesky Formation. In the study area, the most robust, common burrows in quiescent bay deposits of the upper estuary are small *Planolites* and *Palaeophycus*, with diminutive *Thalassinoides* rarely present. In the middle estuary, relatively larger burrows are common, including more robust *Thalassinoides*. In the offshore deposits, robust *Thalassinoides*, *Rosselia*, *Teichichnus* and *Cylindrichnus* are all present in moderate to high abundance (Fig. 4A).

Trophic generalists, or organisms able to adapt their feeding strategy in response to changing environmental conditions, have been considered to be diagnostic of brackish-water settings (Wightman *et al.* 1987; Beynon *et al.* 1988; Pemberton & Wightman 1992). In associated environments, adaptation by an organism generally results in the construction of simple horizontal or vertical burrows. Consequently, in an area transitory between fresh and marine water, burrow morphologies are increasingly complex towards the marine end of the system (e.g. Howard *et al.* 1975). This is the case observed in sediments of the Bluesky Formation. Relatively simple trace fossils, including *Planolites*, *Palaeophycus* and *Skolithos* (Fig. 4A), dominate quiescent bay deposits of the upper estuary. Middle estuary quiescent bay deposits are characterised by a trace fossil assemblage consisting of slightly more complex forms, which include *Gyrolithes*, *Teichichnus* and *Thalassinoides* (Fig. 4A). The most complex feeding and dwelling structures in quiescent deposits in the Bluesky Formation are associated with facies deposited in the offshore, including *Asterosoma*, *Chondrites*, *Helminthopsis*, and *Schaubcylindrichnus* (Fig. 4A).

Tidal flat deposits

The utility of trace fossils from tidal flat deposits in providing an adequate data set for the analysis of salinity stress is negated by their similarity throughout the depositional system, and that in all cases associated facies are pervasively bioturbated (Fig. 4B). The mottled texture does not promote the identification of individual trace fossils and, thus, comparing ichnofossil diversities is not possible. Unfortunately, bioturbation intensity in these deposits does not necessarily aid in the palaeoenvironmental analysis of the fossil estuary. Complete bioturbation of a sediment substrate can result from reworking by a diverse assemblage of organisms in normal marine conditions, or through the work of a highly abundant, low-diversity assemblage of burrowers under extremely stressful conditions (Pemberton & Wightman 1992).

Intertidal flats are shaped by a host of stresses not encountered by subtidal sediment dwellers. These additional stresses include temperature fluctuations, wind and solar desiccation, a fixed oxygen supply, and increased predation. Thus, salinity is not necessarily the principal shaping parameter of the tidal flat deposits. The presence of all of these stresses on intertidal flats throughout the fossil estuary leads to the striking similarity in the ichnological character of the associated units.

Lower estuary and bayhead delta deposits

In environments dominated by sand, sandstone laminae sets are commonly separated by thin mudstone beds reflecting fluctuations in flow strength (Figs. 5C, 6A, 7A, 8E, F). As the sandstones are lacking in features unique to any one depositional environment (e.g. low- or high-angle cross-bedding or trough cross-bedding), palaeoenvironmental interpretation may be greatly enhanced through the careful analysis of the preserved mudstone interbeds (MacEachern *et al.* 1999a).

A comparison of bayhead delta sediments of the upper estuary, flood-tidal delta sediments of the middle estuary, and ebb-tidal delta sediments of the lower estuary is evaluated, focusing on the mudstone interbeds present in each facies (Fig. 4B). In each of the three environments, it is notable that the sandstone facies show no signs of biogenic reworking. The diversity of commonly recurring ichnogenera within mudstone beds across the estuary increases from the upper estuary (bayhead delta – one common ichnogenus) to the middle estuary (flood-tidal delta – two common ichnogenera), and then to the lower estuary (ebb-tidal delta – three common ichnogenera) (Fig. 4B). In each subenvironment, the common trace fossils are simple structures constructed by trophic generalists, probably indicative of salinity-induced stress, which was prominent during periods of waning flow. The increasing diversity of trace fossil forms towards the lower estuary, albeit slight, suggests that salinity-induced stresses were greatest at the landward end of the estuary. The only common trace fossils observed in the upper estuary are diminutive *Planolites*, consistent with observations on maximum burrow size distribution made by Gingras *et al.* (1999) (Fig. 10). The largest burrows observed in all of the deltaic deposits are *Thalassinoides*, which are common only in ebb-tidal delta deposits of the lower estuary.

Summary and conclusions

Estuaries are characterised by numerous stresses on potential trace-making organisms, including low salinities, fluctuating salinity levels, high water turbidity, high current energy, rapid sedimentation rates, and low oxygen levels in bottom and interstitial waters. Although the effects of these stresses are difficult to differentiate, doing so is important in order to understand the palaeoenvironmental conditions in ancient estuarine settings. These factors shape the trace fossil assemblage that is ultimately preserved in the rock record.

Trace fossils in the Cretaceous Bluesky Formation of Alberta provide important palaeoenvironmental information, thereby enhancing subsurface facies mapping and interpretation. Burrowing organisms in the ancient, wave-dominated estuary were particularly influenced by lowered salinities in the middle to upper parts of the system, and by high current energies and rapid sedimentation rates in both the lower and upper extremes of the system. Other factors suspected of contributing to the trace fossil assemblages include high suspended sediment volume in the middle to upper estuary associated with the turbidity maximum, fluctuating salinity levels in intertidal areas, and reduced oxygenation of bottom and interstitial waters locally.

Acknowledgements

Shell Canada Limited is thanked for their financial and technical support of our research on the Peace River Oil Sands property. In particular, we would like to thank Sheri Bechtel, Lorie Cooper, Felix Frey, Bryan Gould, Ed Howard, Ken Lawrence, Patricia Lee, and Mark Thomas for their help and encouragement. Natural Sciences and Engineering Research Council (NSERC) of Canada operating grants to MKG and SGP enabled much of this research. We would also like to recognise contributions from members of the Ichnology Research Group at the University of Alberta, especially Ian Armitage, Jason Frank, Eric Hanson, Jason Lavigne, Tom Saunders, Glen Schmidt, and John-Paul Zonneveld. Reviews of the manuscript by James MacEachern and Duncan McIlroy significantly improved the clarity of the paper, and are much appreciated.

References

Allen, G.P. 1991: Sedimentary processes and facies in the Gironde estuary: a recent model for macrotidal systems. *Canadian Society of Petroleum Geologists Memoir 16*, 29–40.

Beynon, B.M., Pemberton, S.G., Bell, D.D. & Logan, C.A. 1988: Environmental implications of ichnofossils from the lower Grand Rapids Formation, Cold Lake Oil Sands Deposit. *Canadian Society of Petroleum Geologists Memoir 15*, 275–290.

Brownridge, S. & Moslow, T.F. 1991: Tidal estuary and marine facies of the Glauconitic Member, Drayton Valley, central Alberta. *Canadian Society of Petroleum Geologists Memoir 16*, 107–122.

Buatois, L.A. & López Angriman, A.O. 1992: The ichnology of a submarine braided channel complex: the Whiskey Bay Formation of James Ross Island, Antarctica. *Palaeogeography, Palaeoclimatology, Palaeoecology 94*, 119–140.

Buatois, L.A., Mangano, M.G., Alissa, A. & Carr, T.R. 2002: Sequence stratigraphic and sedimentologic significance of biogenic structures from a late Paleozoic marginal- to open-marine reservoir, Morrow Sandstone, subsurface of southwest Kansas, USA. *Sedimentary Geology 152*, 99–132.

Cadée, G.C. 1998: Influence of benthic fauna and microflora. *In* Eisma, D. (ed.): *Intertidal Deposits: River Mouths, Tidal Flats, and Lagoons*, 383–402. CRC Press, Boca Raton.

Clifton, H.E. & Thompson, J.K. 1978: *Macaronichnus segregatis*, a feeding structure of shallow marine polychaetes. *Journal of Sedimentary Petrology 48*, 1293–1302.

Coates, L. & MacEachern, J.A. 1999: The ichnological signature of wave- and river-dominated deltas: Dunvegan and Basal Belly River formations, west-central Alberta. *Canadian Society of Petroleum Geologists and Petroleum Society Core Conference Paper 99-114C*, Calgary.

Curran, H.A. 1985: The trace fossil assemblage of a Cretaceous nearshore environment: Englishtown Formation of Delaware, U.S.A. *Society of Economic Paleontologists and Mineralogists Special Publication 35*, 261–276.

Dalrymple, R.W., Makino, Y. & Zaitlin, B.A. 1991: Temporal and spatial patterns of rhythmite deposition on mud flats in the macrotidal Cobequid Bay–Salmon River estuary, Bay of Fundy, Canada. *Canadian Society of Petroleum Geologists Memoir 16*, 137–160.

Dalrymple, R.W., Zaitlin, B.A. & Boyd, R. 1992: Estuarine facies models: conceptual basis and stratigraphic implications. *Journal of Sedimentary Petrology 62*, 1130–1146.

Dörjes, J. & Howard, J.D. 1975: Estuaries of the Georgia coast, U.S.A.: sedimentology and biology. IV. Fluvial–marine transition indicators in an estuarine environment, Ogeechee River–Ossabaw Sound. *Senckenbergiana Maritima 7*, 137–179.

Edwards, J.M. & Frey, R.W. 1977: Substrate characteristics within a Holocene salt marsh, Sapelo Island, Georgia. *Senckenbergiana Maritima 9*, 215–259.

Ekdale, A.A., Bromley, R.G. & Pemberton, S.G. 1984: Ichnology: the use of trace fossils in sedimentology and stratigraphy. *Society of Economic Paleontologists and Mineralogists Short Course 15*, 1–317.

Ekdale, A.A. & Mason, T.R. 1988: Characteristic trace-fossil associations in oxygen-poor sedimentary environments. *Geology 16*, 720–723.

Finger, K.L. 1983: Observations on the Lower Cretaceous Ostracode Zone of Alberta. *Bulletin of Canadian Petroleum Geology 31*, 326–337.

Frey, R.W. & Pemberton, S.G. 1985: Biogenic structures in outcrops and cores. 1. Approaches to ichnology. *Bulletin of Canadian Petroleum Geology 33*, 72–115.

Gibert, J.M., de & Ekdale, A.A. 1999: Trace fossil assemblages reflecting stressed environments in the Middle Jurassic Carmel Seaway of central Utah. *Journal of Paleontology 73*, 711–720.

Gingras, M.K. 2002: Neoichnology of the Bay of Fundy's north shore. *New England Inter-university Geological Congress Field Guides*, 33–42. Fredericton, New Brunswick.

Gingras, M.K., MacEachern, J.A. & Pemberton, S.G. 1998: A comparative analysis of the ichnology of wave- and river-dominated allomembers of the Upper Cretaceous Dunvegan Formation. *Bulletin of Canadian Petroleum Geology 46*, 51–73.

Gingras, M.K., Pemberton, S.G., Saunders, T. & Clifton, H.E. 1999: The ichnology of modern and Pleistocene brackish-water deposits at Willapa Bay, Washington: variability in estuarine settings. *Palaios 14*, 352–374.

Gingras, M.K., Räsänen, M. & Ranzi, A. 2002: The significance of bioturbated inclined heterolithic stratification in the southern part of the Miocene Solimoes Formation, Rio Acre, Amazonia Brazil. *Palaios 17*, 591–601.

Holmden, C., Creaser, R.A. & Muehlenbachs, K. 1997: Paleosalinities in ancient brackish water systems determined by $^{87}Sr/^{86}Sr$ ratios in carbonate fossils; a case study from the Western Canada Sedimentary Basin. *Geochimica et Cosmochimica 61*, 2105–2118.

Howard, J.D. 1975: The sedimentological significance of trace fossils. *In* Frey, R.W. (ed.): *The Study of Trace Fossils: a Synthesis of Principles, Problems and Procedures in Ichnology*, 131–146. Springer, New York.

Howard, J.D., Elders, C.A. & Heinbokel, J.F. 1975: Estuaries of the Georgia coast, U.S.A.: sedimentology and biology. V. Animal–sediment relationships in estuarine point bar deposits, Ogeechee River–Ossabaw Sound, Georgia. *Senckenbergiana Maritima 7*, 181–203.

Howard, J.D. & Frey, R.W. 1973: Characteristic physical and biogenic sedimentary structures in Georgia estuaries. *American Association of Petroleum Geologists Bulletin 57*, 1169–1184.

Howard, J.D. & Frey, R.W. 1975: Estuaries of the Georgia coast, U.S.A.: sedimentology and biology. I. Introduction. *Senckenbergiana Maritima 7*, 1–31.

Howard, J.D. & Frey, R.W. 1985: Physical and biogenic aspects of backbarrier sedimentary sequences, Georgia Coast, U.S.A. *Marine Geology 63*, 77–127.

Hubbard, S.M. 1999: Sedimentology and ichnology of brackish water deposits in the Bluesky Formation and Ostracode Zone, Peace River Oil Sands, Alberta. Unpublished Master's Thesis, University of Alberta, Edmonton.

Hubbard, S.M., Gingras, M.K., Pemberton, S.G. & Thomas, M.B. 2002: Variability in wave-dominated estuary sandstones: implications on subsurface reservoir development. *Bulletin of Canadian Petroleum Geology 50*, 118–137.

Hubbard, S.M., Pemberton, S.G. & Howard, E.A. 1999: Regional geology and sedimentology of the basal Cretaceous Peace River Oil Sands deposit, north-central Alberta. *Bulletin of Canadian Petroleum Geology 47*, 270–297.

Klein, G. deV. & Ryer, T.A. 1978: Tidal circulation patterns in Precambrian, Paleozoic, and Cretaceous epeiric and mioclinal shelf seas. *Geological Society of America Bulletin 89*, 1050–1058.

Kranck, K. 1981: Particulate matter grain-size characteristics and flocculation in a partially mixed estuary. *Sedimentology 28*, 107–114.

Leithold, E.L. 1989: Depositional processes on an ancient and modern muddy shelf, northern California. *Sedimentology 36*, 179–202.

Longstaffe, F.J., Ayalon, A. & Racki, M.A. 1992: Stable isotope studies of diagenesis in berthierine-bearing oil sands, Clearwater Formation, Alberta, Canada. *In* Kharaka, Y.K. & Maest, A.S. (eds): *Proceedings of the International Symposium on Water–Rock Interaction 7*, 955–958. International Association of Geochemistry and Cosmochemistry, and Alberta Research Council, Park City.

MacEachern, J.A. & Pemberton, S.G. 1992: Ichnological aspects of Cretaceous shoreface successions and shoreface variability in the western interior seaway of North America. *Society of Economic Paleontologists and Mineralogists Core Workshop 17*, 57–84.

MacEachern, J.A. & Pemberton, S.G. 1994: Ichnologic aspects of incised-valley fill systems from the Viking Formation of the Western Canada Sedimentary Basin, Alberta, Canada. *Society of Economic Paleontologists and Mineralogists Special Publication 51*, 129–157.

MacEachern, J.A., Stelck, C.R. & Pemberton, S.G. 1999a: Marine and marginal marine mudstone deposition; paleoenvironmental interpretations based on the integration of ichnology, palynology and foraminiferal paleoecology. *Society for Sedimentary Geology Special Publication 64*, 205–225.

MacEachern, J.A., Zaitlin, B.A. & Pemberton, S.G. 1999b: Coarse-grained, shoreline-attached, marginal marine parasequences of the Viking Formation, Joffre Field, Alberta, Canada. *Society for Sedimentary Geology Special Publication 64*, 273–296.

Martini, I.P. 1991: Sedimentology of subarctic tidal flats of western James Bay and Hudson Bay, Ontario, Canada. *Canadian Society of Petroleum Geologists Memoir 16*, 301–312.

Mattison, B.W. & Wall, J.H. 1993: Early Cretaceous foraminifera from the middle and upper Mannville and lower Colorado subgroups in the Cold Lake Oil Sands area of east-central Alberta; stratigraphic and paleoenvironmental implications. *Canadian Journal of Earth Sciences 30*, 94–102.

McKay, J.L. & Longstaffe, F.J. 1997: Diagenesis of the Lower Cretaceous Clearwater Formation, Primrose area, northeastern Alberta. *Canadian Society of Petroleum Geologists Memoir 18*, 392–412.

Meade, R.H. 1969: Landward transport of bottom sediments in estuaries of the Atlantic coastal plain. *Journal of Sedimentary Petrology 39*, 222–234.

Moslow, T.F. & Pemberton, S.G. 1988: An integrated approach to the sedimentological analysis of some Lower Cretaceous shoreface and delta front sandstone sequences. *Canadian Society of Petroleum Geologists Memoir 15*, 373–386.

Nanson, L.L., Culberson, R.S. & Savrda, C.E. 2000: Sedimentology and ichnology along an axial transect through an Upper Cretaceous Estuary, Eutaw Formation, eastern Gulf Coastal Plain. *Geological Society of America Annual Meeting Abstracts with Programs 32*, A308.

Pemberton, S.G., Flach, P.D. & Mossop, G.D. 1982: Trace fossils from the Athabasca Oil Sands, Alberta, Canada. *Science 217*, 825–827.

Pemberton, S.G., James, D.P. & Wightman, D.M. (eds) 1994: *Mannville Core Conference. Canadian Society of Petroleum Geologists Exploration Update.*

Pemberton, S.G., MacEachern, J.A. & Frey, R.W. 1992a: Trace fossil facies models: environmental and allostratigraphic significance. *In* Walker, R.G. & James, N.P. (eds): *Facies Models: Response to Sea Level Change*, 47–72. Geological Association of Canada, St. John's.

Pemberton, S.G., Van Wagoner, J.C. & Wach, G.D. 1992b: Ichnofacies of a wave-dominated shoreline. *Society of Economic Paleontologists and Mineralogists Core Workshop 17*, 339–382.

Pemberton, S.G. & Wightman, D.M. 1992: Ichnologic characteristics of brackish water deposits. *Society of Economic Paleontologists and Mineralogists Core Workshop 17*, 141–167.

Rahmani, R.A. 1988: Estuarine tidal channel and nearshore sedimentation of a Late Cretaceous epicontinental sea, Drumheller, Alberta. *In* de Boer, P.L., van Gelder, A. & Nio, S.D. (eds): *Tide-influenced Sedimentary Environments and Facies*, 433–471. Reidel Publishing, Dordrecht.

Ranger, M.J. & Pemberton, S.G. 1992: The sedimentology and ichnology of estuarine point bars in the McMurray Formation of the Athabasca Oil Sands deposit, north-eastern Alberta, Canada. *Society of Economic Paleontologists and Mineralogists Core Workshop 17*, 401–421.

Reading, H.G. & Collinson, J.D. 1996: Clastic coasts. *In* Reading, H.G. (ed.): *Sedimentary Environments: Processes, Facies and Stratigraphy*, 154–231. Blackwell Science, London.

Reidiger, C.L., Fowler, M.G. & Snowdon, L.R. 1997: Organic geochemistry of the Lower Cretaceous Ostracode Zone, a brackish/non-marine source for some Lower Mannville oils in southeastern Alberta. *Canadian Society of Petroleum Geologists Memoir 18*, 93–102.

Rhoads, D.C. 1973: The influence of deposit-feeding benthos on water turbidity and nutrient recycling. *American Journal of Science 273*, 1–22.

Rhoads, D.C. 1975: The paleoecological and environmental significance of trace fossils. *In* Frey, R.W. (ed.): *The Study of Trace Fossils: a Synthesis of Principles, Problems and Procedures in Ichnology*, 147–160. Springer, New York.

Rhoads, D.C. & Morse, J.W. 1971: Evolutionary and ecologic significance of oxygen-deficient marine basins. *Lethaia 4*, 413–428.

Rhoads, D.C., Speden, I.G. & Waage, K.M. 1972: Trophic group analysis of Upper Cretaceous (Maestrichtian) bivalve assemblages from South Dakota. *American Association of Petroleum Geologists Bulletin 56*, 1100–1113.

Saunders, T., MacEachern, J.A. & Pemberton, S.G. 1994: Cadotte Member sandstone: progradation in a Boreal basin prone to winter storms. *Canadian Society of Petroleum Geologists Manville Core Conference*, 331–350.

Savrda, C.E. 1992: Trace fossils and benthic oxygenation. *In* Maples, C.G. & West, R.R. (eds): *Trace Fossils. Short Courses in Paleontology 5*, 172–196. The Paleontological Society, Knoxville.

Savrda, C.E. & Bottjer, D.J. 1989: Trace-fossil model for reconstructing oxygenation histories of ancient marine bottom waters: application to Upper Cretaceous Niobrara Formation, Colorado. *Palaeogeography, Palaeoclimatology, Palaeoecology 74*, 49–74.

Savrda, C.E., Locklair, R.E., Hall, J.K., Sadler, M.T., Smith, M.W. & Warren, J.D. 1996: Ichnofabrics in a tidal-inlet sequence, Tombigbee Sand Member, Eutaw Formation (Santonian), central Alabama. *Geological Society of America Annual Meeting (Southeastern Section) Abstracts with Programs 28*, 42.

Smith, D.G. 1988: Tidal bundles and mud couplets in the McMurray Formation, northeastern Alberta, Canada. *Bulletin of Canadian Petroleum Geology 36*, 216–219.

Taylor, A., Goldring, R. & Gowland, S. 2003: Analysis and application of ichnofabrics. *Earth Science Reviews 60*, 227–259.

Weimer, R.J., Howard, J.D. & Lindsay, D.R. 1982: Tidal flats. *American Association of Petroleum Geologists Memoir 31*, 191–246.

Wightman, D.M. & Pemberton, S.G. 1997: The Lower Cretaceous (Aptian) McMurray Formation: an overview of the Fort McMurray area, northeastern Alberta. *Canadian Society of Petroleum Geologists Memoir 18*, 312–344.

Wightman, D.M., Pemberton, S.G. & Singh, C. 1987: Depositional modeling of the Upper Mannville (Lower Cretaceous), East-Central Alberta: implications for the recognition of brackish water deposits. *Society of Economic Paleontologists and Mineralogists Special Publication 40*, 189–220.

Wignall, P.B. 1991: Dysaerobic trace fossils and ichnofabrics in the Upper Jurassic Kimmeridge Clay of southern England. *Palaios 6*, 264–270.

Wilber, C.G. 1983: *Turbidity in the Aquatic Environment: an Environmental Factor in Fresh and Oceanic Waters*. Charles Thomas Publishing, Springfield, IL.

Zaitlin, B.A., Dalrymple, R.W., Boyd, R.A., Leckie, D. & MacEachern, J. 1995: The stratigraphic organization of incised valley systems: implications to hydrocarbon exploration and production. *Canadian Society of Petroleum Geologists Short Course*, 1–189. Canadian Society of Petroleum Geologists, Calgary.

Zaitlin, B.A. & Shultz, B.C. 1990: Wave-influenced estuarine sand body, Senlac heavy oil pool, Saskatchewan, Canada. *In* Barwis, J.H., McPherson, J.G. & Studlick, J.R.J. (eds): *Sandstone Petroleum Reservoirs*, 363–387. Springer, New York.

Zonneveld, J.P., Gingras, M.K. & Pemberton, S.G. 2001: Trace fossil assemblages in a Middle Triassic mixed siliciclastic-carbonate marginal marine depositional system, British Columbia. *Palaeogeography, Palaeoclimatology, Palaeoecology 166*, 249–276.

Compound trace fossils formed by plant and animal interactions: Quaternary of northern New Zealand and Sapelo Island, Georgia (USA)

MURRAY R. GREGORY, ANTHONY J. MARTIN & KATHLEEN A. CAMPBELL

Gregory, M.R., Martin, A.J. & Campbell, K.A. 2004 87 104: Compound trace fossils formed by plant and animal interactions: Quaternary of northern New Zealand and Sapelo Island, Georgia (USA). *Fossils and Strata*, No. 51, pp. 88–105. New Zealand. ISSN 0300-9491.

Quaternary coastal-terrestrial deposits from Aupouri and Karikari Peninsulas, northern North Island (New Zealand), and from Sapelo Island, Georgia (USA), reveal sedimentary structures produced by plant roots that provided habitats suitable for root-sucking, deposit-feeding, burrowing, brooding and other terrestrial invertebrate activities. The structures are typically large (<1 to >2 m in length, with widths from a few to >45 cm), and consist of white sand-filled tubes that are circular to subcircular in cross-section, and cylindrical to downwardly tapering in longitudinal view. Dark brown (humic and ferruginous) haloes demarcate the white sand fill. These structures open upwards, often connecting into overlying palaeosols, and are inferred to reflect the root architecture of several large forest trees that are morphologically comparable with those of some modern conifers, e.g. kauri (*Agathis australis*) in New Zealand and loblolly pine (*Pinus taeda*) in Sapelo Island. The white sand cores of these structures are a passive fill originating from a podzolised horizon, and/or from drifting aeolian sands. The root structures were temporarily open to the Quaternary surface and represent cavities remaining after tree death and/or toppling. The yellowish host sands are frequently mottled by trace fossils, as are boundaries between outer dark brown haloes and inner white sand fill to the root structures. Many of these traces are small, of simple form, and cannot readily be ascribed to any ichnotaxon. Meniscate burrows (*Taenidium*) that have been identified from both localities were probably produced by cicada nymphs. The moist and sheltered tree root-protected environment persisted for some time after tree death and was a desirable microhabitat for a number of invertebrates. Around the margins of these root casts, trace fossils and tiering fabrics may cross one another irregularly, develop oblique to primary bedding surfaces, or can even be inverted. Such stratigraphically disjunct relationships could be misleading in structural and palaeoenvironmental assessment of older or tectonically deformed strata that include palaeosols. These observations add an additional dimension to reconstructions of ancient forest cover and terrestrial continental environments.

Key words: Continental trace fossils; root-insect interactions; compound/complex traces; Quaternary; New Zealand; Sapelo Island.

Murray R. Gregory [m.gregory@auckland.ac.nz] & Kathleen A. Campbell [ka.campbell@ auckland.ac.nz], Department of Geology, The University of Auckland, Private Bag 92019, Auckland, New Zealand

Anthony J. Martin [geoam@learnlink.emory.edu], Department of Environmental Studies, Emory University, Atlanta, GA 30322, USA

Introduction

The casts and moulds of tree trunks, stumps and rooting structures are common in coal measures, continental terrestrial deposits, and marginal-marine sequences. Often they are associated with palaeosols. From the mid-1800s onwards, these occurrences were identified and reported on the basis of botanical taxonomic criteria. For instance, the well-known Carboniferous "form genera" *Stigmaria* and *Lepidodendron* were quickly accepted as casts and moulds of the roots and trunks, respectively, of a tree-sized plant (e.g. Brown 1848). More recent examples are identifications of rooting structures belonging to the palm family in the Pliocene of southern Baja California, Mexico (Fischer & Olivier 1996) and in Quaternary sands from northern New Zealand (Gregory & Campbell 2003).

Where botanical identification of root makers is not possible, these features generally receive but passing mention, other than to signify non-marine environments, with the notable exception of mangrove roots (e.g. Whybrow & MacClure 1981; Galli 1991). Overall, these structures are commonly known as rhizoliths when preserved as calcite-cemented root casts, moulds, tubules and sheaths (e.g. Klappa 1980; Loope 1988; D'Alessandro *et al.* 1992). At the present time, there is an evolving appreciation that plant rooting structures have much to offer habitat reconstructions and palaeoenvironmental interpretations (e.g. Glennie & Evamy 1968; Bown 1982; Cohen 1982; Plaziat & Mahmoudi 1990; Bockelie 1994; White & Curran 1997; Curran & White 2001; Retallack 2001). This may even be the case when rooting structures remain taxonomically unidentified (Pfefferkorn & Fuchs 1991; Bockelie 1994). Numerous invertebrate and vertebrate trace fossils (burrows, borings, trails and trackways) are also known from palaeosols (e.g. Retallack 2001; Hasiotis 2002).

In this contribution, we describe an intimate relationship between small terrestrial invertebrate trace fossils and thallo-ichnomorphs [following the terminology of White & Curran (1997)] that reflects the rooting architecture of large forest trees in Quaternary siliciclastic dune deposits of coastal settings. Identical composite ichnofabrics have been recognised at two widely separated localities – one in the eastern hemisphere and the other in the western, but both lying at similar distances south and north of the equator.

Root-related structures: trace fossils or body fossils?

While structures in the geological record attributable to the activity of plant roots are sometimes identified as such (e.g. Bracken & Picard 1984; Bockelie 1994), they generally present a descriptive and interpretive challenge for ichnologists, compounded by the semantic dilemma of whether they represent trace fossils or body fossils. In the former sense, the mere act of describing suspected root structures is fraught with difficulty, owing to their complex geometries and variability. In some cases, they closely resemble abiogenic structures (e.g. fracture fills, mud cracks) with which they commonly occur, and in others, they mimic trace fossils made by invertebrates or vertebrates (Boyd 1975). Moreover, the vast majority of ichnologists study invertebrate or vertebrate traces; their zoological leanings may result in a lack of botanical training, interest or awareness. For example, the analysis of invertebrate and vertebrate trace fossils has become increasingly sophisticated over the past 30 years, to the point where narrowly defined clades, genera, or even species of trace makers can be reliably attributed to their

traces in the fossil record (e.g. Lockley & Hunt 1994; Hasiotis 2002; Nesbitt & Campbell 2002; Bromley *et al.* 2003; Curran & Martin 2003; Rindsberg & Martin 2003). In contrast, it has been the same three decades since overt recognition of root-related structures as trace fossils (Sarjeant 1975, 1983), yet their assessment typically remains limited to noting presence amidst abundant animal-caused trace fossils, or simply identifying such structures as "root traces" (e.g. Zonneveld *et al.* 2001; Buatois & Mángano 2002; Retallack *et al.* 2002) or "phytoturbation" (e.g. Gregory & Campbell 2000). However, works by Froede (2002), Buatois & Mángano (2002), Traynham & Martin (2003) and Gregory & Campbell (2003) represent exceptions to this generalisation.

Priority in any discussion of the interpretative value of root-related structures must be given to proper descriptions. For example, Gregory & Campbell (2003) provide an account of a *Phoebichnus* look-alike in which the trace maker was identified as the rooting system of the nikau palm (*Rhopalostylis sapida*) in Quaternary coastal dune sediments from New Zealand. *Phoebichnus* is normally associated with an invertebrate trace maker in marine strata (Bromley & Asgaard 1972), thus exemplifying how plants can produce traces similar to those attributed to animals, and underlining the importance of thorough descriptions of suspected rooting structures. Herein, the following descriptive criteria are proposed as a checklist to test whether any given structure is related to plant roots: (1) inconsistent diameters within any given length, which are especially notable if they taper; (2) secondary and tertiary branching that shows the aforementioned distal tapering with each successive branch; (3) dichotomous, Y-shaped branching with junctions that are not noticeably enlarged; (4) downward, near-vertical to oblique orientations (with some exceptions based on responses of a plant to the originally affected substrate); (5) lack of evidence for active fill (or, conversely, evidence favouring passive fill from overlying layers); and (6) carbonised or otherwise preserved plant material in the structures (noting, however, that some invertebrate traces contain plant material that has been "stuffed" into burrows). Although not all of these criteria may be fulfilled in any given description, those that are most diagnostic of roots include both distal tapering and dichotomous branching, without enlargement at branch junctions.

With the establishment of a reasonable interpretation of root-related structures, a point of contention may be whether they should be classified as body fossils or trace fossils or plants. Confusion associated with the terminology of fossil root structures can lead to the equating of root casts with root traces (Driese *et al.* 1997), or the application of numerous terms to root-related structures, such as "rhizolith" (Donovan *et al.* 2002; Froede 2002), "rhizomorph" (Erwin 1984; Rothwell & Erwin 1985), "rhizocretion" (Jones *et al.* 1998), or "ichnorhizomorph"

(Curran & White 1999). Rhizomorph is a particularly problematic term because it is more properly applied to subsurface fungal colonies that mimic the shape and geometry of plant roots (Kwansa 2002; Mihail *et al.* 2002), but has also been appropriated by geologists to describe root-like structures in general (Erwin 1984; Rothwell & Erwin 1985). In some instances, such apparent conflation is resoluble by noting that both body and trace components can compose any given root structure. For example, roots can leave carbonised residues as preserved remnants of a plant body, yet the displacement of sediment grains or geochemical changes imparted to the surrounding substrate as a result of plant movement and respiration, respectively, constitute traces of its behaviour (Ross *et al.* 2001). In the latter sense, root structures are thus traces, and their recognition is grounded in how they represent the products of plant behaviour that impart noticeable, describable, and diagnostic changes to substrates. Accordingly, ancient root structures that are related to substrate changes resulting from plant behaviour should be classified as trace fossils. In contrast, external or internal moulds of roots from plants that were uprooted and subsequently buried are not traces of behaviour. Hence, such structures are more akin to most skin impressions associated with dinosaur bones; both are classified as body fossils. (However, skin impressions in dinosaur footprints are trace fossils because they are associated with the behaviour of living organisms.) A cautionary approach is advised for the description of root-like structures encountered in the field or cores, and palaeontologists should take great care before categorising them as either "trace fossils" or "body fossils". The reality is that root-like structures often result from a blending of trace fossils with body fossils, wherein post-mortem processes, such as roots decaying *in situ* with their traces, blur boundaries between the two types.

Traces of plant behaviour that are most likely to be preserved in the fossil record are associated with plant roots that move downwards as a result of positive gravitropism, also known as geotropism, coupled with their contact with a substrate. These behavioural effects may be preserved more easily because they are typically subsurface, and thus already buried, and occasionally penetrate substrates to great depths. From this perspective, root traces are analogous to "elite" trace fossils defined from marine environments (e.g. *Zoophycos*) that are emplaced far below the zone of active sedimentary reworking (Ekdale & Bromley 1984). However, not all traces of root movement have preservation potential, particularly those that affect lithic substrates in terrestrial settings (Mikuláš 2001), but their visible effect on vertebrate bones supplies an example of how such trace fossils can be preserved in some consolidated substrates (Montalvo 2002).

In both sediment and rock, roots physically shift grains, cause or widen fractures in bedrock, or locally compact

sediments as they extend and grow, whether through tropism, growth, or both. Geochemical changes caused by plant behaviour include local alterations of sediment or rock around roots that produce visible colour changes, evident as "haloes" around the former loci of the roots. These haloes are a result of respiration, ion exchange, water and nutrient flow, and interactions with other organisms such as bacteria, fungi, and termites (Ahonen-Jonnarth *et al.* 2003). Perhaps the most important of these symbiotic relationships are represented by mycorrhizae, fungal colonies that surround roots and improve their nutrient uptake through fine-scale invasion of the soil (Smith & Read 1997; Hibbett *et al.* 2000). Mycorrhizae are also documented in the fossil record. Two Tertiary (Eocene) examples provide compelling evidence of fungal colonies surrounding root structures of conifers such as *Pinus* sp. (LePage *et al.* 1997) and *Metasequoia milleri* (Stockey *et al.* 2001), and are directly comparable with similar colonies associated with modern species. Haloes that result from symbiotic and geochemical processes may or may not contain carbonised remnants of the plants proximal to the zone of original contact with the plant roots. Nonetheless, in these cases, a type of taphonomic overprinting can occur where dead roots may have remained in their traces and geochemically contributed to the surrounding microenvironment through their decay (Jones *et al.* 1998).

Regardless of whether plant remains are present or not, root-related structures in the geological rock record that occur as a direct result of plant behaviour are trace fossils. Moreover, where symbiotic relationships between plant roots and mycorrhizae result in alterations of the surrounding sediment during the lifetimes of those organisms, the resultant structures are compound trace fossils. A further complication arises if the structures served as sites for sediment accumulation and created microenvironments amenable to inhabitation and exploitation by later generations of plants and animals. In such cases, any traces added to those of the plant or fungal traces would constitute a complex trace fossil, a biogenic amalgamation representing behaviours of different taxa and times. This concept is further explored here, with examples from Aupouri and Karikari Peninsulas (northern North Island, New Zealand) and Sapelo Island (Georgia, USA).

Karikari and Aupouri Peninsulas, northern New Zealand

During the Pleistocene and Holocene, two expansive tombolos (Fig. 1) permanently linked to the northern New Zealand mainland an archipelago of small, once-upstanding and erosion-resistant islands of local

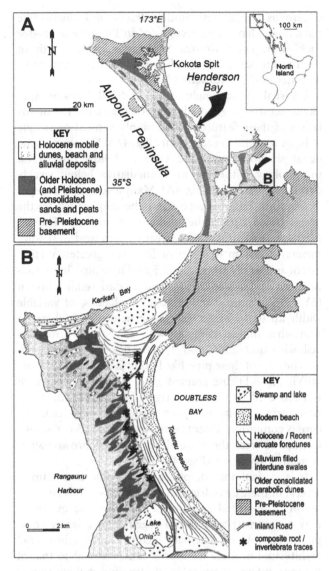

Fig. 1. A: Simplified geological map of northernmost New Zealand (Northland) illustrating the basement archipelago and extensive Quaternary tombolos of Aupouri and Karikari Peninsulas that connect the ancient islands to the mainland of the North Island. Note the inset map of North Island of New Zealand (top right). The Henderson Bay and Karikari Peninsula–Tokerau Beach localities are shown with an arrow, with the location of (B) also indicated. B: This enlarged and more detailed geological map shows the broad zone of arcuate Holocene/Recent beach ridges that back Tokerau Beach and the older (Pleistocene), semi-consolidated, northeast–southwest-orientated, parabolic dunes crossed by the Inland Road. Root trace localities are starred. (After Hay 1975; Newnham *et al.* 1993; Isaac 1996.)

basement rocks (Mesozoic and Tertiary: Isaac 1996; Black & Gregory 2002). These tombolos were built up of a suite of quartz and quartzo-feldspathic shallow-marine, beach and dune sands with associated estuarine, lacustrine and swamp sediments (Isaac 1996). The region was once heavily forested with the giant kauri (*Agathis australis*) dominant, and for which numerous radiocarbon dates from about 40,000 to 2000 years BP have been recorded (e.g. Goldie 1975; Ricketts 1975; Hay 1981; Ogden *et al.*

1992; Newnham *et al.* 1993). Kauri prefers well-drained soils. The local Holocene demise of this cover may be related to heavy leaching, impervious iron pan development, and podzolisation, with ground conditions becoming increasingly boggy. This changing circumstance doomed kauri to local, and episodic, self-destruction, with later human intervention furthering the rate of deforestation. There followed a recent period of destabilisation and, with accompanying deflation, the development of extensive, actively mobile, modern dune fields. More recently, pine plantations have stabilised many dunes and others are now grassed. Dissection of older large dunes through marine and stream erosion, and later road works, reveal excellent cross-sectional profile and deflational plan views of the weakly to moderately consolidated sands. The observations reported below have been made at two localities: Inland Road (Karikari Peninsula) and Henderson Bay (Aupouri Peninsula) (Fig. 1).

Inland Road

Along most of its length, the modern foredune of east-facing Tokerau Beach, Karikari Peninsula is backed by an arcuate system of Holocene/Recent beach-parallel ridges 1000 m or so across. At its inner margin, this beach ridge field abuts a set of older, larger and now fixed parabolic dunes that are strongly aligned northeast–southwest (Fig. 1B) (Hicks 1983). The Inland Road passes over this latter dune field within 200 m of its termination against the most landward beach ridges. Up to 6 m of moderately consolidated, cross-bedded aeolian sands are exposed where the road cuts deeply through dune crests. Interdune swales are commonly occupied by peaty sediments, which may pass laterally into palaeosols and iron-stained or organic-rich sands ("coffee rock"; Fig. 2). Weak palaeosols are developed across some parabolic dune crests. A vague to strong, pervasively mottled fabric (Fig. 3) is often evident in the vertical profiles exposed in roadside cuttings, as well as in views more or less concordant with bedding. The mottles are typically characterised by a dark core and a pale border, and resemble drab haloed root traces, which most likely represent organic matter buried in soil below or near the water table (Retallack 2001). This mottled fabric post-dates depositional cross-bedding and probably originated from plant rooting and growth habits (Gregory & Campbell 2003). Locally and irregularly, the mottling may have destroyed any evidence of the bidirectional cross-bedding that otherwise characterises the cores of these dunes. On exposed surfaces that closely follow bedding or dune relief, a similarly mottled ichnofabric is sometimes replaced by simple traces resembling "*Planolites*". A few of these traces appear to branch and are suggestive of

Fig. 2. Irregularly terraced or benched "coffee rock" and palaeosols (foreground), with a discontinuous cover of drifting sand (middle and far distance) at Henderson Bay. (Viewed to the southeast from GR N03/220 168.)

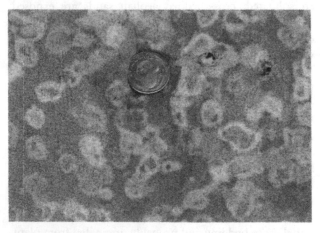

Fig. 3. Drab haloed mottles from <5 to >20 mm across, a common feature of the older, parabolic dune sands, are considered to have formed around decaying roots at, or immediately below, the water table. Coin diameter 23 mm. (Inland Road, GR O03/438 016.)

"*Thalassinoides*". Both traces are here considered to be root related, with the latter bearing similarities to the "branching, horizontally oriented rhizoconcretions" illustrated by Blay & Longman (2001: fig. 5D) in

calcareous aeolianites. Similar traces and dikaka-like phenomena have been recognised in Pleistocene "coffee rock" terraces at Kotoka Spit some 25 km north of Henderson Bay. Here mottling was considered to be of diagenetic origin (Ricketts 1975).

Unusual and strikingly large cylindrical shafts or pipe-like structures are exposed in several road cuts. The orientation of these features is commonly vertical or steeply oblique to dune development (Fig. 4A). However, some lie at shallow angles and tend to follow dune crest morphology, either in or immediately below weakly developed palaeosols (Fig. 4A). Many also appear to taper downwards, but this aspect may be emphasised by the orientation of surfaces being viewed in road cuts. The size of these structures varies greatly from widths (~diameters) >45 cm to 1 cm or less. The greatest vertical extent and/or length reaches 2.5 m. These pipe-like structures are demarcated by conspicuous dark reddish brown (5YR 2.5/2) humic and ferruginous haloes, of variable width up to 10 or 15 cm. These haloes have sharp to somewhat diffuse contacts with the strongly mottled yellowish sand (2.5Y 7/3–7/4) of the hosting dune (Figs. 4, 5). The core of these pipe-like bodies is light grey or white (10YR 7.1–8.1) fine-grained sand, the source of which is passive fill from an overlying, strongly podzolised, near-surface horizon of pure white sand (Fig. 4B). The core is commonly absent from structures <5 cm across. Contact between the white fill and the dark brown halo is typically knife sharp (Figs. 4B, 5).

It is evident that despite their variable orientations, several of these cylindrical shafts and pipe-like structures are in close spatial association and, in some instances, may be geometrically linked together in a larger compound body. Upon schematic reconstructions, the overall architecture of these features is rather similar to that of the root systems of several local, large and shallow-rooted trees. A typical example is the horizontal, broadly spreading root system of kauri (*Agathis australis*) (Fig. 6). There are, however, a number of other local, large (totara: *Podocarpus totara*) and smaller trees (kanuka: *Kunzea ericoides*) that are also potential root cast producers. While most of the pipe-like structures are probably tree root related, there are a few isolated examples suggesting burial of a tree trunk by an advancing sand dune. In these cases, prominent, large-scale, dune cross-bedding is interrupted by untapered pipes through vertical thicknesses of 1 m or more (Fig. 4A).

Associated trace fossils

A great diversity of small, simple, unbranched, "wriggle-like" and unidentified trace fossils is present in the dark haloes that define the margins of these root casts and the immediately adjacent yellowish hosting sand and/or white core fill. Many of these reach lengths of >30 mm

Fig. 4. Inland Road localities: probable tree trunk and root structures with margins sharply defined by dark, humic palaeosols and/or limonitic haloes, and back filled by structureless fine white quartz sand. A: Two vertical, non-branching, shaft-like structures that probably represent casts of tree trunks engulfed by drifting sand; note the fine vertical rooting fabric (*?Skolithos*) between the engulfed tree trunks (length of hammer = 31 cm). B: Rooting structures that tend to follow ancient dune crest morphology; note the partial infilling of fine white sand (arrowed). (Inland Road, GR O03/440 022.)

and widths of > 3 mm, and they often terminate sharply (Fig. 5). These traces display similarities with many of the terrestrial invertebrate ichnofossils described and illustrated by Ratcliff & Fagerstrom (1980). Their fill is derived from either the white, very fine sand of the core, or the slightly coarser yellowish fine to medium sand of the hosting sediment. Dark sediment from the enclosing halo is also worked into both the core fill and the hosting dune sand (Fig. 5). Deeper parts of the core fill are structureless, but upwards and towards the surface in which root casts were formed, the fill may become extensively bioturbated (Fig. 5).

Local heterolithic breccia horizons develop where palaeosol and humic or ferruginous clasts, as well as material derived from consolidated dune sands, have collapsed into tree root and trunk casts (Fig. 7A). Associated with the dark haloes of several of these cavities there are a few unbranched traces which have a finely meniscate character and which are here identified as *Taenidium* (following D'Alessandro & Bromley 1987; D'Alessandro *et al.* 1992). These traces typically open into the core fill, in places exceed 30 mm in length, have widths of 2–4 mm, and exhibit a broadly rounded termination of similar shape to the menisci (Fig. 7B, C). Ricketts (1975: fig. 49)

Fig. 5.   Inland Road (Gr O03/437 011). Cavernous exposure in a fresh road cut Reveals one wall of a vertical, dark organic-rich halo to a root cast structure (centre) which contrasts strongly with the mottled, greyish white sand of the core fill (left) and the hosting yellowish dune sand (right). Note the intensive bioturbation of the core fill (middle and upper) and numerous simple "wriggle-like" traces that reflect reworking of core, halo and host sediments into each other. Towards the visible base of the core fill (centre bottom) bioturbation reduces dramatically to simple isolated burrows with dark fill (length of hammer = 31 cm).

recorded a similar trace fossil in Quaternary "coffee rock" dune sands from Kokota Spit to the north of Henderson Bay. Most of these traces have no defined lining (Fig. 7B, C) and the menisci consist of alternating dark and light bands, the latter having been reworked from fine white sand of root cast cores. A vague to moderately distinct lining, <1 mm across, is recognisable in a few traces that are similar to *Taenidium* in general outline, but whose fill is homogeneous (Fig. 7B). In at least three instances, several similar meniscate traces appear to branch, cluster and/or radiate from an ill-defined central zone (Fig. 8). D'Alessandro & Bromley (1987: text-fig. 11) have ascribed trace fossils of this kind to the ichnogenus *Cladichnus*.

Henderson Bay

Along the sweep of this shore, modern foredunes partially hide an escarpment about 10 m high cut in consolidated Pleistocene shallow-marine and aeolian sands that form the resistant spine of the Aupouri tombolo. An extensive surface behind this beach is devoid of vegetation and subject to periodic cover by modern sand drifts. Across these Pleistocene deposits were cut a flight of terraces between 2 and 60 m above sea level. Extensive exposed surfaces lie behind the modern beach and are subject to irregular cover or exhumation through drifting sand lie behind the modern beach (Fig. 2).

In situ kauri stumps (radiocarbon dated at 32,000 ± 2000 years BP: Goldie 1975), as well as palms and other fossil forest trees (Gregory & Campbell 2003), feature in benches of a prominent terrace at 19–29 m above

Fig. 6.   Stump of a moderate-sized kauri (*Agathis australis*) on display outside the Kauri Museum at Matakohe, Northland, New Zealand. The broadly spreading shallow rooting architecture is clearly revealed (height of child = 1.4 m). The public notice (on the right) reads: "ROOT FORMATION … OF A SMALL KAURI. THIS STUMP WAS IN A SWAMP FOR 1590 YEARS, 100 KM SOUTH AUCKLAND AT ROTONGARO, HUNTLY. PRESENTED BY A.J. FURNESS".

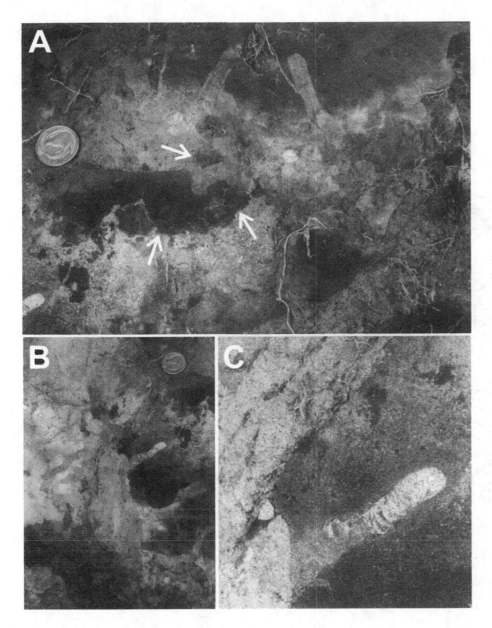

Fig. 7. Inland Road (GR O03/440 022). A: Intense and varied bioturbation that involves a fine white sand fill, a probable palaeosol, organic-rich and limonitic haloes, as well as the pale yellow hosting dune sand. The broken, fragmented and mixed character of this exposure suggests brecciation (arrowed) associated with collapse into a cavity left by rotting roots. Note the clotted bioturbate fabric (centre) and indistinct lining to two burrows with structureless fill (top middle) that penetrate upwards (bottom left, B, C). B: Intense bioturbation (middle left and lower right) and distinctive burrow with backfill menisci strongly suggestive of the ichnogenus *Taenidium*. C: Enlargement of meniscate burrow; note the lack of a wall lining and penetration of brecciated palaeosol (coin diameter shown in A, B = 23 mm).

sea level. Palaeosols and hardened ferruginous crusts are exposed on these deflationary surfaces (Fig. 9). With the exception of occasional clavate burrows suggestive of stiff or firm ground conditions, these horizons are typically marked by a suite of invertebrate terrestrial traces closely similar to those found in the root cast haloes described from Inland Road localities (above) and Sapelo Island (below).

Vertical section views of back-filled root casts similar to those seen along Inland Road (above) were not encountered in the Pleistocene exposures at Henderson Bay. However, two types of structure are considered to be their counterparts. First, several unusual concretion-like, cylindrical structures between 15 and >30 cm in diameter stand proud of deflationary surfaces by 20 cm or

more. These features exhibit a core of variably cemented, white to pale grey sand that is enclosed within a ferruginous halo several centimetres across (Fig. 9). The assemblage of invertebrate trace fossils present in the haloes of Inland Road root casts has not been identified in the halo of these structures. However, similar invertebrate traces occur in nearby Pleistocene palaeosols. Also, next to these vertical bodies, and lying freely on exhumed deflationary surfaces, are broken and separated segments of similar cylindrical character. No ferruginous halo or lining is evident in this material. Reconstructed lengths of these segments may reach 2 m. These cylindrical structures further our appreciation of the geographical extent and three-dimensional aspects to rooting architecture evident at both New Zealand study localities.

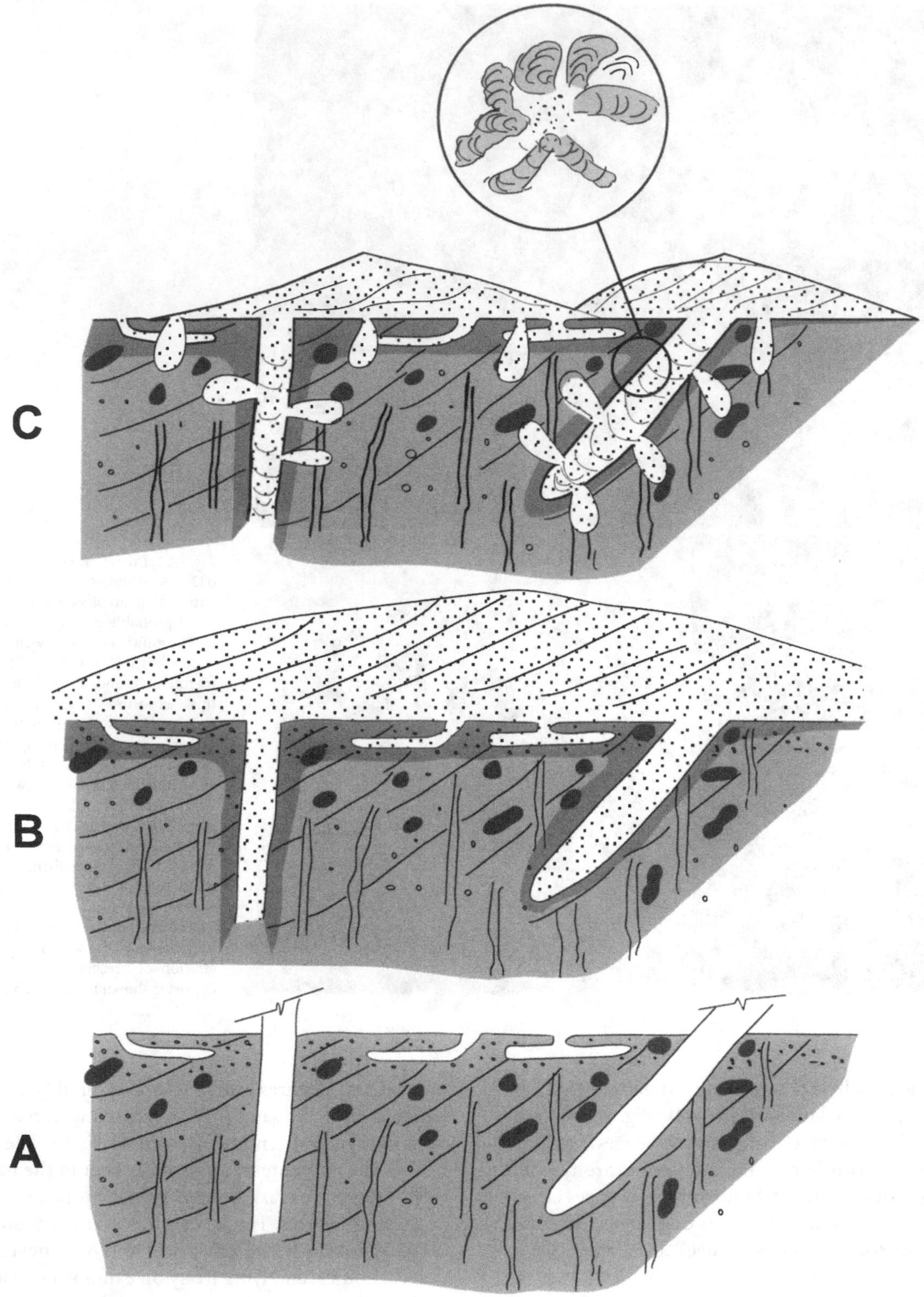

Fig. 8.   Schematic representation of the development of disjunct ichnofabrics, trace fossil orientation and tiering relationships of the composite inver-
tebrate/root structures (A–C). A: Vegetated dunes and swales; development of weak mottling and vertical (rooting) fabrics; trees engulfed by drifting
sand; soil erosion; terrestrial insect traces in soil; burrows follow surface relief. B: Death of trees, decay of tree trunk and roots; development of firm
grounds, iron pan and/or strong humic staining across the surface and as haloes around rotting trunks and roots; exhumation and some back-fill by
drifting sand, invertebrates find a hospitable environment in cavity-fill, moist humic horizons; invertebrate activity and simple trace fossils. C: Harden-
ing of surface crust; clavate burrows orthogonal to exposed surfaces of humic firm grounds and in mould haloes; passive fill of large cavities by drifting
very fine silica sand; complicated trace fossil orientations and cross-cutting relationships; collapse and brecciation; cicada nymphs colonise roots;
meniscate burrow construction, *Taenidium* (inset).

Fig. 9. Henderson Bay (GR N03/224 164). A: Resistant remnants of tree stump and/or root moulds back filled with moderately consolidated fine white sand, emerging from and standing proud on the eroding, irregularly case-hardened, bench surface of an older (Pleistocene) parabolic sand dune at Henderson Bay (arrowed). B: Scattered and fragmented remnants of the weakly cemented siliceous casts of probable tree roots and trunks (arrowed).

Sapelo Island, Georgia (USA)

Sapelo Island (Fig. 10) is a barrier island on the southeastern coast of Georgia (USA), and is one of a series of northeast–southwest trending Pleistocene–Holocene barrier islands marking former positions of the coastline. Georgia barrier islands differ from most others because they formed as islands during the Pleistocene, then were overprinted and supplemented by Holocene sediments (Hoyt *et al.* 1964; Davis 1994). Sapelo Island and the remainder of the Georgia coast constitute what are arguably the best-studied of barrier island and salt marsh

ecosystems. These studies commenced with the founding of the University of Georgia Marine Institute in 1954 by Eugene Odum. Ecological research on Sapelo was followed by intensive sedimentological and ichnological efforts in the 1960s–1980s (Hoyt & Weimer 1964; Frey & Howard 1969, 1988; Basan & Frey 1977). However, these researchers focused on Sapelo littoral environments (e.g. salt marshes, foreshores, beaches, dunes). These environments are affected by a mesotidal regime (2.4–3.4 m range), with wave-dominated beaches on the eastern side and salt marshes on the western (back-barrier) side of the island.

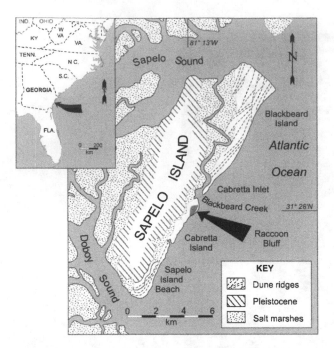

Fig. 10.   Locality map of Sapelo Island, Georgia (USA), with Raccoon Bluff arrowed.

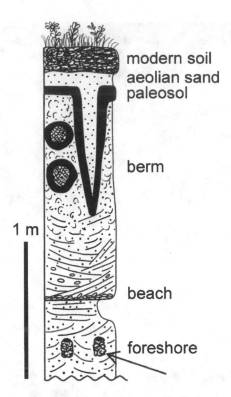

Fig. 11.   Stratigraphic section at Raccoon Bluff, showing vertical succession from foreshore to terrestrial facies. Note *Ophiomorpha nodosa* (arrowed) burrows in foreshore sands. Scale bar = 1 m.

In comparison, studies of maritime forests on Georgia barrier islands are few (Bratton & Miller 1994; Callaway *et al.* 2002; Albers & Albers 2003). Georgia barrier island forests are nearly the same as those on the back-barrier mainland, which contain primary and secondary growth evergreens: slash pine (*Pinus elliottii*) and loblolly pine (*Pinus taeda*), live oak (*Quercus virginiana*), and southern magnolia (*Magnolia grandiflora*). The most striking difference between mainland and barrier island forests is that littoral environments create narrow and abrupt ecotones in the maritime forests, as a result of differences in subsurface pore water. Pleistocene maritime forests were evidently similar in composition to those of modern Georgia barrier islands, as indicated by fossil conifers (*Pinus taeda* and *P. serotina*) and deciduous trees, such as *Quercus* sp. and *Carya* sp., from 37,000 BP on Skidaway Island, 60 km northeast of Sapelo (Booth *et al.* 2003).

As a result of previous research, Pleistocene facies on Sapelo can be reliably interpreted in the context of knowledge about its modern environments and those of other Georgia barrier islands. Such facies are best exposed at Raccoon Bluff (N31°28′, W81°13′) on the eastern shore of Cabretta Island, situated along the east-central coast of Sapelo Island (Fig. 10). The western bank of Blackbeard Creek exposes part of the Raccoon Bluff Formation, which forms a poorly consolidated yet resistant ledge on the Cabretta Island bank, but is absent from Blackbeard (Fig. 10). Blackbeard Island is composed entirely of Holocene sediments, but like Cabretta Island, it was probably connected to the main part of Sapelo earlier in the Holocene before channel incision (Hoyt *et al.* 1966).

Despite the large number of sedimentological and ichnological studies reported from Sapelo, few geological descriptions and interpretations of Raccoon Bluff have been attempted (cf. Hoyt *et al.* 1966 is a notable exception). Thus, a secondary goal of this study is to show how the Raccoon Bluff sedimentary sequence can provide insights on the Pleistocene history of Sapelo Island as a whole.

Fortunately, the 2.5 m thick stratigraphic interval of the Raccoon Bluff Formation exposed along a 200–300 m stretch at Raccoon Bluff provides much sedimentological and ichnological information (Fig. 11). At low tide, the lower part of the sequence reveals a dark brown muddy sandstone containing numerous *Ophiomorpha nodosa*. This bed is succeeded by laterally discontinuous (but as much as 20 cm thick) shell-hash layers within a light brown trough cross-bedded relatively clean sandstone. The shell fragments are most likely those of *Donax variabilis* (coquina clam), a rapid-burrowing bivalve common to sandy intertidal beaches of the southeastern USA. Cross-bedding in this sandstone bed is gradually less defined upward through increased burrow mottling. Bedding is also cross-cut by the large structures that constitute the main focus of this study (described later). These structures are subjacent to and, in some instances, connected to a 0–12 cm thick dark brown (humic) and clayey sand bed interpreted as a palaeosol. While this bed pinches out in places, it is laterally persistent and easily traceable over the outcrop. The palaeosol is overlain by a 12–17 cm thick unconsolidated and structureless sand layer, which is fine grained, well sorted, and with a minor heavy mineral content. The top of this sand layer grades upwards into the 20–25 cm thick modern soil of the maritime forest and modern plant roots pervade the outcrop. Wave-induced bank erosion along Blackbeard Creek causes some mature trees to collapse into the channel, particularly *Quercus virginiana* and *Pinus taeda*.

The stratigraphic sequence at Raccoon Bluff clearly demonstrates a transition from shallow-marine upper foreshore to beach (ridge runnel), berm, and back dune (terrestrial) facies, with soil development in the last of these (Fig. 11). The presence of a former marine foreshore here was first recognised by Hoyt *et al.* (1966) based on the occurrence of *Ophiomorpha*, which Weimer & Hoyt (1964) had linked to the abundant ghost shrimp (*Callichirus major*) traces that occur abundantly in such environments on Sapelo and other Georgia barrier islands (Weimer & Hoyt 1964). The succession of foreshore beach runnel facies is supported by local accumulations of *Donax* and other finely ground molluscan shell fragments within trough crossed-bedded sands. These represent shell lags developed from backwash in runnel systems (Dörjes *et al.* 1986). A change from beach to berm environments is suggested by the gradual erasure of bedding by biogenic sedimentary structures, through increased bioturbation by *Ocypode quadrata* (ghost crabs) and other animals above the high-tide mark (Frey *et al.* 1984; Darrell *et al.* 1993). Missing from the idealised facies sequence are dune sediments, but the humic layer probably represents a palaeosol that developed first as an interdune meadow that was succeeded by maritime forest. Surface depressions in this soil were later filled with aeolian sands, which typically consist of fine-grained quartz sands with heavy mineral-rich laminae (Woolsey *et al.* 1978; Cofer-Shabica 1993).

Structures pertinent to this study can be interpreted with the above palaeoenvironmental context in mind. They are vertically oriented, conical and downwardly tapering bodies. They are also evident horizontally oriented oval to circular cross-sections; a few obliquely oriented examples are also present (Fig. 12). In some places, horizontal and vertical components are clearly interconnected, although isolated examples of each are also present (Fig. 12). Horizontal structures associated with the vertical components are seemingly limited to a 50 cm thick zone in the upper part of the stratigraphic sequence. The vertical structures gradually taper downward and, in rare instances, have thin bifurcations at their ends (Fig. 13A). Vertical extents (depths) are 1–2 m (mean of 1.41 m, $n = 11$) and maximum widths (~diameters) of these same structures are 7–56 cm (mean of 30 cm, $n = 11$). Structures are surrounded by dark brown,

Fig. 12. Schematic diagram of a vertical section at Raccoon Bluff showing the pattern of root-related structures: coarse stipple is muddy sand lithology, black outlines are humic haloes around structures, and fine stipple is sand-filled interiors to the structures. A similar reconstruction can be made for exposures along the Inland Road (Northland). Scale bar = *c.* 1 m (*cf.* Mossa & Schumacher 1993: 711, fig. 8).

humic, and clayey haloes (2–16 cm wide), identical in composition to the aforementioned palaeosol. These haloes (or rims) are filled with white sand probably derived from the aeolian sand layer, providing a colour contrast that helps to define the structures within the consolidated sands. Some structures connect with the palaeosol, whereas others near to it are not directly connected. Although the sand fill probably originated from the sand layer overlying the palaeosol, it has its own distinctive texture caused by burrowing infauna (Fig. 13B). Pervasive and overlapping, 1–2 cm diameter burrows, herein considered *Taenidium*, are an expression of alternating light and dark, 1–2 mm thick menisci differentiated by quartz- and heavy mineral-rich sands, respectively. In most of the examined structures, these burrows transected the brown haloes and effectively mixed the different sediments so that formerly discrete boundaries between the sediment fill and the haloes were blurred in a 5–7 cm thick grey zone similar to those from northern New Zealand (cf. Figs. 7, 14).

Based on the geometry, size, co-occurrence, and internal fill, as well as comparisons with the physiology and rooting architecture of modern trees (e.g. Hasiotis 2002: 108), the large structures and their enclosed small structures are interpreted as root traces from either coniferous or deciduous trees. The most probable candidates for the trace makers are *Pinus taeda* or other coastal species of *Pinus*. Pines, in general, are well known for having robust and deeply penetrating conical taproots that serve as central axes for radiating, near-surface horizontal root systems (Stone & Kalisz 1991; Pallardy *et al.* 1995).

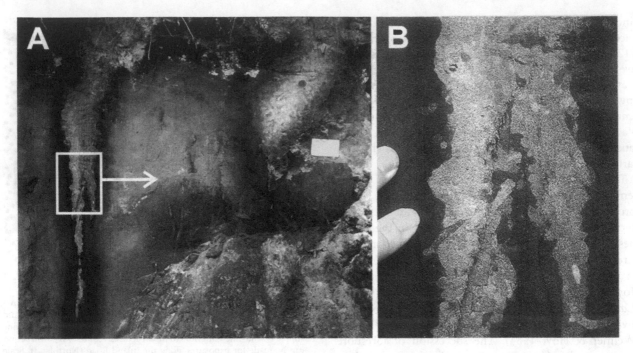

Fig. 13. Root cast structures in the Raccoon Bluff exposure. A: View of a vertical, downwardly tapering and bifurcating structure with a strongly developed, dark humic-rich, halo (left), with an intensely bioturbated, grey to white sandy fill (for detail see B) and, also, in oblique or transverse cross-sectional view to the upper right side, a root cast with a poorly defined halo and white fill that has been worked by burrowing (white scale bar = 10 cm). B: Enlargement of the main downwardly tapering and bifurcated root structure shown in the left middle part of (A) (defined by white-lined rectangle) illustrating the intensely bioturbated, grey to white sandy fill.

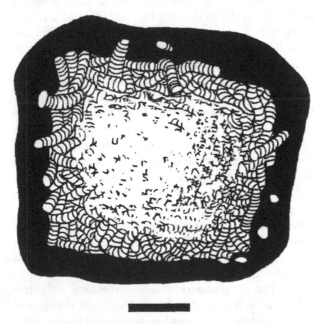

Fig. 14. Cross-sectional schematic view of root structure (horizontal component): black outline is humic halo and meniscate burrows are *Taenidium* that gradually blur into central white sand fill. Scale bar = 10 cm.

Observations of exhumed specimens of *Pinus taeda* on nearby Cabretta Beach (Fig. 15) confirmed root architecture and dimensions similar to those of the Raccoon Bluff structures. Moreover, the exterior surfaces of the root

traces were probably accentuated by mycorrhizae (fungal symbionts) while the trees were still alive (LePage *et al.* 1997; Smith & Read 1997; Hibbett *et al.* 2000). This symbiosis resulted in forming the most visible aspect of the structures, and which qualifies them as both compound and composite trace fossils (*sensu* Pickerill 1994). However, these traces were also later modified through the following steps: (1) death of the trees and their root symbionts; (2) vacancy caused by uprooting and/or decay; (3) passive filling of depressions by overlying aeolian sands; and (4) colonisation of the sands by moisture-seeking infaunal invertebrates which, in turn, intensively burrowed the fill sufficiently to blur the original rims associated with the tree-fungal traces. Because *Taenidium* is a back-filled burrow caused by an invertebrate with appendages (D'Alessandro & Bromley 1987), the trace makers were probably arthropods. In this case, insects are the most likely culprits in the interpreted terrestrial setting. Coleopteran larvae or hemipteran nymphs are considered the most probable trace makers. Sediment filling the root structures was (and still is) more porous and permeable than the root structures themselves, which allowed moistening of sediments filling the structures, an attraction to various insect taxa.

Complicating matters, numerous roots from modern plants in the overlying maritime forest penetrate the outcrop, causing a mixing of ancient and modern root structures, and providing readily observable examples of uniformitarianism. Interestingly, Hoyt *et al.* (1966: 16)

Fig. 15. Roots of *Pinus taeda* (loblolly pine) lying exposed on Cabretta Beach, Sapelo Island. Note distal tapering, strong vertical orientation, and bifurcating ends to some roots.

noticed the commingling of modern and ancient root-related features, although they did not explicitly define the latter: "Root systems are visible on fallen trees along the bluff and *in situ* root patterns are preserved along the banks of Blackbeard Creek. One common type forms almost a solid wall or curtain of roots which radiates out from the central system". These Quaternary structures remain exposed to this day in the Raccoon Bluff outcrop.

Discussion

There are close similarities in morphology between the large structures described from the Quaternary of northern New Zealand and the Holocene of Sapelo Island, USA. Moreover, these structures can be compared with the cylindrical tree root casts, preserved through pedogenesis and diagenesis in late Pliocene and Pleistocene formations of southern Louisiana (Mossa & Schumacher 1993). Further possibly comparable examples are the sand-filled "problematical cavities" figured by Stephenson & Monroe (1940: figs. 37, 38) from the Late

Cretaceous Coffee Sand at Corinth, Mississippi, which were also inferred to represent the infilled casts of decayed pine tree roots. Similar, but much smaller, recent, cylindrical tapering structures (maximum length 3 m and diameters of 4–14 mm) in alluvial sands of the Trent Valley, UK, are considered to have originated through percolating ground waters and iron precipitation about thistle roots (Claxton 1970). Similarities in morphology can also be noted with solution pipes (Herwitz 1993) and other karst phenomena known from calcareous aeolianites, and for which a plant root or trunk cast origin has, in the past, often been contemplated. For many of these features, an abiogenic origin is now generally accepted (Herwitz & Muhs 1995). However, some features of this kind may be difficult to distinguish from the rooting structures we have described, particularly so where they are associated with terra rossa soils. For example, the "palmetto stumps" from Bermuda aeolianites were once identified as the fossilised casts of trunks and/or roots of a palm, *Sabal bermudana* (Verrill 1902), or a cedar, *Juniperus bermudiana* (Livingston 1944). To the best of our knowledge, small invertebrate trace fossils, like those we have recorded herein, await discovery or recognition in solution pipes associated with the terra rossa.

Development of the North Island and Sapelo Island root casts has followed a pattern similar to those outlined by both Claxton (1970) and Mossa & Schumacher (1984: fig. 8). In life, tree roots are sheathed by microbial activity. Upon tree death, roots shrivel and contract away from the soil and sand that they had once penetrated. Cicadas probably desert their host at this time. Decay may leave a dark (black) carbonaceous residue that defines original root margins. Water table fluctuations lead to the precipitation of iron, manganese and humate compounds in the adjacent host sediments. This process creates the haloes described previously. Root decomposition leaves voids that are later, and progressively, filled by drifting aeolian sand (Sapelo Island), or derived from a heavily podzolised soil horizon (Karikari Peninsula). Voids may also be left through the toppling of trees after death, or upon being uprooted by catastrophic events such as cyclones or tsunamis, or simple bank erosion.

The numerous, simple and irregular unidentified trace fossils that are so conspicuous in and immediately adjacent to the root cast haloes suggest varied invertebrate fauna seeking a dark, moist and protected environment that suited burrow-feeding and brooding habits. This fauna probably included solitary wasps and bees, as well as beetles and spiders, crickets, etc. At neither the northern New Zealand nor Sapelo Island localities did we recognise cluster burrows or any other evidence for ant nesting behaviour of the kind reported by Curran & White (2001: figs. 7, 10). An intimate association between the meniscate trace fossil, *Taenidium*, and plant roots has been recognised previously in marine terrace deposits of a Pleistocene regressional setting, from the Ionian coast of southern Italy (D'Alessandro *et al.* 1992). There, numerous *Taenidium*/root cast composite structures occur where the former is clustered around and/or entwined about rhizoliths over vertical lengths that may reach 1.5 m (see D'Alessandro *et al.* 1992: figs. 2, 3, 10, 14). *Taenidium* was formerly considered to be indicative of marine environments. After noting some similarities with *Ancorichnus*, D'Alessandro *et al.* (1992: 509) offered three possibilities for the *Taenidium* trace maker(s) in the composite structure. In summary these were: (1) an animal living contemporaneously with plants in a terrestrial/freshwater environment; (2) marine animals harvesting the products of plant decay during periods of seawater inundation; and (3) the work of two different animals – one a marine deposit feeder and the other a terrestrial herbivore. D'Alessandro *et al.* (1992: 513) concluded that the *Taenidium* trace maker was "most probably a non-marine animal". Bromley (2001, pers. comm.) now recognises that this trace is a cicada nymph burrow. It is likely that the *Taenidium* described herein have a similar origin. Cicadas are common in the forests of the northern New Zealand and Sapelo Island. After hatching, their nymphs drop to earth, and quickly burrow to depths where they feed by sucking on root sap. Of over 20 New Zealand cicada species, one (*Kikihia muta*) is known to spend at least 3 years underground, but some North American species dwell underground for up to 17 years before emerging (Miller 1984).

Fossilised roots and trace fossils are used widely to determine "younging direction", or facing, or "way-up" (e.g. Shrock 1948; Seilacher 1964; Claxton 1970; Basan 1978). In this study, however, invertebrate trace fossils associated with root cast haloes follow boundary margins between cast and fill. In particular, they tend to exhibit two preferred orientations, either orthogonal/steeply inclined to the dark brown halo contacts with the fill and/or hosting background sediment (Figs. 4A, 13A), or more-or-less parallel to them (Figs. 4B, 6). Cross-cutting patterns of trace fossils in dark halo inner and outer margin contacts with the white sand core fill and yellowish host sand imply tiering relationships. Where root casts are more or less vertical, the associated trace patterns may lead to opposite facing directions on either side of a root cast (Fig. 8). Furthermore, with gently inclined root casts, some trace fossils in the hanging wall of a halo may be inverted, including, for example, simple tubular and clavate forms (Figs. 7A, 8). In deformed strata, whether this pattern reflects soft sediment or tectonic processes, and particularly where palaeosols are involved, or surface exposures are limited and/or discontinuous, such disjunct relationships could confuse facing determinations and possibly interfere with palaeoenvironmental interpretations. These aspects could further complicate trace fossil identifications in core analyses.

Conclusions

The composite and compound ichnofabrics described herein, with their intimate relationships between the alteration haloes of large root casts and terrestrial invertebrate trace fossils, including *Taenidium*, bear similarities to composite ichnofabrics and tiering described by Bromley & Ekdale (1986). However, structures of the kind we have described do not appear to have been recognised previously. They are significant because compound and composite root cast/invertebrate trace fossil ichnofabrics may be more common than is appreciated at present. Furthermore, they add an extra dimension to palaeoenvironmental reconstructions of palaeosols and ancient forest ecosystems. However, before this potential can be further exploited, identification of the progenitors of the numerous small and non-diagnostic invertebrate traces will need to be further advanced. Finally, this study shows a need for caution when attempting to establish "younging direction" from disjunct ichnofabrics of this kind, when working in tectonically deformed palaeosol-bearing sequences. Nonetheless, it is clear that terrestrial invertebrates exploit the margins of large rooting galleries, and the cavity-fill left behind upon the death and/or decay of roots. The stratigraphic record of these subterranean communities of organisms is preserved in the complexities of compound and composite trace fossil relationships that reflect plant rooting and invertebrate interactions.

Acknowledgements

Funding support for MRG and KAC came through a grant from the University of Auckland Research Committee. We are indebted to Louise Cotterall for her patience and drafting skills, and to Andrea Alfaro for field assistance and company in Northland. We also thank J. Garbisch, University of Georgia Marine Institute, for help with logistics on Sapelo Island, together with S. Henderson, R. Bromley and A. Uchman for valuable feedback while in the field there. AJM also recognises the field support of R. Schowalter in both New Zealand and Sapelo. While the final product is the authors' responsibility, the paper has been much improved through the critical reviews of A. Rindsberg and G. Cadée, which have been much appreciated.

References

Ahonen-Jonnarth, U., Göransson, A. & Finlay, R.D. 2003: Growth and nutrient uptake of ectomycorrhizal *Pinus sylvestris* seedlings in a natural substrate treated with elevated Al concentrations. *Tree Physiology* 23, 157–167.

Albers, G. & Albers, M. 2003: A vegetative survey of back-barrier islands near Sapelo Island, Georgia. *In* Hatcher, J.K. (ed.): *Proceedings of the Georgia Water Resources Conference, Institute of Ecology, University of Georgia*, 1–4. Athens, Georgia.

Basan, P.B. 1978: Introduction. *Society of Economic Paleontologists and Mineralogists, Short Course 5*, 1–12.

Basan, P.B. & Frey, R.W. 1977: Actual-palaeontology and neoichnology of salt marshes near Sapelo Island, Georgia. *Geological Journal, Special Issue, 9*, 49–90.

Black, P.B. & Gregory, M.R. 2002: Geological gems of the far north. *Geological Society of New Zealand Miscellaneous Publication 112B*, 93–110.

Blay, C.T. & Longman, M.W. 2001: Stratigraphy and sedimentology of Pleistocene and Holocene carbonate eolianites, Kaua'i, Hawai'i, U.S.A. *Society of Economic Paleontologists and Mineralogists, Special Publication 71*, 47–55.

Bockelie, J.F. 1994: Plant roots in core. *In* Donovan, S.K. (ed.): *The Paleobiology of Trace Fossils*, 177–199. Johns Hopkins University Press, Baltimore.

Booth, R.K., Rich, F.J. & Jackson, S.T. 2003: Paleoecology of mid-Wisconsonian peat clasts from Skidaway Island, Georgia. *Palaios 1*, 63–68.

Bown, T.M. 1982: Ichnofossils and rhizoliths of the nearshore fluvial Jebel Quatrani Formation (Oligocene), Fayum Province, Eygpt. *Palaeogeography, Palaeoclimatology, Palaeoecology 40*, 255–309.

Boyd, D.W. 1975: False or misleading traces. *In* Frey, R.W. (ed.): *The Study of Trace Fossils*, 65–83. Springer, New York.

Bracken, S.P. & Picard, M.D. 1984: Trace fossils from Cretaceous/Tertiary North Horn Formation in central Utah. *Journal of Paleontology 58*, 477–487.

Bratton, S.P. & Miller, S.G. 1994: Historic field systems and the structure of maritime oak forests, Cumberland Island. *Bulletin of the Torrey Botanical Club 121*, 1–12.

Bromley, R.G. & Asgaard, U. 1972: A large radiating burrow-system in Jurassic micaceous sandstones of Jameson Land, East Greenland. *Grønlands Geologiske Undersøgelse, Rapport 49*, 23–30.

Bromley, R.G. & Ekdale, A.A. 1986: Composite ichnofabrics and tiering of burrows. *Geological Magazine 123*, 59–65.

Bromley, R.G., Uchman, A., Gregory, M.R. & Martin, A.J. 2003: *Hillichnus lobosensis* igen. et isp nov., a complex trace fossil produced by tellinacean bivalves, Paleocene, Monterey, California, USA. *Palaeogeography, Palaeoclimatology, Palaeoecology 192*, 157–186.

Brown, R. 1848: Description of an upright Lepidodendron with Stigmaria roots in the roof of the Sydney main coal in the Island of Breton. *Quarterly Journal of the Geological Society, London 4*, 46–50.

Buatois, L.A. & Mángano, G.M. 2002: Trace fossils from Carboniferous floodplain deposits in western Argentina: implications for ichnofacies models of continental environments. *Palaeogeography, Palaeoclimatology, Palaeoecology 183*, 71–86.

Callaway, R.M., Reinhart, K.O., Moore, G.W., Moore, D.J. & Pennings, S.C. 2002: Epiphyte host preferences and host traits: mechanisms for species-specific interactions. *Oecologia 132*, 221–230.

Claxton, C.W. 1970: Cylindrical tapering structures in the alluvial sands of the Trent valley. *Mercian Geologist 3*, 265–267.

Cofer-Shabica, S.V. 1993: Ocean shoreline changes. *Georgia Geological Society Guidebook 13*, 12–15.

Cohen, A.S. 1982: Paleoenvironments of root casts from the Koobi Fora Formation, Kenya. *Journal of Sedimentary Petrology 52*, 401–414.

Curran, H.A. & Martin, A.J. 2003: Complex decapod burrows and ecological relationships in modern and Pleistocene intertidal carbonate environments, San Salvador Island, Bahamas. *Palaeogeography, Palaeoclimatology, Palaeoecology 192*, 229–245.

Curran, H.A. & White, B. 1999: Ichnology of Holocene carbonate eolianites on San Salvador Island, Bahamas: diversity and significance. *In* Curran, H.A. & Mylroie, J.E. (eds): *Proceedings of the 9th Symposium on the Geology of the Bahamas and other Carbonate Regions*, 22–35. Bahamian Field Station, San Salvador, Bahamas.

Curran, H.A. & White, B. 2001: Ichnology of Holocene carbonate eolianites of the Bahamas. *Society of Economic Paleontologists and Mineralologists, Special Publication 71*, 47–55.

Darrell, J.H., Brannem, N.A. & Bishop, G.A. 1993: The beach. *Georgia Geological Society Guidebook 13*, 13–16.

D'Alessandro, A. & Bromley, R.G. 1987: Meniscate trace fossils and the *Muensteria-Taenidium* problem. *Palaeontology 30*, 743–763.

D'Alessandro, A., Loiacono, F. & Bromley, R.G. 1992: Marine and nonmarine trace fossils and plant roots in a regressional setting (Pleistocene, Italy). *Rivista Italiana di Paleontologia e Stratigrafia 98*, 495–522.

Davis, R.A., Jr 1994: Barriers of the Florida Gulf. *In* Davis, R.A., Jr (ed.): *Geology of Holocene Barrier Island Systems*, 167–205. Springer, New York.

Donovan, S.K., Pickerill, R.K. & Portell, R.W. 2002: A late Cenozoic "root bed", an unconformity and the tectonic history of Carriacou, The Grenadines, Lesser Antilles. *Proceedings of the Geologists' Association, London 113*, 199–205.

Dörjes, J., Frey, R.W. & Howard, J.D. 1986: Origins of, and mechanisms for, mollusk shell accumulations on Georgia beaches. *Senckenbergiana Maritima 18*, 1–43.

Driese, S.G., Mora, C.I. & Elick, J.M. 1997: Morphology and taphonomy of root and stump casts of the earliest trees (Middle to late Devonian), Pennsylvania and New York, USA. *Palaios 12*, 524–537.

Ekdale, A.A. & Bromley, R.G. 1984: Comparative ichnology of shelf-sea and deep-sea chalk. *Journal of Paleontology 58*, 322–332.

Erwin, D.M. 1984: Growth of the rhizomorph of *Paurodendron* and homologies among the rooting organs of Lycopsida. *American Journal of Botany 71*, 112–113.

Fischer, R. & Olivier, C.G. 1996: Fossilized root systems (?) of land plants in Pliocene sandstones of southern Lower California, Mexico. *Geologica et Palaeontologica 30*, 267–277.

Frey, R.W., Curran, H.A. & Pemberton, S.G. 1984: Tracemaking activities of crabs and their significance: the ichnogenus *Psilonichnus*. *Journal of Paleontology 58*, 333–350.

Frey, R.W. & Howard, J.D. 1969: A profile of biogenic sedimentary structures in a Holocene barrier island–salt marsh complex, Georgia. *Gulf Coast Association of Geological Societies, Transactions 19*, 427–444.

Frey, R.W. & Howard, J.D. 1988: Beaches and beach-related facies. *Geological Magazine 125*, 621–640.

Froede, C.R., Jr. 2002: Rhizolith evidence in support of a late Holocene sea-level highstand at least 0.5 m higher than present at Key Biscayne, Florida. *Geology 30*, 203–206.

Galli, G. 1991: Mangrove-generated structures and depositional model of the Pleistocene Fort Thompson Formation (Florida Plateau). *Facies 2*, 297–314.

Glennie, K.W. & Evamy, B.D. 1968: Dikaka: plants and plant-root structures associated with aeolian sand. *Palaeogeography, Palaeoclimatology, Palaeoecology 4*, 77–87.

Goldie, P.J. 1975: The Quaternary geology of an area north of Houhora. MSc Thesis, The University of Auckland.

Gregory, M.R. & Campbell, K.A. 2000: Towards a phytoturbation index. *Geological Society of New Zealand Miscellaneous Publication 108A*, 57.

Gregory, M.R. & Campbell, K.A. 2003: A '*Phoebichnus* look-alike': a fossilized root system from Quaternary coastal dune sediments, New Zealand. *Palaeogeography, Palaeoclimatology, Palaeoecology 192*, 247–258.

Hasiotis, S.T. 2002: Continental trace fossils. *Society of Economic Paleontologists and Mineralogists, Short Course Notes 51*, 1–132.

Hay, R.F. 1975: Sheet N7 Doubtless Bay. *Geological Map of New Zealand 1:63 360. Map and Notes.* NZ Department of Scientific and Industrial Research.

Hay, R.F. 1981: Sheet N6 Hourora. *Geological Map of New Zealand 1:63 360. Map and Notes.* NZ Department of Scientific and Industrial Research.

Herwitz, S.R. 1993: Stemflow influences in the formation of solution pipes in Bermuda eolianite. *Geomorphology 6*, 253–271.

Herwitz, S.R. & Muhs, D.R. 1995: Bermuda solution pipe soils: a geochemical evaluation of eolian parent materials. *Geological Society of America, Special Paper, 300*, 311–323.

Hibbett, D.S., Gilbert, L.-Z. & Donoghue, M.J. 2000: Evolutionary instability of ectomycorrhizal symbioses in basidiomycetes. *Nature 407*, 506–508.

Hicks, D.L. 1983: Landscape evolution in coastal dunesands. *Zeitschrift für Geomorphologie, Supplement 42*, 245–250.

Hoyt, J.H., Henry, V.J. Jr & Howard, J.D. 1966: Pleistocene and Holocene sediments, Sapelo Island, Georgia and vicinity. *Geological Society of America (Southeastern Section) Guidebook for Field Trip 1*, 1–78.

Hoyt, J.H. & Weimer, R.J. 1964: Comparison of modern and ancient beaches, central Georgia coast. *Bulletin of American Association of Petroleum Geologists 47*, 529–531.

Hoyt, J.H., Weimer, R.J. & Henry, V.J. Jr. 1964: Late Pleistocene and recent sedimentation, central Georgia coast, U.S.A. *In* Van Straaten, M.J.U. (ed.): *Deltaic and Shallow Marine Deposits*, 170–176. Elsevier, Amsterdam.

Isaac, M.J. 1996: *Geology of the Kaitaia: 1:250 000 Geological Map 1*, 1–44. Institute of Geological and Nuclear Sciences, Lower Hutt.

Jones, B., Renaut, R.W., Rosen, M.R. & Klyen, L. 1998: Primary siliceous rhizoliths from Loop Road Hot Springs, North Island, New Zealand. *Journal of Sedimentary Research 68*, 115–123.

Klappa, C.P. 1980: Rhizoliths in terrestrial carbonates: classification, recognition, genesis and significance. *Sedimentology 27*, 613–629.

Kwasna, H. 2002: Changes in microfungal communities in roots of *Quercus robor* stumps and their possible effects on colonization by *Armillaria*. *Journal of Phytopathology 150*, 403–411.

LePage, B.A., Currah, R.S., Stockey, R.A. & Rothwell, G.W. 1997: Fossil ectomycorrhizae in Eocene *Pinus* roots. *American Journal of Botany 84*, 401–412.

Livingston, W. 1944: Observations on the structure of Bermuda. *Geographical Journal 104*, 40–48.

Lockley, M.G. & Hunt, A.P. 1994: A track of the giant theropod dinosaur *Tyrannosaurus* from close to the Cretaceous/Tertiary boundary, northern New Mexico. *Ichnos 3*, 213–218.

Loope, D.R. 1988: Rhizoliths in ancient eolianites. *Sedimentary Geology 56*, 301–304.

Mihail, J.D., Bruhn, J.N. & Leininger, T.D. 2002: The effects of moisture and oxygen availability on rhizomorph generation by *Armillaria tabescens* in comparison to *A. gallica* and *A. mellea*. *Mycological Research 106*, 697–704.

Mikuláš, R. 2001: Modern and fossil traces in terrestrial lithic substrates. *Ichnos 8*, 177–184.

Miller, D. 1984: *Common Insects in New Zealand.* A.H. and A.W. Reed, Wellington.

Montalvo, C.I. 2002: Root traces in fossil bones from the Huayquerian (Late Miocene) faunal assemblage of Telén, La Pampa, Argentina. *Acta Geológica Hispánica 37*, 37–42.

Mossa, J. & Schumacher, B.A. 1993: Fossil tree casts in south Louisiana soils. *Journal of Sedimentary Petrology 63*, 707–713.

Nesbitt, E.A. & Campbell, K.A. 2002: A new *Psilonichnus* ichnospecies attributed to mud shrimp *Upogebia* in estuarine environments. *Journal of Paleontology 76*, 892–901.

Newnham, R., Ogden, J. & Mildenhall, D. 1993: A vegetation history of the far north of New Zealand during the late Otira (last) Glaciation. *Quaternary Research 39*, 361–372.

Ogden, J., Wilson, A. Hendy, C. & Newnham, R. 1992: The late Quaternary history of kauri (*Agathis australis*) in New Zealand and its climatic significance. *Journal of Biogeography 19*, 611–622.

Pallardy, S.G., Cermak, J., Ewers, F.W., Kaufmann, M.R., Parker, W.C. & Sperry, J.S. 1995: Water transport dynamics in trees and stands. *In* Smith, W.K. & Hinckley, T.M. (eds): *Resource Physiology of Conifers: Acquisition, Allocation and Utilization*, 299–387. Academic Press, New York.

Pfefferkorn, H.W. & Fuchs, K. 1991: A field classification of fossil plant substrate interactions. *Neues Jahrbuch für Geologie und Paläontologie Abhandlungen 183*, 17–36.

Pickerill, R.W. 1994: Nomenclature and taxonomy of invertebrate trace fossils. *In* Donovan, S.K. (ed.): *The Palaeobiology of Trace Fossils*, 3–42. Johns Hopkins University Press, Baltimore.

Plaziat, J.C. & Mahmoudi, M. 1990: The role of vegetation in Pleistocene eolianite sedimentation: an example from eastern Tunisia. *Journal of African Earth Sciences 10*, 445–451.

Ratcliff, B.C. & Fagerstrom, J.A. 1980: Invertebrate lebensspuren of Holocene floodplains: their morphology, origin and paleoecological significance. *Journal of Paleontology 54*, 614–630.

Retallack, G.J. 2001: *Soils of the Past.* Blackwell Science, Oxford.

Retallack, G.J., Tanaka, S. & Tate, T. 2002: Late Miocene advent of tall grassland paleosols in Oregon. *Palaeogeography, Palaeoclimatology, Palaeoecology 183*, 329–354.

Ricketts, B.D. 1975: Quaternary geology of the Parengarenga – Te Kao district. MSc Thesis, University of Auckland.

Rindsberg, A.K. & Martin, A.J. 2003: *Arthrophycus* in the Silurian of Alabama (USA) and the problem of compound trace fossils. *Palaeogeography, Palaeoclimatology, Palaeoecology 192*, 187–219.

Ross, D.J., Scott, N.A., Tate, K.R., Rodda, N.J. & Townsend, J.A. 2001: Root effects on soil carbon and nitrogen cycling in a *Pinus radiata* plantation on a coastal sand. *Australian Journal of Soil Research 39*, 1027–1039.

Rothwell, G.W. & Erwin, D.M. 1985: The rhizomorph apex of *Paurodendron*: implications for homologies among the rooting organs of Lycopsida. *American Journal of Botany 72*, 86–98.

Sarjeant, W.A.S. 1975: Plant trace fossils. *In* Frey, R.W. (ed.): *The Study of Trace Fossils: a Synthesis of Principles, Problems, and Procedures in Ichnology*, 163–179. Springer, New York.

Sarjeant, W.A.S. 1983: Editor's comments on Papers 1 and 2. *In* Sarjeant, W.A.S. (ed.): *Terrestrial Trace Fossils*, 4–5. *Benchmark Papers in Geology*. Hutchinson Ross Publishing, Stroudsburg, Pennsylvania.

Seilacher, A. 1964: Biogenic sedimentary structures. *In* Imbrie, J. & Newell, N.D. (eds): *Approaches to Paleoecology*, 296–316. John Wiley, New York.

Shrock, R.R. 1948: *Sequence in Layered Rocks.* McGraw-Hill Book Company, New York.

Smith, S.E. & Read, D.J. 1997: *Mycorrhizal Symbiosis*, 2nd edn. Academic Press, London.

Stephenson, L.W. & Monroe, W.H. 1940: The Upper Cretaceous deposits. *Mississippi State Geological Survey, Bulletin 40*, 1–296.

Stockey, R.A., Rothwell, G.W., Addy, H.D. & Currah, R.S. 2001: Mycorrhizal association of the extinct conifer *Metasequoia milleri*. *Mycological Research 105*, 202–205.

Stone, E.C. & Kalisz, P.J. 1991: On the maximum extent of tree roots. *Forest Ecology and Management 46*, 59–102.

Traynham, B.N. & Martin, A.J. 2003: Comparison of Pleistocene and modern root traces on Man Head Cay, San Salvador Island, Bahamas, and implications for plant biogeography over time. *Geological Society of America Abstracts with Programs, 35(1)*, 16.

Verrill, A.E. 1902: The Bermuda Islands: their scenery, climate, productions, natural history and geology. *Transactions of the Connecticut Academy of Arts and Sciences 11*, 413–916.

Weimer, R.J. & Hoyt, J.H. 1964: Burrows of *Callinassa major* Say, geologic indicators of littoral and shallow neritic environments. *Journal of Paleontology 38*, 761–767.

White, B. & Curran, H.A. 1997: Are the plant-related features in Bahamian Quaternary limestones trace fossils? Discussion, answers, and a new classification system. *In* Curran, H.A. (ed.): *Guide to Bahamian Ichnology: Pleistocene, Holocene and Modern Environments*, 47–54. Bahamian Field Station, San Salvador, Bahamas.

Whybrow, P.J. & MacClure, H.A. 1981: Fossil mangrove roots and palaeoenvironments of the Miocene of the eastern Arabian Peninsula. *Palaeogeography, Palaeoclimatology, Palaeoecology 32*, 213–225.

Woolsey, J.R., Henry, V.J. Jr & Hunt, J.L. 1978: Backshore heavy-mineral concentration on Sapelo Island, Georgia. *Journal of Sedimentary Petrology 45*, 280–284.

Zonneveld, J-P., Gingras, M.K. & Pemberton, S.G. 2001: Trace fossil assemblages in a Middle Triassic mixed siliclastic-carbonate marginal marine depositional system, British Columbia. *Palaeogeography, Palaeoclimatology, Palaeoecology 166*, 249–276.

Permian plant–insect interactions from a Gondwana flora of southern Brazil

KAREN ADAMI-RODRIGUES, ROBERTO IANNUZZI & IRAJÁ DAMIANI PINTO

Adami-Rodrigues, K., Iannuzzi, R. & Pinto, I.D. **2004 10 25**: Permian plant–insect interactions from a Gondwana flora of southern Brazil. *Fossils and Strata*, No. 51, pp. 106–125. Brazil. ISSN 0300-9491.

Preserved foliar compressions and impressions provide evidence for several types of external interaction with insects in taxa from Gondwanan floras of Permian age in the state of Rio Grande do Sul, Brazil. The material analysed originates from the Rio Bonito (Artinskian–Kungurian) and Irati/Serra Alta (Kungurian–early Kazanian) Formations and was collected from horizons interpreted as representing wet palaeoenvironments – marginal accumulations of ancient peat or deposits that formed close to the shoreline. The principal groups of phytophagous insects inferred as the herbivore culprits are the orthopteran-like, homopterous Hemiptera, and holometabolous Coleoptera. A qualitative analysis of plant–insect interactions from these deposits indicate 11 categories of damage inflicted on the vascular plants: continuous and discontinuous external foliage feeding activity of the foliar edge and apex, removal of the foliar lamina, mining, skeletonisation, small incisions related to piercing and sucking, ovipositon scar and galling. The present study concludes that possibly these phytophagous insects had a preference for *Glossopteris* and less of a specific preference for *Cordaites* and *Gangamopteris* foliage.

Key words: Plant-insect interactions; herbivory; qualitative analyses; Permian; Godwana Flora.

Karen Adami-Rodrigues [gerarus@hotmail.com], Roberto Iannuzzi [roberto.iannuzzi@ufrgs.br] & Irajá Damiani Pinto[ipinto@orion.ufrgs.br], Departamento de Paleontologia e Estratigrafia, Instituto de Geociências, Universidade Federal do Rio Grande do Sul, Cx. P. 15.001, Porto Alegre, RS, 91.501-970, Brazil

Introduction

The examination and interpretation of the fossil record of the interactions between plants and insects is an expanding field in palaeobiology, historically ignored by palaeoentomologists as well as palaeobotanists and seldom discussed by biologists, who study a myriad of associations between modern plants and insects. The most continuous record of plant–insect associations in the Late Palaeozoic comprises vegetal remains preserved as compressions and impressions, which includes evidence of herbivory, galls, possible mines and skeletonisation of leaves.

The record of vascular plants, arthropods and their probable interactions is well characterised in a few associations of the Late Palaeozoic. The best examples are in the associations from Mazon Creek, northeastern Illinois (Janssen 1939; Shabica & Hay 1997), at Elmo, in north-central Kansas (Sellards 1909; Moore 1964) and in Chekarda, in the Urals (Martynov 1940; Meyen 1982).

Direct evidence of plant–insect interactions is less common in the Permian than in the Carboniferous, probably because of the extinction of swamp forests, with the associated spectacular preservational mode of coal ball deposits in equatorial Euroamerica, and because terrestrial Permian deposits are less widely distributed. All this documentation has been acquired during the past 25 years and has been summarised in several articles, such as Scott & Taylor (1983), Labandeira & Beall (1990), Labandeira (1998, 2002), Scott *et al.* (1985, 1992), Chaloner *et al.* (1991), Scott (1991, 1992), Stephenson & Scott (1992), Labandeira *et al.* (1994) and Smith (1994).

For Gondwana, Plumstead (1963), Srivastava (1987) and Guerra-Sommer (1995) recorded external feeding on leaves of *Glossopteris* possibly linked to prothorpterid insects. In spite of this, understanding of the evolution of herbivory by insects based on these plants is not well documented in the Late Palaeozoic of Gondwana. Seven articles have discussed aspects of *Glossopteris* damage caused by insects: Anderson & Anderson (1985),

Chaudhan *et al.* (1985), Holmes (1995), Pant & Srivastava (1995), Banerjee & Bera (1998), Srivastava (1994, 1998).

The material studied here includes evidence for several types of external association among various elements of Gondwanan floras from Early Permian strata of the Paraná Basin, in the state of Rio Grande do Sul, Brazil. The analysed material originates from deposits of the Rio Bonito (Artinskian–Kungurian) and the Irati/Serra Alta (Kungurian–early Kazanian) Formations. Because the occurrence of feeding activities may be related to a certain flora, we survey the possible herbivores responsible for the damage and suggest, in addition to the type of diet, the group of insects likely to have caused the damage to the plant.

The hypotheses and observations on the associations between Gondwanan vascular plants and insects are relevant to the evolution of the functional mechanisms associated with diets and trophic structure of modern terrestrial ecosystems. Accordingly, we establish a connection between the data analysed in these Palaeozoic strata and modern feeding activities.

Materials and methods

The specimens of fossil plants studied are preserved as compressions and impressions. The palaeobotanical material has been collected from four localities: in the Faxinal and Morro do Papaléo Mines, from outcrops at Quitéria, and from the highway BR-290, Rio Grande do Sul (Fig. 1). The Faxinal Mine is located in the Arroio dos Ratos municipality, approximately 50 km southwest of the city of Guaíba (coordinates UTM – N 6651.5; E 432.7). The Morro do Papaléo Mine is situated in the Mariana Pimentel municipality (coordinates UTM – N 6620; E 490). The Quitéria outcrop is located in the Pantano Grande municipality (coordinates 52°22'W 30°28'S). The last studied outcrop is situated on a roadcut of highway BR-290, km 90, in the Minas do Leão municipality (coordinates 30°13'S 52°03'W).

In addition, insects have been collected from several localities in southern Brazil, situated in the Minas do Leão, Caçapava do Sul and São Gabriel municipalities (Rio Grande do Sul), the Boituva municipality (state of São Paulo), the Teixeira Soares municipality (state of Paraná), and in the Anitápolis municipality (state of Santa Catarina). They are preserved as impressions.

In total, 786 samples of fossil leaves were analysed. Insect–plant interactions were present in 43 of these specimens, representing a frequency of herbivory of 0.05590. We identified and compared modern angiosperm leaves showing evidence of phytophagy with the types of interaction recorded on the fossil specimens. These angiosperm leaves were assembled at the Universidade Federal do Rio Grande do Sul, Agronomic Campus, and in the "Mata Atlântica" (Brazilian Atlantic

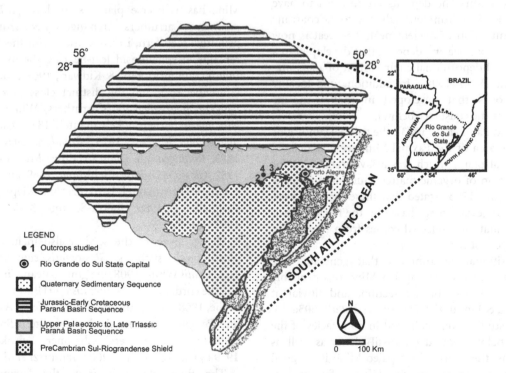

Fig. 1. Geological map showing the regional setting and outcrop distribution of the Morro do Papaléo (1) and Mina do Faxinal (2) Mines, as well as Quitéria (3) and highway BR-290 (4), near the town of Minas do Leão.

rain forest) of the Maquiné district. The parameters used in the qualitative description of palaeophytophagy had been adapted from Beck & Labandeira (1998). The classes of leaf size (i.e. notophyll, mesophyll, etc.) used are based on the Leaf Architecture Working Group (Wing *et al.* 1999).

In the laboratory, specimens were examined using a binocular microscope in order to identify signs of phytophagy, and a camera lucida was employed for making line drawings and measurements. Photography was undertaken in the Photographic Laboratory of the Departamento de Paleontologia e Estratigrafia. The studied material is housed in the collection of the Museu de Paleontologia of the Departamento de Paleontologia e Estratigrafia, Instituto de Geociências, Universidade Federal do Rio Grande do Sul, and has been catalogued under the following accession prefixes: MP-Pb (fossil plants) and MP-I (fossil insects).

Geological and palaeontological background

Geology

The evidence of insect feeding activities on the fossil leaves of coal-bearing palaeofloras is found principally in deposits of the Rio Bonito Formation (Guatá Subgroup, Tubarão Group), of Artinskian–Kungurian age, in the Paraná Basin. The deposits are inferred to have formed in wetland environments adjacent to the coastline – in an accumulation of ancient peat. The peat is now represented by coal-bearing deposits that developed in a localised paludal depositional environment of a lagoon–barrier system (Holz 1998). In addition, one locality includes deposits that developed in shallow-marine conditions near the coast (see below).

According to Paim *et al.* (1983), the paludal conditions present in the area of the Faxinal Mine were associated with an alluvial plain, which was a favourable area for the accumulation of organic matter. Guerra-Sommer & Cazzullo-Klepzig (1993) stated that these deposits were Artinskian–Kungurian in age based on the floristic associations. The material collected comes from roof shales above thin layers of coal.

The depositional environment that generated the sediments from Morro do Papaléo Mine was probably lacustrine at the base of the section, and fluvial at the top (Vieira & Iannuzzi 2000; Iannuzzi *et al.* 2003a, b). The samples studied were collected in both facies of the section (see below). Based on fossil plants as well as palynomorphs, the Morro do Papaléo Mine is assigned a Sakmarian or Artinskian age (Guerra-Sommer & Cazzullo-Klepzig 1993).

Locally, the sequence in the Quitéria outcrop is represented by clastic sediments generated in wet lowlands. The portion of the section from which the analysed foliar material was collected is characterised by a cyclical succession of claystones and thin coal layers in association with diamictites, siltstones and thin yellowish sandstones. The siltstones and sandstones were deposited on a deltaic plain, with a well-developed paludal environment (Picolli *et al.* 1991). An Artinskian–Kungurian age is attributed to the outcrop based on megafloral and palynological records (Picolli *et al.* 1991; Guerra-Sommer *et al.* 1996).

The lithological and sedimentary features of the BR-290 highway outcrop suggest deposition in a shallow-marine environment near the coastline. These deposits are included in the Irati Formation (Bortolluzzi 1975; Backheuser *et al.* 1984) or in the Serra Alta Formation (Adami-Rodrigues & Iannuzzi 2001), both assigned to the Passa Dois Group. Palynological studies indicate a late Kungurian to early Kazanian age for these deposits (Daemon & Quadros 1970; Marques-Toigo 1991; Petri & Souza 1993).

Flora

The floral associations are represented by a number of taphofloras belonging to the major "*Glossopteris* Flora". Below, we analyse these associations in relation to the stratigraphic levels from which they were obtained.

The lower part of the section in the Morro do Papaléo Mine has a diverse plant assemblage of *Botrychiopsis plantiana* (Carruthers) Archangelsky & Arrondo, 1971, a primitive seed fern, in association with the fragmentary remains of sphenopsid leaf-bearing shoots [*Phyllotheca indica* (Bunbury) Pant & Kidwari, 1968], the cordaitean (*Cordaites* sp.), and the distinct glossopterid leaves of *Gangamopteris obovata* (Carruthers) White, 1908, *Gangamopteris buriadica* Feistmantel, 1879, *Gangamopteris angustifolia* McCoy, 1875, *Rubidgea obovata* Maithy, 1965, *Rubidgea lanceolata* (Maithy) Millan & Dolianiti, 1982, *Glossopteris indica* Schimper, 1869, and *Glossopteris communis* Feistmantel, 1876 (Cazzulo-Klepzig *et al.* 1980; Pasquali *et al.* 1986; Guerra-Sommer & Cazzulo-Klepzig 1993) (see Fig. 2).

Higher levels of the section also include records of glossopterids such as *Glossopteris communis* and *G. occidentalis* White, 1908, the ginkgoalean *Ginkgophytopsis* sp., the cordaitean *Cordaites hislopii* (Bunbury) Seward & Leslie, 1908, the compound leaves of ferns *Asterotheca* sp. and *Pecopteris* sp., and the lycopsid stem *Brasilodendron pedroanum* (Carruthers) Chaloner, Leistikow & Hill, 1979 (Vieira & Iannuzzi 2000; Iannuzzi *et al.* 2003a, b).

The flora identified from the Faxinal Mine is composed of leaves, predominantly of glossopterids

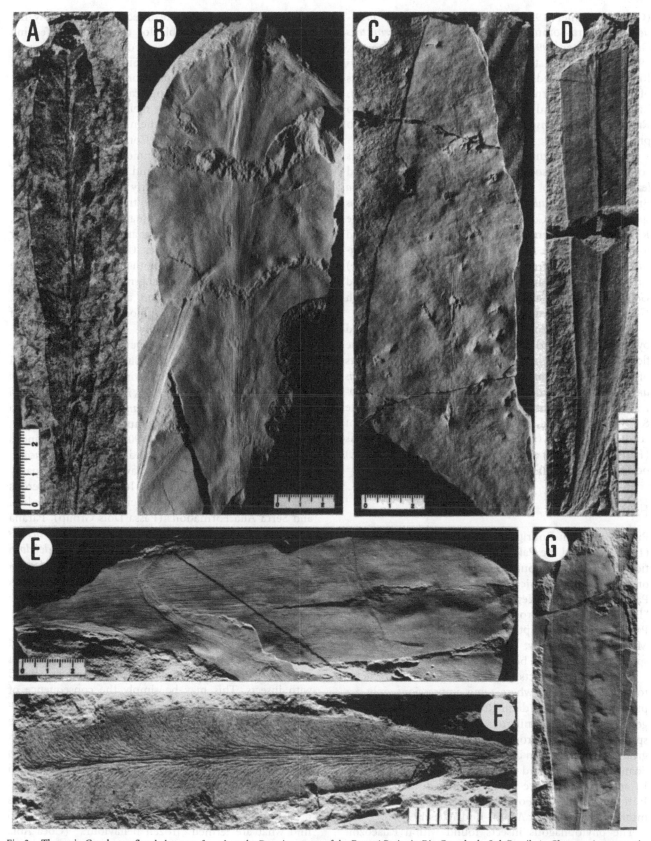

Fig. 2. The main Gondwana floral elements found on the Permian strata of the Paraná Basin, in Rio Grande do Sul, Brazil. A: *Glossopteris communis* (MP-Pb-2863). B: *Gangamopteris obovata* (MP-Pb-3704). C: *Glossopteris* cf. *G. indica* (MP-Pb-3702). D: *Glossopteris angustifolia* (MP-Pb-2719). E: *Cordaites hislopii* (MP-Pb-3671). F: *Glossopteris angustifolia* var. *taeniopteroides* (MP-Pb-3641). G: *Glossopteris occidentalis* (MP-Pb-3757). Scale bars: A–F, in mm; G shows a 5.0 cm scale.

(Guerra-Sommer 1992), including *Glossopteris brasiliensis* Guerra-Sommer, 1992, *G. similis-intermittens* Guerra-Sommer, 1992 and *G. papillosa* Guerra-Sommer, 1992. Reproductive structures such as *Plumsteadia sennes* Rigby, 1978, the seeds of the *Platycardia*-type, fragments of *Sphenopteris* fern-like foliage (cf. *S. ischanovensis* Zalessky, 1934), and the cordaitean leaves assigned to *Rufloria gondwanensis* Guerra-Sommer, 1992, are also present (see Fig. 2).

The flora recognised in the Quitéria outcrop is composed of an association of compressed glossopterid leaves assigned to the taxa *Glossopteris browniana* Brongniart, 1828, *Rubidgea lanceolata* and *Gangamopteris* sp., the cordaitean *Cordaites hislopii*, the fern-like foliage *Rhodea* sp., the sphenopsid leaf-bearing shoots attributed to *Phyllotheca indica*, the seeds *Samaropsis* sp. and *Cordaicarpus* sp., and the fructification *Arberia minasica* (White) Rigby, 1972. Among the glossopterids, *Rubidgea* and *Gangamopteris* are subdominant forms, compared with *Glossopteris*, the most abundant morphogenus (Fig. 2).

The outcrop along highway BR-290 was studied by Bortoluzzi (1975) and Backheuser *et al.* (1984). The flora from this site, according to these authors, consists of dominant *Glossopteris* and subordinate *Glossopteris angustifolia* var. *G. taeniopteroides* Seward, 1908, *Glossopteris* cf. *G. indica*, and *Glossopteris* cf. *G. antartica* (Saporte & Marion) Seward, 1919. In addition, fragments of cordaitean leaves (*Cordaites* sp.), sphenopsid stems (*Paracalamites* sp.), and fern-like foliage (*Pecopteris* sp.) occur rarely (Fig. 2).

Fauna

Pinto & Adami-Rodrigues (1999) and Würdig *et al.* (1999) reviewed the Palaeozoic insects from nine localities in Brazil and four in Argentina. Representatives of these South American insects comprise the orders: Palaeodictyoptera, Megasecoptera, Diaphanopterodea, "Protorthoptera", Protodonata, Perlaria, Blattodea, Paraplecoptera, Hemiptera, Coleoptera, Neuroptera and Mecoptera. Currently, 23 families, 32 genera and 45 species of insects have been recorded.

The record of insects in the Paraná Basin is restricted to few occurrences where wing impressions only are preserved. Pinto & Adami-Rodrigues (1999) have suggested a Late Carboniferous age based on the insect species from fossiliferous localities of the Itararé Subgroup. This succession crops out in the Boituva, Teixeira Soares, Anitápolis, and Caçapava do Sul municipalities. In addition, there are Early Permian localities in the Irati and Serra Alta Formations in the Minas do Leão municipality (Pinto & Adami-Rodrigues 1999).

The phytophagous insects found within Palaeozoic strata of the Paraná Basin consist of the orthopteran-like Prothortoptera, homopterous Hemiptera and coleopterans endopterygotes groups (Pinto & Ornellas 1981; Pinto

1987). The orthopteran-like insects were probably the main organisms responsible for the externally feeding phytophagy on vascular plants during the Palaeozoic (Rohdendorf & Rasnitsyn 1980; Shear & Kukalová-Peck 1990). Having mouthparts used for chewing and triturating food, the modern Orthoptera stand out among all the other orders of insects for their relative lack of host-plant specialisation; 60% of them are polyphagous and 25% are monophagous (Pérez-Contreras 1999).

The Coleoptera or their immediate ancestors arose during the Early Permian, and the first records of "Protocoleoptera" occurred early in the Late Permian (Ponomarenko 1995). Initially they developed several feeding strategies through their mandibulate mouthparts. The recent Coleoptera and their archostematan ancestor are predominantly xylophagous. Currently many homopterous Hemiptera feed on both the epigeous and the hypogeous parts of plants using mouthparts specialised for piercing and sucking. There are three major types: xylem feeders (e.g. cicadas), phloem feeders (most leaf hoppers) and mesophyll feeders.

Orthopteran-like insects have been found in the Palaeozoic strata of the Itararé Subgroup in the Paraná Basin, documented by the following species: *Proedischia mezzalirai* Pinto & Ornellas, 1978, *Narkemina rohdendorfi* Pinto & Ornellas, 1978, and *Paranarkemina kurtzi* Pinto & Ornellas, 1980, from Boituva municipality (São Paulo); *Carpenteroptera onzii* Pinto, 1990, from the Anitápolis municipality (Santa Catarina); and *Narkemina rochacamposi* Pinto, 1978, from the Caçapava do Sul municipality (Rio Grande do Sul) (Fig. 3F–H). In contrast, colepterous and homopterous insects have been recorded for the Irati and Serra Alta Formations (Passa Dois Group), Paraná Basin, being represented by two species (*Kaltanicupes ponomarenkoi* Pinto, 1987, *Protocupoides rohdendorfi* Pinto, 1987), and three species (*Prosbolecicada gondwanica* Pinto, 1987, *Fungoringruo kukalovae* Pinto, 1990, *Gondwanoptera capsii* Pinto & Ornellas, 1981), respectively, all collected from the Minas do Leão municipality of Rio Grande do Sul (Fig. 3A–E).

Among the most abundant evidence of fossil record in the Paraná Basin are the leaves of vascular plants. It is very rare to find evidence of plants directly associated with insects. This may be simply the consequence of taphonomic processes (Baxendale 1979). The composition and structure of these two organismic groups are so different that they preserve under particular and often exclusive fossilisation conditions. Of the many localities where the fauna of Late Palaeozoic insects of the Paraná Basin are recorded, only one outcrop, near Minas do Leão (highway BR-290, km 90), has insects and plants occurring in association. Consequently, it is difficult to find what are termed "frozen behaviours" (Coca-Abia *et al.* 1999), which record insects interacting with the plant in *flagrante delicto*, or even insect mouthparts, preserved in the Paraná Basin strata (cf. Labandeira 2002).

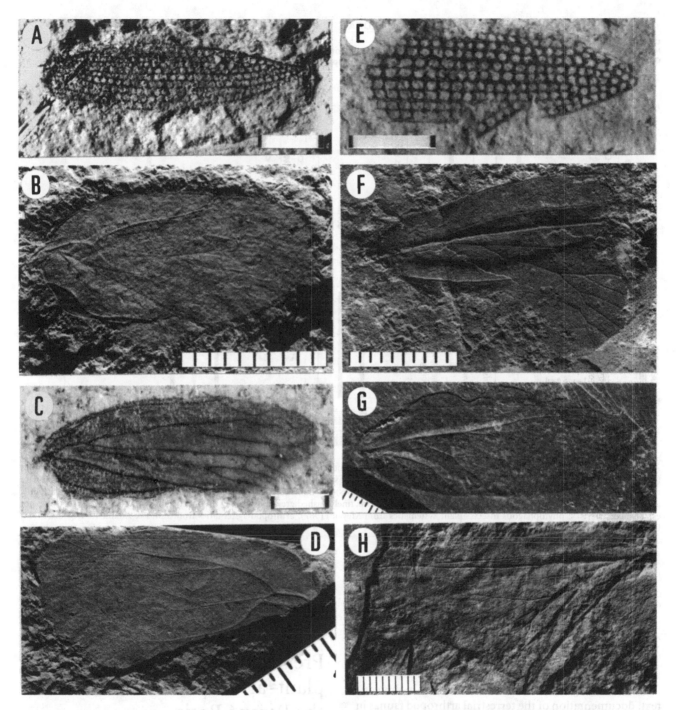

Fig. 3.   Some of the phytophagous insects found in the Palaeozoic strata of the Paraná Basin. A, E: Coleopteran group. A: *Kaltanicupes ponamorenkoi* (MP-I-5269). E: *Protocupoides rohdenforfi* (MP-I-5266). B–D: Hemipteran group. B: *Prosbolecicada gondwanica* (MP-I-5263). C: *Fulgoringruo kukalovae* (MP-I-5273). D: *Gondwanaptera capsii* (MP-I-5259). F–H: Orthopteroid group. F: *Narkemina rohdendorfi* (MP-I-5283). G: *Carpenteroptera onzii* (MP-I-6608). H: *Narkemina rochacamposi* (MP-I-5286). Scale bars: B, D, F, G, in mm; A, C, E show a 1.0 mm scale.

Faunal and floral succession in the Paraná Basin

The chronostratigraphic and biostratigraphic framework of the Paraná Basin is based on palynological evidence as well as on the marine and terrestrial macrofaunas spanning the Late Carboniferous to Late Permian interval (Daemon & Quadros 1970; Rocha-Campos & Rösler 1978; Rösler 1978; Barberena *et al.* 1985; Marques-Toigo 1991; Guerra-Sommer & Cazzulo-Klepzig 1993; Petri & Souza 1993; Pinto 1995; Langer 2000). Megafossil plants are abundant and geographically widespread within the

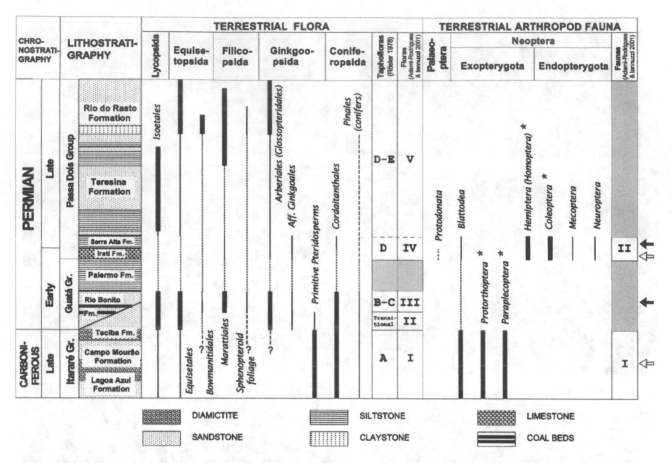

Fig. 4. Stratigraphic distribution of the major fossil plant and arthropod groups in Paraná Basin during Late Carboniferous to Permian times (modified from Adami-Rodrigues & Iannuzzi 2001). The floral and faunal successions are shown in the vertical columns, while stratigraphic gaps are represented as grey areas. Arrows at the right indicate faunas (white arrows) and floras (black arrows), as analysed herein. Continuous thick dark line: abundant; continuous thin dark line: common; broken dark line: rare; finely dashed grey line: absent (no direct evidence); question mark: range not established; asterisks: phytophagous insects [plant systematics from Meyen (1987); arthropod systematics from Pinto & Adami-Rodrigues (1999); lithostratigraphy based on Milani *et al.* (1994)].

Palaeozoic deposits of the Paraná Basin. The floral associations were subdivided into six stratigraphically distinct "taphofloras", A–E (Fig. 4), which was proposed as an informal phytostratigraphic scheme for the entire basin (Rösler 1978). The other formal biozones refer to specific time intervals and they are applied only to restricted portions of the basin (Millan 1987; Guerra-Sommer & Cazzulo-Klepzig 1993; Rohn & Rösler 2000). In this context, documentation of the terrestrial arthropod faunas in the Paraná Basin is restricted to a few occurrences and currently there are no proposed biostratigraphic units. The faunal associations have been largely revised by Pinto & Adami-Rodrigues (1999).

Recently, Adami-Rodrigues & Iannuzzi (2001) completed a revision and reinterpretation of the available data, in order to analyse the evidence of plant–arthropod interactions in Palaeozoic strata of Paraná Basin. In an attempt to informally reconcile the various formal biostratigraphic units with the general framework of Rösler (1978), Adami-Rodrigues & Iannuzzi (2001) defined a

palaeontological succession divided into five floras (I–V) and two faunas (I, II), thus adjusting the record for the entire basinal area (Fig. 4).

Previous record of plant–arthropod interactions in the Paraná Basin

Indirect evidence

Fauna I of Adami-Rodrigues & Iannuzzi (2001) is related to the typical Carboniferous flora succession of the Paraná Basin (Pre-*Glossopteris* Flora, or Flora I of Fig. 4), although similar associations are related to more ancient elements of the "*Glossopteris* Flora" in Argentina, such as in the Bajo de Veliz locality (Pinto & Adami-Rodrigues 1997, 1999). Consequently, the insects of Fauna I may have coexisted, also, in part, with elements of

"*Glossopteris* Flora" of the Paraná Basin. Among the insects of Fauna I, Prothortoptera and Plecoptera (primitive orthoperoids) are considered the major consumers of external foliage. Contemporaneously, the Blattidae were the likely detritivores (Shear & Kukalová-Peck 1990; Labandeira 1998). Unfortunately, no direct evidence of herbivory has been recognised in the associations of Flora I which can be potentially related to the insects of Fauna I.

The insects belonging to Fauna II of Adami-Rodrigues & Iannuzzi (2001) are directly associated with elements of the "*Glossopteris* Flora" (Flora IV, see Fig. 4), principally based on the evidence of the fauna and flora found in the BR-290 outcrop near the town of Minas do Leão (Bortolluzzi 1975; Backheuser *et al.* 1984; Pinto & Adami-Rodrigues 1999). The majority of phytophagous insects recorded in Fauna II are undoubtedly Hemiptera and Coleoptera (Shear & Kukalová-Peck 1990; Labandeira 1998). Hemiptera were probably piercing–sucking insects (Shear & Kukalová-Peck 1990; Kukalová-Peck 1991; Labandeira 1998, 2002).

Direct evidence

The first direct evidence of plant–anthropod interactions in the Paraná Basin was reported by Guerra-Sommer (1995). This evidence originates from compressed leaves of *Glossopteris* from the Faxinal Mine, the same locality and strata described and characterised in this study (see elsewhere).

Guerra-Sommer (1995) described foliar damage represented by external foliage feeding. Two categories of damage from external foliage feeding were present: continuous feeding activity of the foliar edge and top, based on the terminology of Scott & Titchener (1999). Guerra-Sommer (1995) attributed foliar damage to orthopteroid insects, considering that this type of feeding activity characterised this group during the Late Palaeozoic (Beck & Labandeira 1998; Labandeira 1998). Qualitative analyses of folivory have indicated that damage by insect herbivory was preferentially targeting megaphyllous taxa, especially glossopterids (Guerra-Sommer 1995). In contrast to the data by Beck & Labandeira (1998), which indicated quantitatively much higher percentages for gigantopterid taxa, our results show that quantitatively only 5% of the glossopterid leaves were attacked.

The floral association from the Rio Bonito Formation occupies a position stratigraphically between Fauna I and Fauna II of Adami-Rodrigues & Iannuzzi (2001) (Fig. 4). However, there are no records of insect body fossils preserved in these horizons. On the other hand, signs of marginal feeding are also present on glossopterid foliage in the upper units of the Paraná Basin, such as in the Irati and Serra Alta Formations, where the insects of Fauna II are directly associated (e.g. BR-290 outcrop; see previous description of insect fauna).

Results of qualitative analysis

The precise recognition of the identity of insects that damage plants in foliar compressions is difficult to determine. According to Labandeira (1997), it is only possible to recognise the patterns of plant–insect interactions of some characteristic herbivorous insect lineages.

Damage to leaves caused by insects has been classified into five categories of feeding types for modern insects by Coulson & Witter (1984), and the types defined and identified by Beck & Labandeira (1998) for the Palaeozoic record. These damage types are: excisions of the leaf margin, holes of the leaf blade, skeletonisation, possible mines and possible galls. Margin feeding, hole feeding, and skeletonisation are damage types that are grouped under external foliage feeding. In this study, the kind of evidence is based on the types identified by Beck & Labandeira (1998).

Because the occurrence of this evidence is restricted to Palaeozoic floras, we found it useful to indicate and compare the fossil taxa with recent herbivorous insects responsible for similar damage, whenever possible. The most significant evidence for the Permian strata of the Paraná Basin is presented below.

Evidence of foliar margin feeding

Three kinds of damage have been recorded for this kind of interaction, as follows.

Discontinuous activity at the foliar margin [Fig. 5D(b), C(c)]

Specimens studied
Faxinal Mine: MP-Pb 2811A – *Glossopteris* sp. (notophyll); MP-Pb 2803 – *Glossopteris* sp. (notophyll); MP-Pb 2810 – *Glossopteris* sp. (notophyll); MP-Pb 2796A – *Glossopteris* cf. *communis* (notophyll); MP-Pb 3130A, B – *Glossopteris brasiliensis* (notophyll); MP-Pb 2796B – bract of *Glossopteris*? (microphyll); MP-Pb 2796C – *Cordaites hislopii* (= *Rufloria gondwanensis*) (mesophyll). Morro do Papaléo Mine: MP-Pb 3699 – *Glossopteris* cf. *communis* (mesophyll); MP-Pb 3700 – *Glossopteris* cf. *communis* (mesophyll). Minas do Leão locality (BR-290, km 90): MP-Pb 2238 – *Glossopteris angustifolia* (notophyll); MP-Pb 2223 – *Glossopteris angustifolia* (nanophyll).

Plant hosts
Glossopteris sp., *G. angustifolia*, *G. brasiliensis*, *G.* cf. *communis*, bract of *Glossopteris*?, *Cordaites hislopii*.

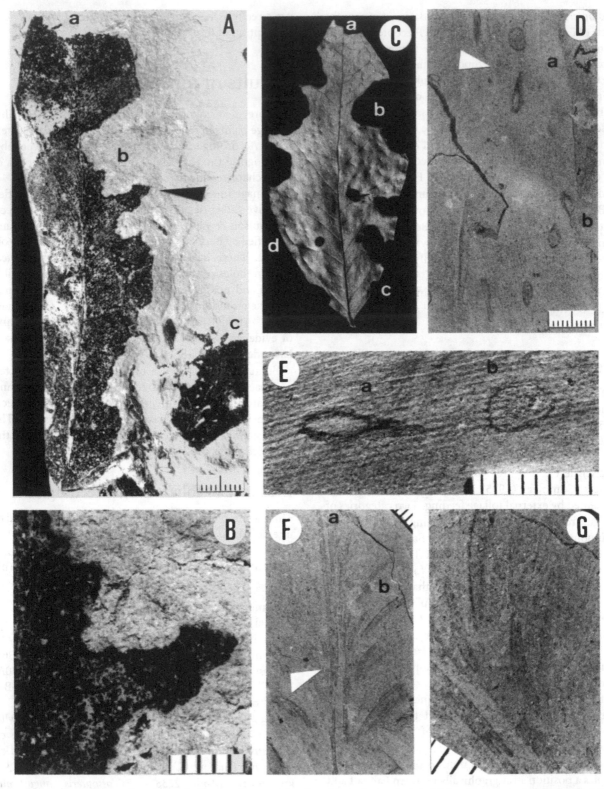

Fig. 5.   Evidence of foliar margin and hole feeding on glossopterid and cordaitean leaves. A: Apical (a) and continuous margin feeding (b) on *Glossopteris* sp. (Pb-2796A), and hole feeding (c) on a probable glossopterid bract, and also showing a reaction tissue rim (arrow). B: Enlargement of the margin feeding area indicated by the arrow in (A). C: Examples of plant–insect interactions on a modern angiosperm leaf (Mirtaceae family); (a) feeding activity of foliar apex, (b) continuous activity of the foliar edge, (c) discontinuous activity of the foliar margin, (d) hole feeding. D: Hole (a) and discontinuous margin feeding (b) on *Cordaites hislopii* (Pb-3655A), showing a developed reaction tissue rim (arrow). E: Enlargement of the hole feeding area indicated by the arrow in (D), showing the ellipsoidal (a) to ovoidal (b) shape of feeding holes. F: Intervenal hole feeding on *Glossopteris* cf. *communis* (Pb-3655E) showing feeding holes between major (a) and secondary (b) veins. G: Enlargement of the hole feeding area indicated by the arrow in (F). Scale bars: A–G, in mm.

Description
Marginal traces of foliar excision that are interrupted. The presence of a plant defence mechanism with clear reaction tissue rims, evidenced by a thickened or coloured contour surrounding the leaf edge where phytophagy has taken place.

Dimensions
Small excisions from 1.3 to 2.0 mm in chord length.

Remarks
Cuspate excisions on foliar impressions from the Mina do Morro do Papaléo Mine bear edges that match the reaction tissue marked by a reddish colour due to the presence of iron oxide. As for foliar compressions from the Faxinal Mine, the reaction tissue is present as a more noticeable accumulation of organic matter, indicated by a dark contour, which corresponds to callus tissue.

Some plants have developed defence mechanisms that generate toxins. Insects respond to this chemical environment by producing discontinuous feeding activity on the foliar edge, in order to avoid the toxic response of the plant. Most of the attacked leaves of *Glossopteris*, classified as notophylls to mesophylls, have reaction tissue. The record of this activity in *Cordaites*-like leaves is very sparse. In the glossopterid species, this type of activity is more intensive on the leaves of *Glossopteris* cf. *communis* (Guerra-Sommer 1995).

Feeding activity at the foliar apex with reaction tissue [Fig. 5A(a), A(b), B, C(a)]

Specimens studied
Faxinal Mine: MP-Pb 3130A, B – *Glossopteris brasiliensis* (notophyll); MP-Pb 2796A – *Glossopteris* cf. *communis* (notophyll).

Plant hosts
Glossopteris brasiliensis, G. cf. *communis.*

Description
Continuous extraction of entire foliar leaf blade at the apex with reaction tissue rims clearly defined by a darkened and thickened contour.

Dimensions
Small cuspate excisions of 1.9–2.0 mm, with transverse truncation of the foliar apex.

Remarks
This type of interaction has been documented in notophylls of *Glossopteris* cf. *communis* and *Glossopteris brasiliensis* from the Faxinal Mine locality. This damage to the plant was probably caused by small orthopteroid insects with flexible mandibulate mouthparts, different from the other insects that would be involved in an extensive or continuous herbivory of the foliar edge (Gangwere 1966; Edwards & Wratten 1980).

Feeding activity at the foliar apex without reaction tissue

Specimen studied
Minas do Leão locality (BR-290, km 90): MP-Pb2241 – *Glossopteris angustifolia* (notophyll).

Plant host
Glossopteris angustifolia.

Description
Continuous extraction of the entire leaf blade at the foliar apex without evident reaction tissue rims.

Remarks
This type of interaction has been recorded in one notophyll of *Glossopteris angustifolia* from the Minas do Leão locality. The assignment of this damage is complicated due to the absence of the clearly defined reaction tissue.

Continuous activity at the foliar edge [Fig. 5A(b), B, C(b)]

Specimens studied
Quitéria outcrop: MP-Pb 2863 – *Glossopteris* cf. *communis* (notophyll); MP-Pb 3138 – *Glossopteris* cf. *communis* (notophyll). Faxinal Mine: MP-Pb 2796A, B – *Glossopteris communis* (notophyll); MP-Pb 3124 – *Glossopteris browniana* (mesophyll); MP-Pb 2811 – *Glossopteris* sp. (notophyll).

Plant hosts
Glossopteris sp., *G. browniana, G. communis.*

Description
Large-scale removal of the foliar edge, placed in an uninterrupted series of marginal excisions, and provided with a strongly darkened contour defined as a reaction tissue rim.

Dimensions
Small cuspate excisions of 2–3 mm around the foliar edge damaged.

Remarks
The continuous activity of the foliar edge is found only in leaves of *Glossopteris* sp., *G. browniana* and *G. communis* from Faxinal Mine and Quitéria outcrop, respectively. No evidence was found on the leaves of *Cordaites.*

General discussion
Guerra-Sommer (1995) attributed the foliar damage described above to orthopteroid insects, considering that

these kinds of feeding habit are mainly related to this group during the Late Palaeozoic time, a view shared by others (Beck & Labandeira 1998; Labandeira 1998). Primitive Coleoptera may also be candidates for the consumption of glossopterid external foliage such as that found in the Irati and Serra Alta Formations (Minas do Leão locality). Although there is apparently a lack of records for orthopteroids in the associated fauna (Fauna II of Adami-Rodrigues & Iannuzzi 2001) (Fig. 6), these results are not conclusive. Further studies are required to establish whether orthopteroid insects are totally absent from these levels in the Paraná Basin.

Small removals of the foliar lamina

Three types of excavation within the leaves are recognised in this category.

Ovoid type [Fig. 5D(a), E]

Specimens studied
Faxinal Mine: MP-Pb 2796B – bract of *Glossopteris?*. Morro do Papaléo Mine: MP-Pb 3668 – *Cordaites hislopii* (megaphyll); MP-Pb 3678B – *Cordaites hislopii* (megaphyll); MP-Pb 3655A, C, D – *Cordaites hislopii* (megaphyll); MP-Pb3678A – *Cordaites hislopii* (mesophyll).

Plant hosts
Glossopteris? sp., *Cordaites hislopii*.

Description
Ovoid to elongated ellipsoidal holes identified by a reddish reaction tissue where the greater cavity axis aligns venationwise with the removal of the foliar lamina.

Dimensions
Elongated ellipsoids in the shape of drops, ranging from 3 to 6 mm in length.

Linear type 1 [Fig. 5F(a–b), G]

Specimens studied
Quitéria outcrop: MP-Pb 2863 – *Glossopteris communis* (notophyll). Faxinal Mine: MP-Pb 2796A – *Glossopteris* cf. *communis* (notophyll). Morro do Papaléo Mine: MP-Pb 3699 – *Glossopteris* cf. *communis* (notophyll); MP-Pb 3700 – *Glossopteris* cf. *communis* (notophyll); MP-Pb 3655B, E – *Glossopteris* cf. *communis* (notophyll).

Plant host
Glossopteris cf. *communis*.

Description
Removal of linear slots of foliar tissue between veins, starting from the midrib veins and thickening between the midrib and secondary venation, surrounded by rims of reaction tissue.

Linear type 2 [Fig. 6C(a)]

Specimens studied
Morro do Papaléo Mine: MP-Pb 3704 – *Gangamopteris obovata* (mesophyll); MP-Pb 3703 – *Gangamopteris obovata* (mesophyll); MP-Pb 3678A – *Cordaites hislopii* (megaphyll).

Plant hosts
Gangamopteris obovata, Cordaites hislopii.

Description
Elongate structures parallel to venation with increase in width, surrounded by thickened tissue that consists of two parallel and similar ridges.

Dimensions
Linear traces 3–5 cm long and 1–3 mm wide.

General discussion
Leaf removal of the ovoid and elongated ellipsoidal type is only found on *Cordaites hislopii* and a small-scale leaf attributed to glossopterids, possibly a bract of *Glossopteris*, and extraction between veins on *Glossopteris* cf. *communis* (= linear type 1). The foliar impressions from the Morro do Papaléo Mine show these two kinds of evidence, lack contrast with the matrix, but the preservation of the contour where the attack took place is marked by the presence of iron oxide and an irregular border that indicates the presence of a reaction tissue rim. On other leaves of *Cordaites hislopii* and *Gangamopteris obovata*, linear traces (= linear type 2) are considered as a narrow type of slot feeding surrounded by thickened tissue interpreted as reaction tissue rims. The slot feedings identified are always found at the base and towards the middle portion of the leaf.

Beetles produce modern evidence of this type of feeding activity, especially from the families Chysomelidae and Curculionidae (Johnson & Lyon 1993). However, these two groups of beetles have only been recorded from the Late Triassic, and the Late Jurassic to the present, respectively (Medvedev 1968; Zherikin & Gratshev 1993). Thus, the present evidence probably relates to the activity of unknown primitive insects related to the coleopteran group.

Possible leaf mines (Figs. 6A, B, 7)

Specimens studied

Morro do Papaléo Mine: MP-Pb 3702a, b – *Glossopteris* cf. *indica* (megaphyll).

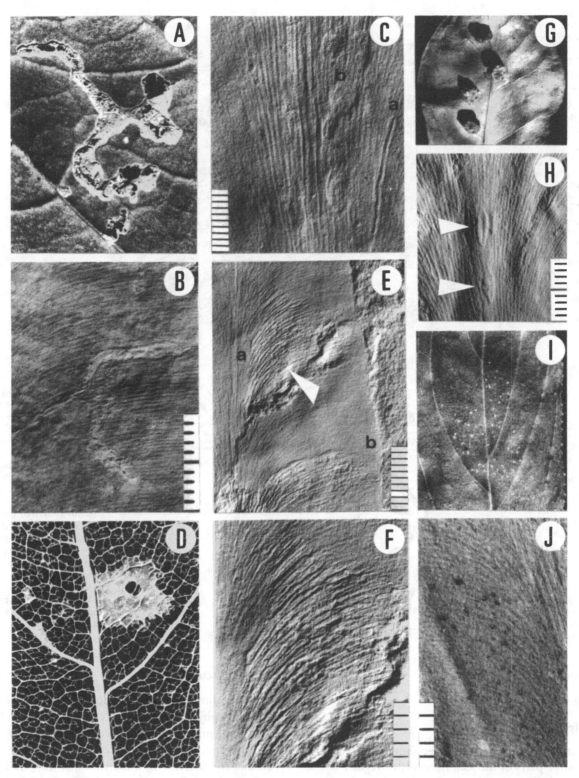

Fig. 6. Evidence of possible leaf mines and galls, as well as characteristic skeletonisation and piercing and sucking feeding on glossopterid leaves. A: Example of a mine on a modern angiosperm leaf. B: A possible bifurcating leaf mine on *Glossopteris* cf. *G. indica* (Pb-3702A); note the similarity to the modern mine shown in (A). C: Linear slot feeding (a) and ellipsoidal to spherical gall-like structure (b) on *Gangamopteris obovata* (Pb-3704). D: Skeletonisation on a modern angiosperm leaf. E: Skeletonisation on *Glossopteris communis* (Pb-3705), displaying two damaged areas (a, b). F: Enlargement of the skeletonised area (a) indicated by the arrow in (E). G: Gall in modern angiosperm leaf (*Eugenia uniflora* L., Myrtaceae family). H: Lenticular-shaped oviposition scars by protodonatan dragonflies on *Gangamopteris obovata* (Pb-3703). I: Piercing and sucking feeding area on a modern angiosperm leaf [*Trema micrantha* (L.) Blume, Ulmaceae family]. J: A possible piercing and sucking feeding area on *Glossopteris communis* (Pb-3655B), exhibiting small spheroidal punctures along the middle part of the leaf and on the secondary venation, indicating piercing and sucking activity; note the similarity to punctures on a modern leaf area in (I). Scale bars: A–J, in mm.

A
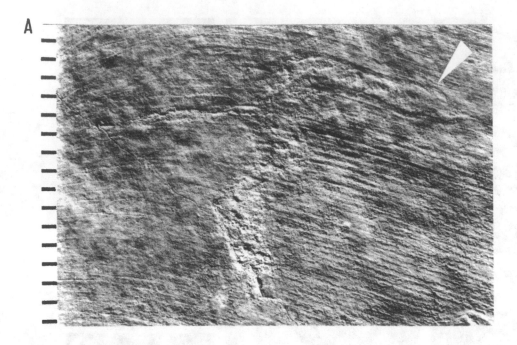

B
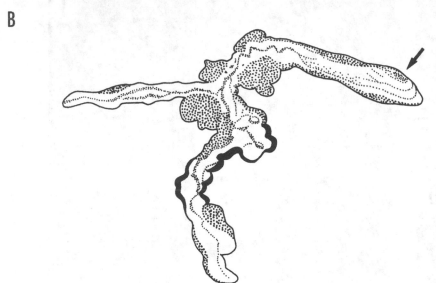

Fig. 7. Detail of a possible bifurcating leaf mine on *Glossopteris* cf. *G. indica* (Pb-3702A). A: Enlargement of the mine illustrated in Fig. 6B. B: Camera lucida drawing of the mine in (A) at the same scale, displaying mine-like structures; the continuous line defines the limits of the mine; the dotted fine area represents a frass trail; the thicker dotted area comprises the irregular area of affected tissue; the inner white area represents a smooth area of affected tissue; the narrow black rim includes the reaction tissue rim. Arrows in (A), (B) define the probable terminal chamber. Scale bar, in mm.

Plant host

Glossopteris cf. *indica*.

Description

Bifurcating curvilinear structures found on the basal and middle parts of the leaf, filled with irregular coarse and smooth tissues, and surrounded by rims of reaction tissue.

Dimensions

Sinusoidal lines 1 cm long by 2 mm wide in the initial phase, and 3–4 mm wide in the final phase of development.

General discussion

Evidence showing foliar mining has been documented on a megaphyll of *Glossopteris* cf. *indica*, from only one bifurcating mine. The mine trace can be identified within the venation, forming a space containing frass and a distinct size increase in mine width. A frass trail and a contrast in colour and texture among unaffected foliar surface, surrounding affected foliar surface (= reaction tissue rims), and an inner mine trace, indicate mining activity. Besides, an expanded termination is interpreted as a possible terminal chamber that the larva, pupa or adult has left. The mine identified is found in the middle portion of the leaf.

Organisms producing mines show special behaviour like the ovipositional penetration of leaf tissue in order to introduce larvae into the leaves. This has been suggested for occurrences in the Lower Permian deposits of Euroamerican localities, in spite of diagnostic enquiries over their real origin (Labandeira 1998).

Modern mines are found on leaves of angiosperms and are generally attributed to larvae of Coleoptera, Diptera, Hymenoptera and Lepidoptera (Connor & Taverner 1997; Coca-Abia *et al.* 1999). However, the earliest Lepidoptera arose during Jurassic times and basal Hymenoptera first appeared in the Middle Triassic. Only the Coleoptera has a record in the Permian (Ollerton 1999). Therefore, it is suggested that this type of structure on a glossopterid leaf should possibly be related to coleopteran larvae, a group that has been recorded in the Permian strata of the Paraná Basin.

Skeletonisation (Fig. 6D–F)

Specimens studied

Morro do Papaléo Mine: MP-Pb 3705A – *Glossopteris communis* (mesophyll); MP-Pb 3705B – *Glossopteris occidentalis* (megaphyll).

Plant hosts

Glossopteris communis, Glossopteris occidentalis.

Description

Lack of mesophyll tissue within a venal network, apparently occurring at or adjacent to an intercoastal region. The spherical perforations occur among cells of intervenal tissue, displaying a meshwork of venules. No reaction tissue rims have been detected.

Dimensions

Small, spherical excavations among veins ranging from 1 to 2 mm in diameter. The average area affected is 1.8–4.0 cm and is located 1 cm from the foliar edge.

Remarks

The total removal of the foliar lamina preserved three dimensionally occurs on glossopterid leaves, and two morphospecies are known as hosts: *Glossopteris communis* and *Glossopteris occidentalis*. These leaves have been preserved as authigenic replicas, which produce three-dimensional moulds or casts of the specimens and allow identification of this type of interaction. However, this kind of preservation does not allow verification of the presence of interpretable reaction tissues surrounding the skeletonised areas of specimens analysed. Determination of true skeletonisation is uncertain without the presence of a recognisable reaction rim which indicates a herbivory

while the leaf was on the plant and alive. The absence of a reaction rim suggests that detritivorous skeletonisation occurred after the leaf died.

Current records of this type of activity are normally related to modern Orthoptera and larval Lepidoptera (Kazakova 1985; Johnson & Lyon 1993). Therefore, if a real skeletonisation process occurred on the *Glossopteris* leaves, it could be related to the interaction with orthopteroids that are known to have existed in the Paraná Basin. Otherwise, if it is not true, this evidence should be interpreted as a result of detritivorous activity on already dead leaves, probably the activity of aquatic scavenging insects, which in the Paraná Basin could have been represented by blattoids.

Possible insect galls [Fig. 6C(b), G]

Specimen studied

Morro do Papaléo Mine: MP-Pb 3704 – *Gangamopteris obovata* (mesophyll).

Plant host

Gangamopteris obovata.

Description

Hypertrophied or hyperplasic tissue organised into an ellipsoid to spherical shape, and placed in the central part of the leaf.

Dimensions

Ellipsoid to spherical tissues from 0.5 to 1.0 mm in width, and 4 to 7 mm in length.

Remarks

The structures found and identified as possible galls occur only on *Gangamopteris obovata* leaves from the Morro do Papaléo Mine, where the same species also displays the possible presence of oviposition insertions.

Galls are the result of a physiological reaction that in most instances is not well understood, although there are a few system models where it is well documented (Shorthouse & Rohfritsch 1992). Galls result from the production of hypertrophied tissue as a reaction to the attack or the presence of a juvenile or ovipositing insect. About 80% of modern galls are found on leaves (Mani 1964). In the fossil record, the oldest known galls are found on stalks of *Rhynia* from the Devonian, and they proliferated on parenchymatic tissues of petioles during the Late Carboniferous (Coca-Abia *et al.* 1999). These latter were derived most probably from endophytic, tissue-boring insects (Roskam 1992; Labandeira & Philips 1996). The oldest records on leaves are found in the

Carboniferous, but a greater diversity occurs in the Tertiary (Diéguez *et al.* 1996). During Permian times, records are sparse. Galls on gymnosperms from the Late Triassic deposits of the southwestern USA (Labandeira 2002) are similar to the ones studied in this paper.

The identification of the gall maker, based only on the shape of galls, is a questionable endeavour. In modern galls, the form often has no relationship to the producer; commonly, the identification of the originator is only recognised from an opened gall. However, considering that the gall phenotype resulted from the interaction between the insect and plant genotypes, and the insect physiologically and developmentally altered the plant tissues, it is likely that these galls would only result from insects that had this ability. Even so, it remains impossible to determine the type of insect that may have been responsible for producing the galls.

Modern fauna of gall-making phytophagous insects include representatives of the Hemiptera, Thysanoptera, Hymenoptera, Diptera, Coleoptera and Lepidoptera. Given the lack of Hymenoptera, Diptera and Lepidoptera during Palaeozoic times possibly leaves the coleopterans as the only likely group responsible for the galls.

Possible oviposition scars (Fig. 6H)

Specimen studied

Morro do Papaléo Mine: MP-Pb 3703 – *Gangamopteris obovata* (mesophyll).

Plant host

Gangamopteris obovata.

Description

Evidence of hypertrophied or hyperplasic tissue organised into a lenticular shape, and sited on the central part of the leaf.

Dimensions

Lenticular tissues from 3 mm in width, by 5 mm in length.

Remarks

These lenticular structures are interpreted as oviposition scars produced by protodonatan (or odonatan?) dragonflies. This kind of structure is quite common on sphenophyte stems, midribs of riparian plants, and even on leaf blades. It has been documented in western Europe (Grauvogel-Stänn & Kelber 1996) and it occurs abundantly in Triassic deposits of the Molteno Formation in South Africa (C. Labandeira 2003, pers. comm.). However, it is very rare in the assemblages under study here, being represented so far by only one specimen. This seems

to be related to protodonatan dragonflies, a group found in Palaeozoic deposits of the Paraná Basin.

Small punctures (Fig. 6I, J)

Specimens studied

Morro do Papaléo Mine: MP-Pb3655F – *Glossopteris communis* (mesophyll), MP-Pb3655G – *Cordaites hislopii* (megaphyll).

Plant hosts

Glossopteris communis, Cordaites hislopii.

Description

Small spheroidal point-source incisions placed along midrib veins or along the central blade region of the leaf.

Dimensions

Spheroidal perforations 0.5–1.0 mm in diameter.

Remarks

The fossil record indicates that the feeding activity of piercing–sucking insects is rare. In our material, small spheroidal incisions are found along the midveins on *Glossopteris communis* and on the middle portion of the leaf blade of *Cordaites hislopii*. The identification of this feeding damage was easily recognised because the perforations are filled with iron oxide, improving the contrast between the foliar impressions and their insect-mediated damage.

The groups of insects responsible for feeding through piercing–suction have changed significantly throughout the history of this functional feeding group. Earliest known perpetrators of the damage are unknown micro-arthropods of the earlier Devonian, and four orders of paleodictyopteroids in the Late Palaeozoic (Carpenter 1971; Rohdendorf & Rasnitsyn 1980; Labandeira & Phillips 1996). These were replaced by basal lineages of Hemiptera and Thysanoptera during the Middle Permian (Becker-Migdisova 1940; Vishiniakova 1981). The Hemiptera and Thysanoptera clades had extended their ecological distribution through several major groups of vascular plants, after the extinction of the paleodictiopteroids during the major Permo-Triassic mass extinction. Currently, these two groups are the predominant piercing and sucking herbivores, even though several families of Coleoptera are included in this functional feeding group (Labandeira & Beall 1990). In the Paraná Basin, the probable phytophagous producers belong to the Hemiptera and include homopterous groups such as the Auchenorryncha, Sternorrhyncha and Coelorhyncha.

Discussion and conclusions

The analysis of collections from the fossiliferous localities defined herein allows the identification and definition of 11 categories of plant–insect interactions. These 11 defined categories indicate damage to the leaves of *Glossopteris*, *Gangamopteris* and *Cordaites*. In addition, based on the number and frequency of specificity for herbivorised leaves, we identified a phytophagous preference for glossopterid leaves, and particularly for the morphospecies *Glossopteris communis* (Fig. 8). For these and other plant hosts, external foliage feeding activity was recognised by continuous and discontinuous consumption of the foliar edge and apex. The total removal of foliar laminae was typically found in *Cordaites*-like leaves, whereas the leaves of *Glossopteris* and *Gangamopteris* were seldom so intensively consumed. Herbivore preference for the removal of the foliar lamina in *Cordaites*, as opposed to that of the glossopterids, is probably based on the foliar thickness. The megaphylls of *Glossopteris* are apparently thicker than those of *Cordaites*. The thin lamina of *Cordaites* is more easily abraded by the mouthparts of the insect perpetrator. This hypothesis is valid only if the damage is recorded for mature

TYPES OF EVIDENCE	PLANT TAXA - MORPHOGENERA			INSECT MOUTHPARTS	POSSIBLE INDUCER INSECT GROUP
	Glossopteris	*Gangamopteris*	*Cordaites*		
Discontinuous margin-feeding	*G. angustifolia;* *G. brasiliensis;* *G.* cf. *communis; Glossopteris* sp.; bract of *Glossopteris*		*Cordaites hislopii* (=*Rufloria gondwanensis*)		orthopteroid / coleopterous
Apical margin-feeding	*G. angustifolia;* *G. brasiliensis,* *G.* cf. *communis.*			Mandibulate type	orthopteroid
Continuous marginal-feeding	*G. browniana;* *G. communis;* *Glossopteris* sp.				orthopteroid / coleopterous
Total removal of foliar lamina	bract of *Glossopteris* *G. communis*	*Gangamopteris obovata*	*Cordaites hislopii*	Piercing-and-Chewing	coleopterous
Leaf mine	*G.* cf. *indica*				coleopterous (?)
Skeletonization	*G. communis;* *G. occidentalis*			Mandibulate type	orthopteroid / blattoid
Galls		*Gangamopteris obovata*			coleopterous (?)
Ovoposition scar		*Gangamopteris obovata*			protodonatan
Small punctures	*G. communis*		*Cordaites hislopii*	Piercing-and-sucking	homopterous

Fig. 8. Relationship between different types of plant–insect interaction and plant hosts based on the present study. The potential functional feeding groups are represented by possible predator insect groups and distinct mouthparts in columns on the right.

megaphylls of *Cordaites* and for nanophylls and
notophylls of *Glossopteris*.

Skeletonisation is a rare and difficult process to identify
and could only be determined in this study through the
three-dimensional preservation of the *Glossopteris* leaves
by authigenic taphonomic processes. In these leaves,
the total foliar removal with venule exposure occurs in
particular regions of the leaf blade for those analysed
mesophylls and megaphylls. More specialised than
skeletonisation is the possible presence of mining insects
recorded on leaves of *Glossopteris* cf. *G. indica*. However,
this is a very low frequency type of damage.

The most intriguing evidence of plant–insect interaction on *Gangamopteris obovata* is the presence of galls,
where one can detect the preference of a host-specialist
inducer insect. As gall production or phenotype depends
on both the plant genotype and the insect, we conclude
that there was a certain species of gall-producing insect
inhabiting the lacustrine palaeocommunity. This community is known from the lower section of the Morro do
Papaléo Mine, and is directly linked to *Gangamopteris
obovata*. Other less common evidence, recorded in one
leaf of *Gangamopteris obovata*, is the presence of oviposition scars. They are attributable to ancient protodonatan
dragonflies.

There is a very sparse record of damage caused
by piercing and sucking insects on material preserved as
foliar impressions of *Glossopteris* and *Cordaites*. No preference for one or the other of the above-mentioned plant
taxa has been identified.

The evidence recorded on external foliage either as
feeding continuous or discontinuous feeding activity of
the foliar edge and apex in glossopterids may be attributed to orthopteroid insects such as the Protorthoptera.
This group would also be responsible for the skeletonisation on glossopterid leaves by virtue of the fact
that they possessed a strong mandibular system, theoretically capable of drilling the thickest edges of the
leaves of *Glossopteris* spp. However, as already noted,
orthopteroid insects do not have any directly related
connection to floristic associations (Fig. 9). Moreover,
there are doubts as to whether primitive Coleoptera – the
only phytophagous group present in the "*Glossopteris*
Flora"-bearing stratigraphic levels of the Paraná Basin
(Adami-Rodrigues & Iannuzzi 2001) – could have
been responsible for the damage (Figs. 8, 9). On the
other hand, hemipteroids such as Sternorrhyncha or
Auchenorrhyncha could have been responsible for the
small perforations on *Cordaites* and *Glossopteris* leaves.
The total removal of foliar lamina, often involving
Cordaites, is attributed to taxa other than Coleoptera,
which performs this type of feeding activity, but only has
a fossil record from the Triassic onwards (Medvedev
1968; Zherikin & Gratshev 1993).

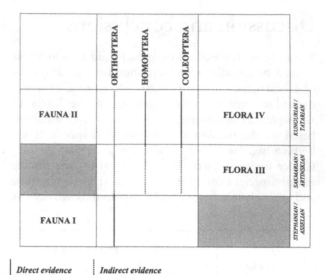

Fig. 9. Stratigraphic distribution of faunas and floras studied, indicating the presence of phytophagous insects, based on direct and indirect
evidence after Adami-Rodrigues & Iannuzzi (2001) and this contribution. See Fig. 4 for a complete list of faunal and floral assemblages. Grey
areas indicate biostratigraphic gaps.

The presence of a putative mining insect on a leaf of
Glossopteris may be attributed to coleopteran larvae.
This assignment is made based on the knowledge that
other modern mining groups, such as the Hymenoptera
and Lepidoptera, were not present during the Permian
(Coca-Abia *et al.*1999; Ollerton 1999), and the Diptera
arguably lacks a Palaeozoic record (Pinto & Adami-
Rodrigues 1999; Krzeminski & Evenhuis 2000). Like the
mining insects, gall-producing insects often have poorly
known life cycles. Some mining groups may have been
candidates for the production of galls and, given that the
Diptera and Lepidoptera orders were absent during the
Permian, it is conceivable that coleopteran larvae served
also as gall producers in these Palaeozoic associations.

The hypotheses and observations derived from this
qualitative analysis of plant–insect interactions are of
particular relevance and importance to an understanding
of the development and structure of food webs in the
terrestrial ecosystems of the Permian strata in the Paraná
Basin. Apparently, the glossopterid and cordaitean leaves
were sufficiently abundant and nutritious to attract a
varied fauna of coleopteran, hemipteran and orthopteroid insects, and allowed them to choose their own
particular feeding preferences. However, other Permian
groups of extinct phytophagous insects were probably
present, and may have been responsible for some of
the patterns of feeding activity recognised here. The
lack of a greater knowledge of the mouthpart anatomy
and habits of these Palaeozoic phytophagous insects
(Labandeira 1997), however, given that their fossil record
is based mainly on preservation of their wings, presently

limits the further analysis of these complex plant–insect interrelationships.

Acknowledgements

Suggestions for improvements of the manuscript from Conrad Labandeira, Barry Webby and one anonymous reviewer are gratefully acknowledged. Many thanks to PhD student Carlos Eduardo Lucas Vieira for his co-operation in the preparation of this paper and to Luiz Flávio Lopes for the photographs. This report is part of a PhD thesis by KAR at the Universidade Federal do Rio Grande do Sul (UFRGS), and is a contribution to IGCP Project 471.

References

Adami-Rodrigues, K. & Iannuzzi, R. 2001: Late Paleozoic terrestrial arthropod faunal and floral successions in the Paraná Basin: a preliminary synthesis. *Acta Geologica Leopoldensia 52/53*, 165–179.

Anderson, J.M. & Anderson, H.M. 1985: *Palaeoflora of Southern Africa: Prodromus of South African Megafloras - Devonian to Lower Cretaceous.* Balkema, Rotterdam.

Backheuser, Y., Silveira, J.B.R. & Guerra-Sommer, M. 1984: Revisão da tafoflora do Afloramento do Km 89–90 da rodovia BR 290, RS, Brasil (Formação Irati?). In: *Anais do XXXIII Congresso Brasileiro de Geologia, 2*, 1062–1074. Sociedade Brasileira de Geologia, Rio de Janeiro.

Banerjee, M. & Bera, S. 1998: Record of Zoocecidia on leaves of *Glossopteris browniana* Brongn. from Mohuda Basin, Upper Permian, Indian Lower Gondwana. *Indian Biologist 30*, 58–61.

Barberena, M.C., Araújo, D.C. & Lavina, E.L. 1985: Late Permian and Triassic tetrapods of southern Brazil. *National Geographic Research 1*, 5–20.

Baxendale, R.W. 1979: Plant-bearing coprolites from North American Pennsylvanian coal balls. *Paleontology 22*, 537–548.

Beck, A.L. & Labandeira, C.C. 1998: Early Permian insect folivory on a gigantopterid-dominated riparian flora from north-central Texas. *Palaeogeography, Palaeoclimatology, Palaeoecology 142*, 139–173.

Becker-Migdisova, E.E. 1940: Fossil Permian cicadas of the family Prosbolidae from the Soyana River. *Transactions of Paleontological Institute 11*, 1–79.

Bortoluzzi, C.A. 1975: Étude de quelques empreintes de la Flore Gondwaienne du Brésil. In: *Actes des 95 Congrés National des Sociétés Savantes, 1970, 3*, 171–187. Reims.

Carpenter, F.M. 1971: Adaptations among Paleozoic insects. *Proceedings of the 1st North American Paleontological Convention*, 1236–1251. Chicago.

Cazzulo-Klepzig, M., Guerra-Sommer, M. & Bossi, G.E. 1980: Revisão fitoestratigráfica do Grupo Itararé no Rio Grande do Sul. I Acampamento Velho, Cambaí Grande, Budó e Morro Papaléo. *Boletim IG-USP 11*, 31–189.

Chaloner, W.G., Scott, A.C. & Stephenson, J. 1991: Fossil evidence for plant-arthropod interactions in the Palaeozoic and Mesozoic. *Philosophical Transactions of the Royal Society of London 333*, 177–186.

Chaudhan, D.K., Tiwari, S.P. & Misra, D.R. 1985: Animal and plant relationships during Carboniferous–Permian period of India. *Bionature 5*, 5–8.

Coca-Abia, M.M., Alonso, P.D. & Ratcliffe, B.C. 1999: Evidencias de actividad biológica producidas por artrópodos terrestres a lo largo del tiempo geológico. *Boletín de la SEA 26*, 213–221.

Connor, E.F. & Taverner, M.P. 1997: The evolution and adaptive significance of the leaf-mining habit. *Oikos 79*, 6–25.

Coulson, R.N. & Witter, J.A. 1984: *Forest Entomology, Ecology and Management.* Wiley Interscience, New York.

Daemon, R.F. & Quadros, L.P. 1970: Bioestratigrafia do Neopaleozóico da Bacia do Paraná. In: *Anais do XXIV Congresso Brasileiro de Geologia*, 359–412. Sociedade Brasileira de Geologia, Brasília.

Diéguez, C., Nievez-Aldrey, J.L. & Barrón, E. 1996: Fossil galls (zoocecids from the Upper Miocene of La Cerdaña, Lérida, Spain). *Review of Palaeobotany and Palynology 94*, 329–343.

Edwards, P.J. & Wratten, S.D. 1980: Ecology of insect–plant interactions. *Studies in Biology 121*, 1–60.

Gangwere, S.K. 1966: Relationships between the mandibles, feeding behavior, and damage inflicted on plants by the feeding of certain acridids (Orthoptera). *Michigan Entomology 1*, 13–16.

Grauvogel-Stänn, L. & Kelber, K.P. 1996: Plant–insect interactions and coevolution during the Triassic in Western Europe. *Palaeontologia Lombarda N.S. 5*, 5–23.

Guerra-Sommer, M. 1992: Padrões Epidérmicos de Glossopteridales da Tafoflora do Faxinal (Formação Rio Bonito, Artinskiano-Kunguriano, Bacia do Paraná, Brasil). *Pesquisas 19*, 26–40.

Guerra-Sommer, M. 1995: Fitofagia em Glossopterídeas na Paleoflora da Mina do Faxinal (Formação Rio Bonito, Artinskiano, Bacia do Paraná). *Pesquisas 22*, 58–63.

Guerra-Sommer, M. & Cazzulo-Klepzig, M. 1993: Biostratigraphy of the southern Brazilian Neopaleozoic Gondwana sequence: a preliminary palaeobotanical approach. In: *Comptes Rendus des XII International Congrès de la Stratigrafie et Géologie du Carbonifère et Permien, 1991, 2*, 61–72. Buenos Aires.

Guerra-Sommer, M., Cazzulo-Klepzig, M. & Marques-Toigo, M. 1996: Gondwanostachyaceae (Equisetopsida) no Gondwana Sul-Brasileiro (Formação Rio Bonito) com Mega e Microflora Associadas. *Pesquisas 22*, 64–73.

Holmes, W.B.K. 1995: The Late Permian megafossil flora from coal, New South Wales, Australia In Pant, D.D. *et al.* (eds): *Proceedings of the International Conference on Global Environment and Diversification of Plants through Geological Time*, 123–152. Society of Indian Plant Taxonomists, Allahabad.

Holz, M. 1998: The Eo-permian coals of the Paraná Basin in southernmost Brazil: an analysis of the depositional conditions using sequence stratigraphy concepts. *International Journal of Coal Geology 36*, 141–163.

Iannuzzi, R., Marques-Toigo, M., Scherer, C.M.S., Caravaca, G., Vieira, C.E.L. & Pereira, L.S. 2003a: Reavaliação da fitobioestratigrafia da seqüência gondvânica sul-riograndense: estudo de caso do afloramento Morro do Papaléo (Bacia do Paraná, Permiano Inferior). In: *Atas I Encontro sobre a Estratigrafia do Rio Grande do Sul: Escudo e Bacias*, 182–185. SBG, Porto Alegre.

Iannuzzi, R., Marques-Toigo, M., Scherer, C.M.S., Caravaca, G., Vieira, C.E.L. & Pereira, L.S. 2003b: Phytobiostratigraphical revaluation of the southern Brazilian Gondwana sequence (Paraná Basin, Lower Permian). In: *Abstracts XV International Congress on Carboniferous and Permian Stratigraphy*, 240–242. Utrecht.

Janssen, R.E. 1939: Leaves and stems from fossil forests. *Illinois State Museum Popular Science Ser. 1*, 1–190.

Johnson, T.L. & Lyon, J.E. 1993: *Insects on Trees and Shrubs*, 2nd edn. Cornell University Press, Ithaca.

Kazakova, I.G. 1985: The character of damage to plants by Orthoptera (Insecta) linked to the structure of their mouthparts (on the example of Novosibirsk Akademgorodok fauna). *In* Zolotarenko, G.S. (ed.): *Anthropogenic Influences on Insect Communities*, 122–127. Nauka, Novosibirsk.

Krzeminski, W. & Evenhuis, N. 2000: Review of Diptera palaeontological records. *In* Papp, L. & Darvas, B. (eds): *Contributions to Manual of Palaearctic Diptera*, 1, 535–564. Science Herald, Budapest.

Kukalova-Peck, J. 1991: Fossil history and the evolution of hexapod structures. *In* Naumann, I.D., Carne, P.B., Lawrence, J.F., Nielsen, E.S. & Spradbery, J.P. (eds.): *The Insects of Australia*, 2nd edn, 141–179. Cornell University Press, Ithaca.

Labandeira, C.C. 1997: Insect mouthparts: ascertaining the paleobiology of insect feeding strategies. *Annual Reviews of Ecology Systematics 28*, 153–193.

Labandeira, C.C. 1998: Early history of arthropod and vascular plant associations. *Annual Reviews of Earth & Planetary Sciences 26*, 329–377.

Labandeira, C.C. 2002: The history of associations between plants and animals. *In* Herrera, C. & Pellmyr, O. (eds): *Plant–Animal Interactions*, 26–74, 248–261. Blackwell Science, Oxford.

Labandeira, C.C. & Beall, B.S. 1990: Arthropod terrestriality. *In* Mikulic, C.D. (ed.): *Arthropods: Notes for a Short Course*, 214–256. University of Tennessee Press, Knoxville.

Labandeira, C.C., Dilcher, D.L., Davis, D.R. & Wagner, D.L. 1994: Ninety-seven million years of angiosperm–insect association: paleobiological insights into the meaning of coevolution. *Proceedings of the National Academy Sciences USA 91*, 12278–12282.

Labandeira, C.C. & Phillips, T.L. 1996: Insect fluid-feeding on Upper Pennsylvanian tree ferns (Palaeodictyoptera, Marattiales) and the early history of the piercing-and-sucking functional feeding group. *Annals of the Entomological Society of America 89*, 157–183.

Langer, M.C. 2000: The first record of dinocephalians in South America: Late Permian (Rio do Rasto Formation) of the Paraná Basin, Brazil. *Neues Jahrbuch für Geologie und Paläontologie Abh. 215*, 69–95.

Mani, M.S. 1964: *Ecology of Plant Galls*. W. Junk, The Hague.

Marques-Toigo, M. 1991: Palynobiostratigraphy of the southern Brazilian Neopalaeozoic Gondwana Sequence. *In* Ulbrich, H. & Rocha-Campos, A.C. (eds): *Proceedings of Gondwana Seven. International Gondwana Symposium, 1988*, 503–515. Instituto de Geociências, Universidade de São Paulo, São Paulo.

Martynov, A.V. 1940: Permian fossil insects from Tschekarda. *Transactions of the Paleontological Institute 11*, 1–63.

Medvedev, L.N. 1968: Polyfagous beetles from the Jurassic of Karatau. *In* Rohdendorf, B.B. (ed.): *Jurassic Insects of Karatau*, 155–165. Akademiya Nauk, Moscow.

Meyen, S.V. 1982: The Carboniferous and Permian floras of Angaraland (a synthesis). *Biological Memoir 7*, 1–109.

Meyen, S.V. 1987: *Fundamentals of Palaeobotany*. Chapman & Hall, London.

Milani, E.J., França, A.B. & Schneider, R.L. 1994: Bacia do Paraná. *Boletim de Geociências 8*, 69–82.

Millan, J.H. 1987: Os pisos florísticos do carvão do Subgrupo Itararé do Estado de São Paulo e suas implicações. In: *Anais X Congresso Brasileiro de Paleontologia*, 2, 832–857. SBP, Rio de Janeiro.

Moore, R.C. 1964: Paleoecological aspects of Kansas Pennsylvanian and Permian cyclothems. *Kansas State Geological Survey Bulletin 169*, 287–380.

Ollerton, J. 1999: La evolución de las relaciones polinizador-planta en los artrópodos. *Boletín de la SEA 26*, 741–758.

Paim, P.S.G., Piccoli, A.E.M., Sartur, J.A.D., Munaro, P., Holz, M. & Granitoff, W. 1983: Evolução paleogeográfica do Super Grupo Tubarão na área de Mariana Pimentel-Faxinal, Guaíba, RS. In: *Atas I Simpósio Sul-Brasileiro de Geologia*, 140–159. SBG, Porto Alegre.

Pant, D.D. & Srivastava, P.C. 1995: Lower Gondwanan insect remains and evidence of insect–plant interaction. *In* Pant, D.D. *et al.* (eds): *Proceedings of the International Conference on Global Environment and Diversification of Plants through Geological Time*, 317–326. Society of Indian Plant Taxonomists, Allahabad.

Pasqualini, M., Cunha, A.S., Guerra-Sommer, M. & Piccoli, A.E.M. 1986: Análise paleoecológica de seqüências paleoflorísticas na Área de Mariana Pimentel, Guaíba, RS. In: *Anais XXXIV Congresso Brasileiro de Geologia*, 1, 556–569. SBG, Goiânia.

Pérez-Contreras, T. 1999: La especialización en los insectos fitófagos: una regla más que una excepción. *Boletín de la SEA 26*, 759–776.

Petri, S. & Souza, P.A. 1993: Síntese dos conhecimentos e novas concepções sobre a bioestratigrafia do Subgrupo Itararé, Bacia do Paraná, Brasil. *Revista do Instituto Geológico 14*, 7–18.

Piccoli, A.E.M., Menegat, T., Guerra-Sommer, M., Marques-Toigo, M. & Porcher, C.C. 1991: Faciologia da seqüência sedimentar gonduânica nas Folhas de Quitéria e Várzea do Capivarita. *Pesquisas 18*, 31–43.

Pinto, I.D. 1987: Permian insects from Paraná Basin, South Brazil. IV Coleoptera. *Pesquisas 19*, 5–12.

Pinto, I.D. 1995: Paleobotanical and paleozoological age divergences in South American strata. *Pesquisas 22*, 46–52.

Pinto, I.D. & Adami-Rodrigues, K. 1997: *Velisoptera taschi* nov. gen., nov. sp. A Carboniferous insect from Bajo de Veliz, Argentina. *Pesquisas 20*, 47–52.

Pinto, I.D. & Adami-Rodrigues, K. 1999: A revision of South American Paleozoic insects. In: *Proceedings I International Palaeontological Conference, 1998, PFIMC98/1*, 99, 117–124. AMBA/AM, Moscow.

Pinto, I.D. & Ornellas, L.P. 1978: Carboniferous insects Protorthoptera and Paraplecoptera from the Gondwana (South America, Africa, Asia). *Pesquisas 11*, 305–321.

Pinto, I.D. & Ornellas, L.P. 1981: Permian insects from Paraná Basin, South Brazil. III Homoptera 1 – Pereboridae. In: *Anais II Congresso Latino-americano de Paleontologia*, 1, 209–219. UFRGS, Porto Alegre.

Plumstead, E.P. 1963: The influence of plants and environment on the developing animal life of Karoo times. *South African Journal of Science 59*, 147–152.

Ponomarenko, A.G. 1995: The geological history of beetles. *In* Palaluk, J. & Slipinski, S.A. (eds): *Biology, Phylogeny, and Classification of Coleoptera. Papers Celebrating the 80th Birthday of Roy A. Crowson*, 155–171. Museum I Instytut Zoologii PAN, Warszawa.

Rocha-Campos, A.C. & Rösler, O. 1978: Late Paleozoic faunal and floral successions in the Paraná Basin, southeastern Brazil. *Boletim IG-USP 9*, 1–16.

Rohdendorf, B.B. & Rasnitsyn, A.P. 1980: Historical development of the Class Insecta. *Transactions of the Paleontological Institute 175*, 1–270.

Rohn, R. & Rösler, O. 2000: Middle to Upper Permian phytostratigraphy of the eastern Paraná Basin. *Revista Universidade de Guarulhos, Geociências V*, 69–73.

Roskam, J.C. 1992: Evolution of the gall-inducing guild. *In* Shorthouse, J.C. & Rohfritsch, O. (eds): *Biology of Insect-induced Galls*, 34–49. Oxford University Press, New York.

Rösler, O. 1978: The Brazilian Eogondwanic floral succession. *Boletim IG-USP 9*, 85–91.

Scott, A.C. 1991: Evidence for plant–arthropod interactions in the fossil record. *Geological Today 7*, 58–61.

Scott, A.C. 1992: Trace fossils of plant–arthropod interactions. *In* Maples, C.G. & West, R.R. (eds): *Trace Fossils: Notes for a Short Course*, 197–223. University of Tennessee Press, Knoxville.

Scott, A.C., Chaloner, W.G. & Paterson, S. 1985: Evidence of pteridophyte–arthropod interactions in the fossil record. *Proceedings of the Royal Society of Edinburgh B 86*, 133–140.

Scott, A.C., Stephenson, J. & Chaloner, W.G. 1992: Interaction and coevolution of plants and arthropods during the Palaeozoic and Mesozoic. *Philosophical Transactions of the Royal Society of London B 335*, 129–165.

Scott, A.C. & Taylor, T.N. 1983: Plant/animal interactions during the Upper Carboniferous. *Botanical Review 49*, 259–307.

Scott, A.C. & Titchener, F.R. 1999: Techniques in the study of plant–arthropod interactions. *In* Jones, T.P. & Rowe, N.P. (eds): *Fossil Plants and Spores: Modern Techniques*, 310–315. Geological Society, London.

Sellards, E.H. 1909: The Permian flora of Kansas. *Kansas Geological Survey Bulletin 9*, 434–467.

Shabica, C.W. & Hay, A.A. (eds) 1997: *Richardson's Guide to the Fossil Fauna of Mazon Creek*. Northeast Illinois University Press, Chicago.

Shear, W.A. & Kukalova-Peck, J. 1990: The ecology of Paleozoic terrestrial arthropods: the fossil evidence. *Canadian Zoological Journal 68*, 1807–1834.

Shorthouse, J.D. & Rohfritsch, O. 1992: *Biology of Insect-induced Galls*. Oxford University Press, New York.

Smith, D.M. 1994: Plant–insect interactions in the fossil record. *Phytophaga 6*, 35–49.

Srivastava, A.K. 1987: Lower Barakar flora of Raniganj coalfield and insect/plant relationship. *Palaeobotanist 36*, 138–142.

Srivastava, A.K. 1994: Plant/animal relationship in Lower Gondwana of India. In: *Ninth International Gondwana Symposium*, 549–555. Hyderabad.

Srivastava, A.K. 1998: Fossil records of insect and insect-related plant damage in India. *Zoos' Print 5*, 5–9.

Stephenson, J. & Scott, A.C. 1992: The geological history of insect-related plant damage. *Terra Nova 4*, 542–552.

Vieira, C.E.L. & Iannuzzi, R. 2000: Presença de *Pecopteris* e *Asterotheca* no afloramento Morro do Papaléo, Município de Mariana Pimentel, Rio Grande do Sul (Formação Rio Bonito, Eopermiano da Bacia do Paraná). *Pesquisa 27*, 49–64.

Vishniakova, V.N. 1981: New Paleozoic and Mesozoic lophioneurids (Thripida, Lophioneuridae). *Transactions of Paleontological Institute 183*, 43–63.

Wing, S., Ash, A., Ellis, B., Hickey, L.J., Johnson, K. & Wilf, P. 1999: Leaf Architecture Working Group. *Manual of Leaf Architecture – Morphological Description and Categorization of Dicotyledonous and Net-veined Monocotyledonous Angiosperms.* Smithsonian Institution, Washington, DC.

Würdig, N.L., Pinto, I.D. & Adami-Rodrigues, K. 1999: South American Paleozoic faunulae and two new insects: chronological, paleogeographical and systematic interpretation. In: *Proceedings I International Palaentomological Conference, 1998, PFICM98/1.*, 99, 177–184. AMBA/AM, Moscow.

Zherikin, V.V. & Gratshev, V.G. 1993: Obrieniidae, fam. nov. the oldest Mesozoic weevils (Coleoptera, Curculionoidea). *Paleontological Journal 27*, 50–69.

The stratigraphic record of microborings

INGRID GLAUB & KLAUS VOGEL

Glaub, I. & Vogel, K. **2004 10 25**: The stratigraphic record of microborings. *Fossils and Strata*, No. 51, pp. 126–135. Germany. ISSN 0300-9491.

Approximately 50 fossil endolith species (biotaxa and ichnotaxa) were screened for their stratigraphic ranges. The study indicates (i) the first occurrence of microborings in the Proterozoic and of some others in the Ordovician and Silurian, (ii) diversity increases at the base of the Mesozoic, (iii) and a change in taxa composition at the base of the Cenozoic. Remarkably, 35% of the endolith taxa are known from the Proterozoic or Palaeozoic until now and can be considered as "living fossils". The protected microenvironment in omnipresent calcareous substrates, the speed of endolith spore settlement, and the ability of photoautotrophic endoliths to tolerate declining light levels are regarded as important reasons for their longevity. The first borings were made by Cyanobacteria in the Proterozoic, whereas the endolithic niche became established by Chlorophyta by the Ordovician, by Rhodophyta by the Silurian (perhaps already the Ordovician), and by fungi by the Ordovician.

Key words: Endolith; evolution; stratigraphy; Cyanobacteria; algae; fungi.

Ingrid Glaub [I.Glaub@em.uni-frankfurt.de] & Klaus Vogel, Geologisch-Paläontologisches Institut, Johann Wolfgang Goethe-Universität Frankfurt, Senckenberganlage 32-34, D-60325 Frankfurt, Germany

Introduction

Microborings are characterized by tunnel diameters of less than 100 μm. They are observed in high abundance from fossil and modern calcareous substrates. The boring patterns include simple spherical holes, bunch-like clusters, ramified networks, as well as dendrite arrangements. A few fossil examples exist where organic remains of the endolithic organism had been preserved within the boring system. As for the majority of microborings, they are found without organic remains of their producers. However, most microborings match exactly the body outline of the endoliths. That is the essential fact, that even empty fossil microborings can be assigned to their trace makers with a high degree of precision.

Modern microborings are mainly produced by Cyanobacteria, red algae, green algae, and fungi. The preservation potential of soft parts is low for these organism groups. This holds true for their endolithic as well as non-endolithic representatives. As for Cyanobacteria and the algae mentioned above, fossil documentation so far mainly focuses on calcified and calcifying species. Knowledge of the fossil record of fungi is meagre. In contrast, microborings are often well preserved, displaying tiny details even in Palaeozoic samples. Based on their species-specific features, they offer information on taxa so far not taken into account in the reconstruction of the evolutionary history of Cyanobacteria, algae and fungi.

Early studies of microborings go back to the 19th and early 20th century (e.g. Quenstedt 1849; Pratje 1922; Mägdefrau 1937; Pia 1937). After the development of the casting embedding technique (Golubic *et al.* 1970), research activity on fossil and modern microborings was intensified (e.g. Golubic *et al.* 1975; Budd & Perkins 1980; Glaub 1994; Bundschuh 2000; Vogel *et al.* 2000). Studies covered different stratigraphic units and different environments. The role of microborers in the stratigraphy of bioerosional trace fossils was examined by Bromley (2004). He gives a comprehensive survey of micro- and macroborings through time at the genus level.

The photoautotrophic producers of microborings (Cyanobacteria and algae) (Table 1) are indicative of the optical properties of ancient marine basins. Characteristic microendolith assemblages for distinctive water depths, represented by photic zones, were defined by Glaub (1994) and Vogel *et al.* (1995). This was confirmed for modern oceans of different latitudes as well as for ancient basins back to the Ordovician (Vogel *et al.* 1999).

Studies on microboring assemblages through time displayed great conservatism in both taxonomic composition and ecological requirements (e.g. Vogel *et al.* 1995, 1999; Glaub & Bundschuh 1997; Glaub *et al.* 1999;

Table 1. Stratigraphic ranges of microborers and microborings (body fossils and trace fossils), sorted by their first occurrences. See right column for organism group of supposed producer.

	Proterozoic	Cambrian	Ordovician	Silurian	Devonian	Carbonifer.	Permian	Triassic	Jurassic	Cretaceous	Palaeogene	Neogene	Quaternary	Organism group
Number of taxa	7	5	8	8	4	9	9	21	21	21	24	12	27	
Eohyella campbellii	■													Cyanobacteria
Graviglomus incrustus	■													?
Parenchymodiscus endolithicus	■													?
Cunicularius halleri	■	■												Cyanobacteria
Eohyella elongata	■	■												Cyanobacteria
Eohyella dichotoma	■		■								■	■	■	Cyanobacteria
Eohyella rectoclada	■		■	■			■		■	■	■		■	Cyanobacteria
Fasciculus frutex		■												
Cunicularius isodiametrus		■												Cyanobacteria
Endoconchia angusta		■												Cyanobacteria
Eohyella lata		■												Cyanobacteria
Fasciculus dactylus			■				■	■			■	■	■	Cyanobacteria
Orthogonum fusiferum			■	■		■		■	■		■		■	Fungi
Orthogonum tripartitum			■								■		■	Rhodophyta ?
Reticulina elegans			■	■	■			■	■		■	■	■	Chlorophyta
Saccomorpha clava			■		■			■	■		■		■	Fungi
Scolecia filosa			■	■	■						■	■	■	Cyanobacteria
Fasciculus rogus				■			■	■			■		■	Cyanobacteria
Palaeoconchocelis starmachii				■			■	■	■		■	■	■	Rhodophyta
Planobola macrogota				■		■			■		■		■	?
Polyactina araneola				■		■	■	■	■		■	■	■	Fungi
Hyellomorpha microdendritica					■	■				■			■	Cyanobacteria
Globodendrina monile						■			■		■	■	■	Foraminifera
Orthogonum lineare						■					■		■	?
Orthogonum spinosum						■		■			■		■	?

Table 1. (Continued)

Number of taxa	Proterozoic 7	Cambrian 5	Ordovician 8	Silurian 8	Devonian 4	Carbonifer. 9	Permian 9	Triassic 21	Jurassic 21	Cretaceous 21	Palaeogene 24	Neogene 12	Quaternary 27	Organism group
Eurygonum nodosum							■	■	■				■	Cyanobacteria
Fasciculus acinosus							■	■	■				■	Cyanobacteria
Cavernula pediculata								■	■		■	■	■	Chlorophyta
Cavernula zancobola								■						Fungi
Planobola cebolla								■						?
Planobola microgota								■						Fungi
Planobola radicatus								■		■				Fungi
Rhopalia catenata								■	■	■	■	■	■	Chlorophyta
Saccomorpha terminalis								■	■		■			Fungi
Scolecia maeandria								■			■			Bacteria ?
Cavernula coccid									■					?
Orthogonum appendiculatum									■					?
Orthogonum giganteum									■				■	?
Orthogonum tubulare									■	■	■		■	Fungi ?
Dendrina anomala										■				?
Dendrina belemniticola										■				?
Dendrina brachiopodicola										■				?
Dendrina constans										■				?
Dendrina fluensis										■				?
Dendrina lacerata										■				?
Dendrina orbiculata										■				?
Fasciculus grandis											■		■	Chlorophyta
Fasciculus parvus											■		■	Cyanobacteria ?
Polyactina fastigata											■			?
Saccomorpha sphaerula											■	■	■	Fungi
Scolecia botulifera											■			Rhodophyta ?
Scolecia serrata											■	■	■	Bacteria ?

Bundschuh 2000). This prompted a closer look at the known stratigraphic record of about 50 endolith taxa. The study gives a survey and an interpretation of the geological record of microendoliths at species level, considering trace fossils as well as body fossils. The use of microborings as stratigraphic key fossils and their contribution to the reconstruction of evolutionary history are addressed. Furthermore, the survival strategies of microborers having a long geological record are discussed.

Database

The selected taxa are restricted to microborings. We excluded bryozoans and sponges, as far as identity is known, because adult sponges usually produce boring structures of more than 100 µm tunnel diameter. As for bryozoans, some are in the microboring size range, but a detailed stratigraphic report has been given by Pohowsky (1978), and hence the data should not be repeated here.

The presentation mainly refers to empty boreholes, which were prepared using the casting embedding technique (Golubic *et al.* 1970). Some data result from natural casts or from boreholes with organic remains. As a result, the survey is based on biotaxa and on ichnotaxa, simply called "taxa" in the following.

The evaluation is based on own research (published and unpublished data) as well as on data obtained from the literature. The stratigraphic ranges given in Table 1 for the endolith taxa under consideration are based on the following data. Our research group screened about 200–500 substrates in each of the following geological periods, considering different localities and environments, and the data set is completed by descriptions of individual findings [summaries are to be found in Vogel *et al.* (1995, 1999) and Glaub *et al.* (1999)]: Silurian (Campbell *et al.* 1979; Campbell 1980; Glaub & Bundschuh 1997; Bundschuh 2000), Devonian (Vogel *et al.* 1987), Triassic (Schmidt 1992; Glaub & Schmidt 1994; Balog 1996, 1997), Jurassic (Glaub 1994; Plewes 1996; Plewes *et al.* 1993; Glaub & Bundschuh 1997), Cretaceous (Hofmann 1996; Schnick 1992; Glaub 1994), and Palaeogene (Radtke 1991; Radtke *et al.* 1997b; Vogel & Marincovich 2004). For other stratigraphic units, reports are so far restricted to local areas: Proterozoic (Knoll *et al.* 1986, 1989; Zhang & Golubic 1987; Green *et al.* 1988; Golubic & Seong-Joo 1999; Golubic *et al.* 1999), Cambrian (Runnegar 1985; Li 1997), and Permian (Balog 1996, 1997). For the Ordovician (Olempska 1986), the Carboniferous (Vogel 1991; Glaub *et al.* 1999, 2001), and the Neogene (Vogel & Marincovich 2004) data are still preliminary, because investigations have not been completed. The literature on the modern equivalents of the ancient endoliths is

numerous and references can be found in the individual papers mentioned above or in the bibliographic overview catalogued by Radtke *et al.* (1997a). The survey presented here refers to 54 taxa (Table 1). It is not considered to be complete nor has additional work been carried out concerning synonymy, etc.

Taxonomic questions

Nododendrina nodosa Vogel, Golubic & Brett 1987, *Platydendrina platycentrum* Vogel, Golubic & Brett 1987, *Platydendrina convexa* Vogel, Golubic & Brett 1987, *Ramodendrina alcicornis* Vogel, Golubic & Brett 1987, and *Ramodendrina cervicornis* Vogel, Golubic & Brett 1987 are not considered, because these taxa are discussed as junior synonyms of *Clionolithes* (Plewes 1996). "*Dodgella*-like borings" (Vogel *et al.* 1987) seem to contain borings which belong to *Sacccomorpha clava* Radtke 1991 and were registered under the latter name. "Semidendrina-Form" (Glaub 1994) is regarded to be identical to *Globodendrina monile* Plewes, Palmer & Haynes 1993. The geological record of the modern *Hyella gigas* is reconstructed by the occurrence of the body fossil *Eohyella rectoclada* Green, Knoll & Swett 1988 and the trace fossil *Fasciculus frutex* Radtke 1991. The taxa identification for the Cretaceous is based mainly on the observations of Hofmann (1996). In agreement with K. Hofmann (pers. comm., and "errata" page in Hofmann 1996), we see the following correspondence between the forms described by him and known ichnotaxa: "Zickzack-Form" is *Reticulina elegans* Radtke 1991, "Lagenoid-Rizoid-Form" is *Rhopalia catenata* Radtke 1991, "Kleine Büschel-Form" is *Fasciculus dactylus* Radtke 1991, "Große Büschel-Form" is *Fasciculus grandis* Radtke 1991, "Faden-Form" is *Scolecia filosa* Radtke 1991, "Säulen-Form" is *Cavernula pediculata* Radtke 1991, "Tubular-Form" is *Orthogonum tubulare* Radtke 1991, "Fungoid-Form A" is *Saccomorpha clava*, "Fungoid-Form B" is *Saccomorpha terminalis* Radtke 1991, "Fungoid-Form D" is *Polyactina araneola* Radtke 1991, and "Fungoid-Form E" is *Planobola radicatus* Schmidt 1992.

Stratigraphic ranges

Table 1 gives an overview of the stratigraphic ranges of microendolith species. The data result from the references cited in the previous section entitled "Database". According to this evaluation, *Eohyella campbellii* Zhang & Golubic 1987 is considered to be the oldest cyanobacterial endolith (1500 or 1700 Ma) as well as the oldest endolith and the oldest bioeroder at all. It has no modern counterpart and is present exclusively in the Proterozoic. Two microendoliths occur from the Proterozoic to the Recent (*Eohyella dichotoma* Green, Knoll & Swett 1988, *Eohyella rectoclada/Fasciculus frutex*). However, *Eohyella dichotoma* is not known from the Mesozoic so far.

Some taxa are known from the Proterozoic as well as from the Cambrian (*Cunicularius halleri, Eohyella elongata* Knoll, Swett & Burkhardt 1989). About 35% (18 species) of the taxa evaluated persist from the Palaeozoic to the Quaternary. They include all key ichnotaxa used for bathymetrical reconstructions by photic-related ichnocoenoses (*sensu* Glaub 1994; Vogel *et al.* 1995): *Fasciculus acinosus* Glaub 1994, *Fasciculus dactylus, Palaeoconchocelis starmachii* Campbell, Kazmierczak & Golubic 1979, *Reticulina elegans, Orthogonum lineare* Glaub 1994, and *Saccomorpha clava*. In general, the taxa with records starting in the Palaeozoic are dominated by those attributed to the Cyanobacteria (eight species) (*Eohyella dichotoma, Eohyella rectoclada/Fasciculus frutex, Fasciculus acinosus, Fasciculus dactylus, Fasciculus rogus* Bundschuh & Balog 1999, *Hyellomorpha microdendritica* Vogel, Golubic & Brett 1987, *Eurygonum nodosum* Schmidt 1992, and *Scolecia filosa*). Others belong to green algae (*Reticulina elegans*), red algae (e.g. *Palaeoconchocelis starmachii*), fungi (e.g. *Saccomorpha clava, Orthogonum fusiferum* Radtke 1991, *Polyactina araneola*), Foraminifera (*Globodendrina monile*) and unknown producers (e.g. *Orthogonum lineare*). That means that all organism groups creating microborings have already appeared with boring representatives by the Palaeozoic.

We discuss the diversity of microendolith taxa through Earth history on the basis of eras. The base of the Mesozoic is characterised by a remarkable increase in diversity. In total, 19 taxa appear in Mesozoic strata for the first time. This observation is supported by the average number of taxa in the Palaeozoic (approximately eight taxa per period) compared with that in the Mesozoic (approximately 20 taxa per period). In total, 11 taxa are restricted to Mesozoic periods: *Cavernula coccidia* Glaub 1994, *Dendrina anomala* Mägdefrau 1937, *Dendrina brachiopodicola* Hofmann 1996, *Dendrina constans* Hofmann 1996, *Dendrina fluensis* Hofmann 1996, *Dendrina lacerata* Hofmann 1996, *Dendrina orbiculata* Hofmann 1996, *Orthogonum appendiculatum* Glaub 1994, *Planobola cebolla* Schmidt 1992, *Planobola microgota* Schmidt 1992, and *Planobola radicatus*. For most of these microborings the producers are unknown. The following taxa started in the Mesozoic and continue to the Cenozoic. *Cavernula pediculata* and *Rhopalia catenata* are affiliated to green algae. Others are supposed to be produced by fungi or unknown heterotrophs: *Saccomorpha terminalis, Cavernula zancobola* Schmidt 1992, *Orthogonum tubulare, Orthogonum giganteum* Glaub 1994, and *Scolecia maeandria* Radtke 1991. The

increase in numbers of taxa at the base of the Mesozoic is mainly caused by microborings of unknown origin.

In contrast to the remarkable diversity increase at the base of the Mesozoic, the base of the Cenozoic displays nearly the same average number of taxa as counted for the Mesozoic, but an essential change in the composition of taxa took place. The 12 specific taxa of the Mesozoic mentioned above as well as *Orthogonum tripartitum*, known from the Palaeozoic and the Mesozoic, do not persist into the Cenozoic. That equals a taxa loss of about 24% (13 taxa) from Mesozoic to Cenozoic times. On the other hand, the following six taxa have their oldest known record in the Palaeogene: *Fasciculus parvus* Radtke 1991, *Fasciculus grandis, Saccomorpha sphaerula* Radtke 1991, *Scolecia serrata* Radtke 1991, *Polyactina fastigata* Radtke 1991, and *Scolecia botulifera* Radtke 1991.

As demonstrated in Table 1, some taxa are restricted to one period or some to a few periods. They hold the potential to serve as stratigraphic key fossils, but this observation demands further investigation to confirm the limitations of the taxa. Those taxa that are bound to Mesozoic strata seem to characterize this era.

Relationship between ichnotaxa and their producers

The interpretation of the microboring stratigraphic record is more valuable if the trace maker's identity is taken into account. We consider most microboring ichnotaxa to be produced by one and the same organism (or a group of closely related organisms) through time. The following arguments make it very probable that this has actually been the case.

The first argument concerns borings that were found to contain organic remnants of the producing organisms due to a very early silicification. For instance, the features observed in two Precambrian *Eohyella* species equal those of their modern equivalent *Hyella* species in detail (Green *et al.* 1988). This also holds for *Palaeoconchocelis starmachii* and its modern counterpart *Porphyra nereocystis* (red alga). As for *Palaeoconchocelis starmachii*, Campbell (1980) found characteristic red algal cells preserved in Silurian boring systems. Even the minute pit connections (2–3 μm in diameter) between cell walls, which clearly indicate a red algal relationship, are visible in the fossil material. The arrangement of endolithic conchosporangial branches, spores and filaments shows a close similarity to the conchocelis phase of the modern *Porphyra nereocystis*.

The second observation supporting the constancy in the relationship between microendoliths and their traces is the ecological stability of boring traces. As already mentioned, not only their morphology remained unchanged through time, but they have also retained their environmental preferences, i.e. especially their close relationship to light properties as is shown by their connection to the same bathymetric key ichnocoenoses through the geological history.

Finally, even special traits of behaviour can be traced far back into the past. For example, the borings of the modern cyanobacterium *Hyella caespitosa* trace *Fasciculus dactylus* are oriented perpendicularly to the substrate surface in well-illuminated environments, whereas they change to a parallel orientation in deeper water to use the remaining light more efficiently. This change in orientation was also observed in *Fasciculus dactylus* from Jurassic sediments.

Implications for the interpretation of evolutionary history

The microendolithic representatives of prokaryotes and eukaryotes belong mainly to other classes and orders than the calcified and calcifying species. This means that the geological record of microborings provides additional information on the phylogeny of Cyanobacteria, algae, and fungi.

The oldest record for Cyanobacteria is in the Early Archean, possibly 3500 Ma ago (see synopsis in Golubic & Seong-Joo 1999). Their contribution to the oxygen content in the atmosphere is well known. It seems that it takes quite a long time before the first cyanobacterial boring appears. The first borings affiliated to Cyanobacteria are 1500 or 1700 Ma old (Proterozoic), called *Eohyella campbellii* (Zhang & Golubic 1987).

Studies of cyanobacterial borings display details of cell size, arrangement of heterocysts, ramification patterns, and ecological requirements (e.g. Glaub *et al.* 1999; Golubic *et al.* 1999). *Eohyella campbellii* as well as many other taxa (e.g. *Eohyella dichotoma*, *Fasciculus dactylus*, *Fasciculus acinosus*) are similar to modern *Hyella* species, which belong to the baeocyte-forming Cyanobacteria. For all *Hyella* species known from the fossil record, findings reach back to the Proterozoic or Palaeozoic, except for *Fasciculus parvus*. The first endolithic representative of heterocyst-bearing Cyanobacteria is given by findings of *Endoconchia angusta* Li (Runnegar) 1997 in Cambrian samples. The first endolithic representative of the hormogonia-forming Cyanobacteria is clearly indicated by the occurrence of *Scolecia filosa* in the Ordovician.

As for red algae, the oldest record is from Proterozoic samples. Butterfield (2000) described in detail preserved organic remains of a non-endolithic rhodophyte belonging to the class Bangiaceae. Red algae with endolithic conchocelis stages are also classified with the Bangiaceae.

The oldest well-documented report for them is given by the finding of *Palaeoconchocelis starmachii* in Silurian samples (Campbell 1980; Glaub & Bundschuh 1997; Bundschuh 2000). *Palaeoconchocelis starmachii* was also found in many younger periods and therefore gives a good documentation of the Bangiacean evolution for periods so far mainly represented by the calcifying taxa of Floridophycean red algae. It is dependent on further investigations to prove whether *Orthogonum tripartitum* (since the Ordovician) and *Scolecia botulifera* (since the Palaeogene) may also be affiliated to red algae. In general, a remarkable increase in taxa related to fossil endolithic red algae is expected from further research. This assumption is based on initial studies by Bundschuh & Glaub, which show that the species-rich modern red alga *Porphyra* (more than 100 species) has fossil equivalents in Silurian samples, so far described only using open nomenclature (Bundschuh 2000).

Discoveries in 600 Ma old samples are considered to be the oldest records of green algae (Tappan 1980). The first endolithic green alga is reported from the Ordovician, represented by the ichnotaxon *Reticulina elegans*, today produced by *Ostreobium quekettii*. This green alga is widely distributed in modern reefs and many other environments, but belongs to the small group of siphonally organised green algae. The observation of *Ostreobium quekettii* through time may clarify the evolution of this type of organisation. *Cavernula pediculata* (since the Triassic) is nowadays produced by endolithic stages of *Gomontia polyrhiza* (class Ulvophyceae). Today, this class is widely distributed as "sea lettuce", but has poor preservation potential, except for the endolithic part, which is preserved as a microboring. Another microboring, *Rhopalia catenata*, that is affiliated to green algae also started in the Triassic. The modern counterparts of this boring are *Phaeophila* sp. and *Eugomontia sacculata*. These green algae are classified with the Chaetophorales, which is not an important group today. Thus, *Rhopalia catenata* may shed light on the Chaetophorales lineage of green algae.

The modern equivalent of *Saccomorpha terminalis* (first occurrence in the Triassic) is *Phythophthora* sp. This genus belongs to the class Oomycetes, which was earlier classified with the lower fungi (e.g. Schlegel 1992) and later with the algal division Heterokontophyta as stated by van den Hoek *et al.* (1993). Heterokontophyta are well known by their most prominent members, the diatoms. Diatom phylogeny is not yet clearly understood. The geological record is short (since the Cretaceous, according to van den Hoek *et al.* 1993), possibly because the opaline silica cell wall has a low resistance against diagenetic alterations. Members of the Oomycetes are believed to be in the ancestral lineage of diatoms (Falciatore & Bowler 2002). Thus, microborings that have supposedly been produced by Oomycetes may aid in understanding the early evolution of diatoms and related heterokont taxa.

As for fungal evolution, based on rRNA sequencing the first fungi are estimated to have arisen in the Proterozoic (Blackwell 2000; Redecker *et al.* 2000). The oldest record of endolithic fungi is from the Palaeozoic. As reported in Müller & Loeffler (1992: 5), Chytridiomycetes were observed in Cambrian shell fragments. On the basis of the evaluation presented here, the occurrence of *Orthogonum fusiferum* and *Saccomorpha clava* in the Ordovician are indicated as the oldest fungal endoliths. The producer of *Orthogonum fusiferum* is *Ostracoblabe implexa*. Its systematic position within the lower fungi is not clear (Zeff & Perkins 1979). The modern producers of *Saccomorpha clava*, *Polyactina araneola* (since the Silurian) and *Cavernula zancobola* (since the Triassic) belong to the Chytridiomycetes. Thus, the evolution of the class Chytridiomycetes is well documented by microborings since the Palaeozoic.

Mass extinctions

As demonstrated, the geological record of microborings displays different time spans for individual taxa. Some taxa have been on Earth since Proterozoic or Palaeozoic times, whereas others show restriction to one or sometimes a few stratigraphic units. The brevity of taxa occurrence as well as the changes in the diversity and composition of taxa through Earth history are not yet known sufficiently to allow interpretation. However, we can discuss the longevity of the 18 taxa known from the Proterozoic or Palaeozoic to the Recent. They represent approximately 35% of the taxa under consideration. These taxa have been on Earth for more than 300 Ma, that is, much longer than the average geological age span of species of 4–5 Ma (Skelton 1993). These biotaxa and the producers of the ichnotaxa can be considered as "living fossils" if our assumption of the relationship between microborings and microborers is conclusive.

The taxa known from Proterozoic/Palaeozoic to modern times are supposed to have been produced by Cyanobacteria (eight taxa), Chlorophyta (one), Rhodophyta (one), fungi (three), Foraminifera (one), and unknown producers (four). Even the incisive extinction events in Earth history had no impact on them. The studies of microendoliths in Permian and Triassic reefs (Balog 1996) clearly show their survival in typical environments in spite of the most influential catastrophe of the biosphere during the Permian–Triassic transition. This provokes the question of the possible reasons for such a surprising stability of microendolithic taxa.

Microendoliths colonise diverse calcareous substrates (e.g. molluscan shells, coral skeletons, hardgrounds, and ooids). Very early in Earth history, microendoliths adapted to this as yet unused space that has been abundant since then, and available in every ocean, even the smallest marine and brackish basins. The endolithic mode of life opened a third dimension for them, so that they did not have to compete for living space with other benthic organisms. It implies protection from grazers (e.g. sea urchins, molluscs) and hydrodynamic desiccation. For heterotrophic endoliths (e.g. fungi), the organic compounds in shells, corals, etc. serve as food resources. Living within such a substrate affords a further means of benefit in a quite stable microenvironment. To a certain degree, the substrate is a closed and well-buffered system (Campbell 1980). This reduces the effect of external fluctuations in temperature, humidity, salinity, turbidity, and other ecological factors.

Studies on One Tree Island (Australia) demonstrated that within these substrates the nutrient supply seems not to be essentially affected by changes in the overlying water body (Kiene 1997; Vogel *et al.* 2000). For experiments, small lagoons of mini-atolls were fertilised with nitrogen and phosphorus 10–15 times the natural concentration. The introduction of these chemicals had no apparent impact on the microborer community after 5 months of treatment.

Thus, the first important explanation for the longevity of the taxa mentioned and their survival even through catastrophes is their protected and buffered habitat within calcareous substrates, which have been abundant since early Earth history.

Even if their survival was endangered, their very mobile reproductive cells helped them to colonise niches with better ecological conditions. They can do so very efficiently, because after touching the substrate in the new environment, their ability to bore helps them to rest there and to resist water turbulence. Especially the baeocytes of Cyanobacteria are known for their quick settlement in new habitats (Wilkinson & Burrows 1972). Thus, the speed of recolonisation seems to be the second essential reason in favour of their longevity.

One environmental factor that may have changed essentially during Earth history is light. Especially at mass extinction events, a high particle content in the atmosphere induced by meteorites or volcanoes is supposed. This seems even more probable given that calcifying red Corallinales and green Dasycladales suffered extinction of about two-thirds of their species at the Cretaceous/Palaeogene boundary (Aguirre *et al.* 2000). There was only a slight difference observed between deeper-water species and shallow-water species. The authors interpreted these observations as resulting from a large-scale reduction in incident light at the Maastrichtian–Danian transition.

As for endoliths, light was identified as the main determining factor for their distribution patterns. Accordingly, it became the basis for the characterisation of photic-related microboring assemblages, which are usable for palaeodepth reconstructions (Glaub 1994; Vogel *et al.* 1995).

Many microendoliths are obligate photoautotrophs. Their life depends on oxygenic photosynthesis. Out of the 18 most conservative taxa mentioned above, 10 taxa are considered obligate photoautotrophs. These are the taxa attributed to Cyanobacteria and algae, except *Scolecia filosa*. The producer of *Scolecia filosa*, the cyanobacterium *Plectonema terebrans*, is probably facultatively heterotroph [see the discussions in Glaub (1994) and Glaub *et al.* (2001)]. How, then, can we explain their longevity in spite of the supposed dramatic light-changing events in Earth history?

The following discussion is based on the data available on endolith light requirement and their physiological capabilities to react in changing light conditions. Unfortunately, for microendoliths no experiments have been performed on the minimum of light required. Thus, we mainly base the discussion on their known bathymetrical distribution. There is evidence that *Ostreobium quekettii*, the only green alga among the conservative taxa under consideration, can live under low light conditions. It was detected several times in about 200 m water depth (e.g. near the Bahamas). For the habitats studied, this water depth corresponds to the dysphotic zone with a light supply of 0.01–0.001% of surface light (see references in Glaub 1994). The characterisation of *Ostreobium quekettii* as a low light alga is corroborated by findings of *Reticulina elegans* in ancient dysphotic environments. On the basis of these data, it would appear that *Ostreobium quekettii* could have stayed in place, even if only with a small amount of light remaining at its disposal.

No other conservative taxa – Cyanobacteria and red algae – are found in the dysphotic zone, either in the present or in the past. That means that they are restricted to the euphotic zone, and at least their recent representatives require 1% or more of surface light. In case of dramatic light decline to less than 1% of surface light they were forced to emigrate to better-illuminated areas.

Cyanobacteria have multiple ways of reacting to light fluctuations within the euphotic zone (down to 1% of surface light). In the case of too much light they can produce black pigments in their sheath to avoid sun burn damage. Production of the red light harvesting pigment phycocyanin occurs higher in the upper water column, whereas the synthesis of phycoerythrin is enhanced in deeper water (chromatic adaptation). As for some *Hyella* species, the growth form changes from perpendicular to the substrate surface under well-illuminated conditions to a parallel orientation if less light is available.

As for red algae, they are also equipped with pigments that allow red light harvesting in the upper and blue light harvesting in the deeper sections of the water column. A non-endolithic red alga holds the record (268 m) for the deepest multicellular alga (Littler *et al.* 1985). In darkness,

Porphyra leucosticta retreats into a resting state. After re-illumination the chloroplasts are quickly reorganised (Sheath *et al.* 1977).

In consequence, all photoautotrophs in question can react to changing light conditions within the euphotic zone, that means for the present at more than 1% of surface light. Only the green alga *Ostreobium queketti* can cope with dysphotic zone conditions of about 0.01 or 0.001% of surface light.

Summarising, we consider that (i) the abundance of protected, buffered hard substrates since early Earth history, (ii) the high mobility of endolith reproduction cells, and (iii) the ability of microborers to react to changing light conditions, are the essential factors that favour their long stratigraphic ranges.

Conclusions

The survey of approximately 50 fossil endolith species (biotaxa and ichnotaxa) provides the following information: microborings first occurred in the Proterozoic, and some others in the Ordovician and Silurian; the diversity increased at the base of the Mesozoic; and the composition of taxa changed remarkably at the base of the Cenozoic.

A surprisingly high number of the taxa evaluated (about 35%) have survived from the Proterozoic or Palaeozoic to the present day. As "living fossils" they represent an appreciated insight into the evolution of Cyanobacteria, Rhodophyta, Chlorophyta, and fungi. The following reasons may have favoured their longevity. The calcareous substrate colonised by microendoliths has been abundantly present since early Earth history and is a well-protected and buffered habitat, which helps in overcoming unfavourable environmental conditions. The high mobility of endolith reproductive spores allowed them to emigrate from poor environmental conditions and colonise better habitats. As for photoautotrophs, the ability to react to changing light conditions at least to a certain degree promotes survival in decreasing light intensity. Thus, essential members of photic-related ichnocoenoses have survived some hundreds of millions of years of Earth history, including several extinction events, without apparent changes in morphology, behaviour or ecological requirements.

The endolithic niche became established by Cyanobacteria in the Proterozoic. The stratigraphic record of microborings gives detailed information on the oldest known baeocyte-forming Cyanobacteria (*Eohyella campbellii*, Proterozoic), and those characterized by heterocysts (*Endoconchia angusta*, Cambrian) and hormogonia (*Scolecia filosa*, Ordovician). *Ostreobium quekettii* (order Siphonales) is the first species of green algae to become endolithic (in the Ordovician). Red algae with endolithic stages (class Bangiaceae) first appear in the Silurian. Data are also provided on the earliest endolithic representative of Oomycetes (Heterocontophyta) in the Triassic and the Chytridiomycetes (lower fungi) in the Ordovician.

Stratigraphic key fossils for individual periods were not identified. It is up to further investigations to prove if taxa so far only known from one or a few periods (e.g. many dendritic traces) are actually restricted to these time intervals. Many taxa are linked to the Mesozoic and may serve as an indication for this era. Further studies are required to explain the increase in taxonomic diversity at the base of the Mesozoic and the change in taxonomic composition at the base of the Cenozoic.

Acknowledgements

We thank our colleagues Marcos Gektidis (Frankfurt) and Gudrun Radtke (Wiesbaden) for valuable comments. Preparation of samples was kindly performed by Olga Sagert (Frankfurt). Financial and technical support is gratefully acknowledged from the German Research Foundation (DFG, Vo 90/21, Vo 90/23). For essential support in the field and important information, Klaus Vogel extends thanks to Carl Brett (Cincinnati) regarding Ordovician samples, and Fritz Steininger (Frankfurt), Oleg Mandic (Wien), Mathias Harzhauser (Wien) and Martin Zuchin (Wien) regarding Neogene samples. Additionally, we thank Gabriela Mángano and Luis Buatois, for their patience and friendly co-operation. We gratefully acknowledge the competent suggestions of Richard G. Bromley and Mark A. Wilson.

References

Aguirre, J., Riding, R. & Braga, J.C. 2000: Late Cretaceous incident light reduction: evidence from benthic algae. *Lethaia 33*, 205–213.

Balog, S.-J. 1996: Boring thallophytes in some Permian and Triassic reefs: bathymetry and bioerosion. *Research Reports, Göttinger Arbeiten für Geologie und Paläontologie Sb2*, 305–309.

Balog, S.-J. 1997: Mikroendolithen im Capitan Reef Komplex (New Mexico, USA). *Courier Forschungsinstitut Senckenberg 20*, 47–55.

Blackwell, M. 2000: Terrestrial life – fungal from the start? *Science 289*, 1884.

Bromley, R.G. 2004: A stratigraphy of bioerosion. *In* McIlroy, D. (ed.): *The Application of Ichnology to Palaeoenvironmental and Stratigraphic Analysis. Geological Society of London Special Publication 228.*

Budd, D.A. & Perkins, R.D. 1980: Bathymetric zonation and paleoecological significance of microborings in Puerto Rican shelf and slope sediments. *Journal for Sedimentary Petrology 50*, 881–903.

Bundschuh, M. 2000: Silurische Mikrobohrspuren – ihre Beschreibung und Verteilung in verschiedenen Faziesräumen (Schweden, Litauen, Großbritannien und U.S.A.). Doctoral thesis, Johann Wolfgang Goethe University, Frankfurt am Main.

Bundschuh, M. & Balog, S.-J. 1999: *Fasciculus rogus* nov. isp., an endolithic trace fossil. *Ichnos 7*, 149–152.

Butterfield, J.N. 2000: *Bangiomorpha pubescens* n. gen., n. sp.: implications for the evolution of sex, multicellularity, and the Mesoproterozoic/Neoproterozoic radiation of eukaryotes. *Paleobiology 26*, 386–404.

Campbell, S.E. 1980: *Palaeoconchocelis starmachii*, a carbonate boring microfossil from the Upper Silurian of Poland (425 million years old): implications for the evolution of the Bangiaceae (Rhodophyta). *Phycologia 19*, 25–36.

Campbell, S.E., Kazmierczak, L. & Golubic, S. 1979: *Palaeoconchocelis starmachii* gen.n., sp.n., an endolithic Rhodophyte (Bangiaceae) from the Silurian of Poland. *Acta Palaeontologica Polonica 24*, 405–408.

Falciatore, A. & Bowler, C. 2002: Revealing the molecular secrets of marine diatoms. *Annual Review of Plant Biology 53*, 109–130.

Glaub, I. 1994: Mikrobohrspuren in ausgewählten Ablagerungsräumen des europäischen Jura und der Unterkreide (Klassifikation und Palökologie). *Courier Forschungsinstitut Senckenberg 174*, 1–324.

Glaub, I., Balog, S.-J., Bundschuh, M., Gektidis, M., Hofmann, K., Radtke, G., Schmidt, H. & Vogel, K. 1999: Euendolithic Cyanobacteria/-phyta and their traces in earth history. *Bulletin de l'Institut Oceanographique, Monaco 19*, 135–142.

Glaub, I. & Bundschuh, M. 1997: Comparative study on Silurian and Jurassic/Lower Cretaceous microborings. *Courier Forschungsinstitut Senckenberg 201*, 123–135.

Glaub, I. & Schmidt, H. 1994: Traces of endolithic microboring organisms in Triassic and Jurassic bioherms. *Kaupia, Darmstädter Beiträge zur Naturgeschichte 4*, 103–112.

Glaub, I., Vogel, K. & Gektidis, M. 2001: The role of modern and fossil cyanobacterial borings in bioerosion and bathymetry. *Ichnos 8*, 185–195.

Golubic, S., Brent, G. & LeCampion-Alsumard, T. 1970: Scanning electron microscopy of endolithic algae and fungi using a multipurpose casting embedding technique. *Lethaia 3*, 203–209.

Golubic, S., LeCampion-Alsumard, T. & Campbell, S.E. 1999: Diversity of marine Cyanobacteria. *Bulletin de l'Institut Oceanographique, Monaco 19*, 53–76.

Golubic, S., Perkins, R.D. & Lukas, K.J. 1975: Boring microorganisms and microborings in carbonate substrates. *In* Frey, R.W. (ed.): *The Study of Trace Fossils*, 229–259. Springer, New York.

Golubic, S. & Seong-Joo, L. 1999: Early cyanobacterial fossil record: preservation, paleoenvironment and identification. *European Journal of Phycology 34*, 339–348.

Green, J.W., Knoll, A.H. & Swett, K. 1988: Microfossils from oolites and pisolites of the Upper Proterozoic Eleonore Bay Group, Central East Greenland. *Journal of Paleontology 62*, 835–852.

Hoek, C. van den, Jahns, H.M. & Mann, D.G. 1993: *Algen*. Thieme, Stuttgart.

Hofmann, K. 1996: Die mikro-endolithischen Spurenfossilien der borealen Oberkreide Nordwest-Europas und ihre Faziesbeziehungen. *Geologisches Jahrbuch, A 136*.

Kiene, W.E. 1997: Enriched nutrients and their impact on bioerosion: results from ENCORE. *Proceedings of the 8th International Coral Reef Symposium, Balboa (Panama) 1*, 897–902.

Knoll, A.H., Golubic, S., Green, J. & Swett, K. 1986: Organically preserved microbial endoliths from the late Proterozoic of East Greenland. *Nature 321*, 856–857.

Knoll, A.H., Swett, K. & Burkhardt, E. 1989: Paleoenvironmental distribution of microfossils and stromatolites in the Upper Proterozoic Backlundtoppen Formation, Spitsbergen. *Journal of Paleontology 63*, 129–145.

Li, G.X. 1997: Early Cambrian phosphate-replicated endolithic algae from Emei, Sichuan, SW China. *Bulletin of the National Museum of Natural Science, Taiwan 10*, 193–216.

Littler, M.M., Littler, D.S., Blair, S.M. & Norris, L.N. 1985: Deepest known plant life discovered on an uncharted seamount. *Science 227*, 57–59.

Mägdefrau, K. 1937: Lebensspuren fossiler „Bohr" Organismen. *Beiträge zur naturkundlichen Forschung in Südwestdeutschland 2*, 54–67.

Müller, E. & Loeffler, W. 1992: *Mykologie*. Thieme, Stuttgart.

Olempska, E. 1986: Endolithic microorganisms in Ordovician ostracode valves. *Acta Palaeontologica Polonica 31*, 229–236.

Pia, J. 1937: Die kalklösenden Thallophyten. *Archiv Hydrobiologie 31*, 264–328, 341–398.

Plewes, C.R. 1996: Ichnotaxonomic studies of Jurassic endoliths. PhD thesis, Institute of Earth Studies, University of Wales, Aberystwyth.

Plewes, C.R., Palmer, T.J. & Haynes, J.R. 1993: A boring foraminiferan from the Upper Jurassic of England and northern France. *Journal of Micropalaeontology 12*, 83–89.

Pohowsky, R.A. 1978: The boring ctenostomate Bryozoa: taxonomy and paleobiology based on cavities in calcareous substrata. *Bulletin of American Paleontology 73(301)*, 1–192.

Pratje, O. 1922: Fossile kalkbohrende Algen (*Chaetophorites gomontoides*) in Liaskalken. *Zentralblatt für Mineralogie*, 299–301.

Quenstedt, F.A. 1849: Petrefaktenkunde Deutschlands. I. Abt., 1. Die Cephalopoden: 580 pp.

Radtke, G. 1991: Die mikroendolithischen Spurenfossilien im Alt-Tertiär West-Europas und ihre paläokologische Bedeutung. *Courier Forschungsinstitut Senckenberg 138*, 1–185.

Radtke, G., Gektidis, M., Golubic, S., Hofmann, K., Kiene, W.E. & Le Campion-Alsumard, T. 1997b: The identity of an endolithic alga: *Ostreobium brabantium* Weber-van Bosse is recognized as carbonate-penetrating rhyzoids of *Acetabularia* (Chlorophyta, Dasycladales). *Courier Forschungsinstitut Senckenberg 201*, 341–347.

Radtke, G., Hofmann, K. & Golubic, S. 1997a: A bibliographic overview of micro- and macroscopic bioerosion. *Courier Forschungsinstitut Senckenberg 201*, 307–340.

Redecker, D., Kodner, R. & Graham, L.E. 2000: Glomalean fungi from the Ordovician. *Science 289*, 1920–1921.

Runnegar, B. 1985: Early Cambrian endolithic algae. *Alcheringa 9*, 179–182.

Schlegel, H.G. 1992: *Allgemeine Mikrobiologie*. Thieme, Stuttgart.

Schmidt, H. 1992: Mikrobohrspuren ausgewählter Faziesbereiche der tethyalen und germanischen Trias (Beschreibung, Vergleich und bathymetrische Interpretation). *Frankfurter geowissenschaftliche Arbeiten A, Geologie-Paläontologie 12*, 228 pp.

Schnick, H. 1992: Zum Vorkommen der Bohrspur *Hyellomorpha microdendritica* Vogel, Golubic & Brett im oberen Obermaastricht Mittelpolens. *Zeitschrift für geologische Wissenschaften 20*, 109–124.

Sheath, R.G., Hellebust, J.A. & Sawa, T. 1977: Changes in plastid structure, pigmentation and photosynthesis of the conchocelis stage of *Porphyra leucosticta* (Rhodophyta, Bangiaceae) in response to low light and darkness. *Phycologia 16*, 265–275.

Skelton, P. 1993 (ed.): *Evolution*. Addison-Wesley, Wokingham.

Tappan, H. 1980: *The Paleobiology of Plant Protists*. Freeman, San Francisco.

Vogel, K. 1991: Comment on: Delle Phosphatic Member: An anomalous phosphatic interval in the Mississipian (Osagean-Meramecian) shelf sequence of central Utah. *Newsletter Stratigraphy 24*, 109–110.

Vogel, K., Balog, J., Bundschuh, M., Gektidis, M., Glaub, I., Krutschinna, J. & Radtke, G. 1999: Bathymetrical studies in fossil reefs, with microendoliths as paleoecological indicators. *Profil 16*, 181–191.

Vogel, K., Bundschuh, M., Glaub, I., Hofmann, K., Radtke, G. & Schmidt, H. 1995: Hard substrate ichnocoenoses and their relations to light intensity and marine bathymetry. *Neues Jahrbuch für Geologie und Paläontologie, Abhandlungen 193*, 49–61.

Vogel, K., Gektidis, M., Golubic, S., Kiene, W.E. & Radtke, G. 2000: Experimental studies on microbial bioerosion at Lee Stocking Island, Bahamas and One Tree Island, Great Barrier Reef, Australia: implications for paleoecological reconstructions. *Lethaia 33*, 190–204.

Vogel, K., Golubic, S. & Brett, C.E. 1987: Endolith associations and their relation to facies distribution in the Middle Devonian of New York State, U.S.A. *Lethaia 20*, 263–290.

Vogel, K. & Marincovich, L. 2004: Paleobathymetric implications of microborings in Tertiary strata of Alaska, USA. *Palaeogeography, Palaeoclimatology, Palaeoecology 206*, 1–20

Wilkinson, M. & Burrows, E.M. 1972: The distribution of marine shell-boring green algae. *Journal of the Marine Biological Association of the United Kingdom 52*, 59–65.

Zeff, M.L. & Perkins, R.D. 1979: Microbial alteration of Bahamian deepsea carbonates. *Sedimentology 26*, 175–201.

Zhang, Y. & Golubic, S. 1987: Endolithic microfossils (Cyanophyta) from early Proterozoic stromatolites, Hebei, China. *Acta Micropalaeontologica Sinensis 4*, 1–12.

Ichnological evidence for the arthropod invasion of land

SIMON J. BRADDY

Braddy, S.J. **2004 10 25**: Ichnological evidence for the arthropod invasion of land. *Fossils and Strata*, No. 51, pp. 136–140. UK. ISSN 0300-9491.

Palaeozoic terrestrial trace fossils (particularly arthropod trackways and trails) provide valuable data on the landfall, and the subsequent diversification of early arthropods on land. This ichnological evidence indicates that the earliest invasion of land, evident from trackways from the Late Cambrian of Ontario, occurred around 90 million years before the earliest reliable terrestrial body fossils. Terrestrial trace fossils are generally rare in the Ordovician. Eurypterid trackways from New York State indicate that this group was capable of amphibious excursions (via marine routes) from the Late Ordovician. Narrow myriapod trails from north-west England, and burrows from Pennsylvania, indicate that this group occupied the early bryophyte soils (via freshwater margin routes) in the Late Ordovician. Terrestrial trace fossil diversity and distribution increase in the Early Devonian, indicating the major phase of colonisation of coastal and fluvial settings. The widespread colonisation of land continued throughout the Devonian, until all non-marine habitats were colonised by the Carboniferous.

Explanations for the arthropod invasion of land are traditionally linked to the exploitation of under-utilised ecospace, or aquatic predator evasion. The evolution of land plants in the Ordovician represents a major ecological shift and is probably associated with the terrestrialisation of the myriapods. Other groups probably had different reasons, possibly associated with their life cycles; a "mass-moult-mate" hypothesis for eurypterid reproductory behaviour is supported by abundant accumulations of their exuviae in marginal settings, the functional morphology of their reproduction and respiration, trackway occurrence, and modern analogues (e.g. *Limulus*).

Key words: Palaeoecology; Palaeozoic; Terrestrial; Trace fossil; Trackway.

Simon J. Braddy [S.J.Braddy@bris.ac.uk], Department of Earth Sciences, University of Bristol, Wills Memorial Building, Queen's Road, Bristol BS8 1RJ, UK

Introduction

The conquest of land by arthropods was a major milestone in the evolution of life. Body fossils of early terrestrial arthropods, however, are based on only a handful of fragmentary remains from a few localities; fragments of early arachnids, myriapods and insects are known from the Early Devonian Rhynie Chert in Scotland, Alken-an-der-Mosel in Germany, and Gilboa in New York State (USA), and the Late Silurian of Shropshire in England [see Selden (2001) for a review]. Wilson & Anderson (2004) have recently described three new millipedes from the Mid-Silurian of Scotland, including *Pneumodesmus newmani*, which shows the oldest known spiracles, demonstrating that this form was fully terrestrial.

The morphology of these early arthropods indicates that they were well adapted for life on land, implying that the earliest terrestrial ecosystems occurred much earlier. The evidence from body fossils also indicates that the major groups of terrestrial arthropods colonised the land separately (Selden 2001). However, this record is severely limited, particularly in the Lower Palaeozoic, and is inadequate to determine the actual timing and environmental distribution of the earliest arthropods on land.

Trace fossils, particularly arthropod trackways and trails, provide crucial data on the temporal and environmental distribution of these earliest terrestrial arthropods. Trace fossils reliably record their distribution as they are preserved *in situ*, extend the stratigraphic range of key taxa (e.g. myriapods and euthycarcinoids), and provide direct evidence for the walking techniques used by the early land arthropods as they crossed the functional threshold from in-phase (aquatic) to more stable out-of-phase (terrestrial) gaits (Braddy 2001a).

This paper presents a review of the early terrestrial trace fossil record and summarises what this reveals about the different arthropod groups making the transition to life on land. Devonian non-marine ichnofaunas are highly diverse and both environmentally and geographically widespread, known from lacustrine, fluvial, estuarine, intertidal, littoral, and aeolian coastal dune settings, from Antarctica, northern India, New York State (USA), Canada, Norway, south Wales (Smith *et al.* 2003; Morrissey & Braddy in press), and the Midland Valley of Scotland. These ichnofaunas are reviewed elsewhere (e.g. Pollard 1985; Buatois *et al.* 1998; Draganits *et al.* 2001; Selden 2001). Lower Palaeozoic non-marine ichnofaunas, on the other hand, have received limited attention, although they are much more significant in terms of what they reveal about the earliest phases of the colonisation of land by arthropods, as they pre-date their body fossil record. These earlier records are therefore the focus of this review.

The Lower Palaeozoic trace fossil record

Cambrian

In the Early Cambrian, rare "non-marine" trace fossils are known from Spain (Crimes *et al.* 1977), Canada (MacNaughton & Narbonne 1999), and the Czech Republic (Mikuláš 1995) (Table 1). These ichnofaunas, together with a Late Cambrian ichnofauna from Jordan (Selley 1970), contain arthropod trackways, trails and rare, low-diversity worm burrows that are more typical of shallow-marine settings. Thus, these assemblages do not represent the establishment of a true non-marine biota, but instead represent temporary amphidromous migrations of marine animals (Maples & Archer 1989) (see below).

The Late Cambrian Nepean Formation of southeastern Ontario includes non-marine depositional environments; one distinctive facies is characteristic of an aeolian dune deposit (MacNaughton *et al.* 2002). Large

arthropod trackways, up to 12 cm wide, with series of up to eight tracks (some specimens also show a well-preserved medial impression), occur on subaerial foreset surfaces. These trackways are assigned to *Diplichnites* (i.e. without a medial impression) and *Protichnites* (i.e. with a medial impression), although the latter ichnogenus is poorly defined, and in need of revision. The producer of these trackways was a large homopodous arthropod with at least eight pairs of walking legs and a tail spine (MacNaughton *et al.* 2002). The morphology of euthycarcinoids is consistent with these trackways. This group was proposed by Trewin & McNamara (1995) as the producer of similar trackways in the Tumblagooda ichnofauna of Western Australia.

Ordovician

In the Ordovician, non-marine trace fossils are generally rare (Table 2). Early Ordovician "non-marine" assemblages are known from Spain (Baldwin 1977) and Antarctica (Weber & Braddy 2004). The Antarctic occurrence represents a very shallow, restricted marginal-marine environment, probably a tide-dominated estuary; mudcracks and rain prints on surfaces intercalated with the trace fossil horizons indicate intermittent subaerial exposure (e.g. intertidal conditions), although trace fossils are not preserved on these surfaces. This ichnofauna is dominated by crustacean-produced walking, foraging and resting traces (Weber & Braddy 2004).

Sharpe (1932) described a eurypterid trackway from the "Caradoc" of New York State (USA), preserved on a mud-cracked surface, representing a mudflat, indicating that eurypterids were capable of amphibious excursions (via marine routes) from the Late Ordovician. Eurypterid and other arthropod trackways are also known from the Graafwater Formation (Table Mountain Group) of the western Cape in South Africa (Braddy & Almond 1999).

Table 1. Cambrian non-marine ichnoassemblages, inferred environment (bold text denotes evidence for subaerial exposure), and ichnotaxa present (asterisks denote arthropod-produced trackways, trails and resting traces).

Age	Stratigraphy; locality; reference	Setting	Ichnotaxa present
Late	Disi Group; Jordan; Selley (1970)	Fluvial, tidal and deltaic	*Cruziana*, Diplichnites*, Merostomichnites*, Rouaultia, Rusophycus*, Sabellarifex*
	Nepean Formation; Ontario, Canada; MacNaughton *et al.* (2002)	**Aeolian dune to sublittoral**	***Bergaueria, Diplichnites*, Protichnites*,* "horizontal burrows"**
Early	Cándona Quartzite and Herrería Sandstone; Spain; Crimes *et al.* (1977)	Tidal channel and intertidal sand–mud flats	*Arenicolites, Astropolithon, Bergaueria, Cruziana*, Diplocraterion, Planolites, Rusophycus*, Skolithos*
	Backbone Ranges Formation; northwest Canada; MacNaughton & Narbonne (1999)	Interdistributary bar and mouth bar	*Bergaueria, Didymaulichnus, Heminthoidichnites, Monomorphichnus*, Palaeophycus, Planolites, Rusophycus*, Teichichnus, Treptichnus*
	Paseky Shale; Czech Republic; Mikuláš (1995)	Restricted brackish lagoon	*Bergaueria, Dimorphichnus*, Diplichnites*, Monomorphichnus*, Rusophycus**

Table 2. Ordovician non-marine ichnoassemblages, inferred environment (bold text denotes evidence for subaerial exposure), and ichnotaxa present (asterisks denote arthropod-produced trackways, trails and resting traces).

Age	Stratigraphy; locality; reference	Setting	Ichnotaxa present
Late	**Tumblagooda Formation;** *Heimdallia–Diplichnites* **association; Western Australia; Trewin & McNamara (1995)**	**Mixed waterlain aeolian sandsheet and flooded interdune facies**	*Beaconites, Cruziana*, Didymaulichnus, Didymaulyponomos, Diplichnites*, Diplocraterion, Heimdallia, Lunatubichnus, Paleohelcura** [= *Palmichnium**], *Rusophycus, Skolithos, Tigillites, Tumblagoodichnus**, "meander-loop trails"*, "paired burrows"
	Juniata Formation; Pennsylvania; Retallack (2001)	**Palaeosol**	*Scoyenia*
	Borrowdale Group; Cumbria; Johnson *et al.* (1994)	**Freshwater margin**	*Diplichnites*, Diplopodichnus**
	Hudson River Shales; New York State; Sharpe (1932)	**Tidal mudflat**	*Protichnites** [= *Palmichnium**]
Early	Graafwater Formation; Cape Province, South Africa; Braddy & Almond (1999)	Estuarine and shallow subtidal	*Merostomichnites*, Monomorphichnus*, Petalichnus*, Palmichnium*, Rusophycus* Cruziana*, Planolites, Skolithos*
	Cabos Series; Spain; Baldwin (1977)	Littoral, subtidal	*Asaphoidichnus*, Beaconites,*
	Blaiklock Glacier Group; Shackelton Range, Antarctica; Weber & Braddy (2004)	Shallow, restricted marginal marine, subtidal	*Didymaulichnus, Diplichnites*, Gordia, Laevicyclus, Merostomichnites*, Monomorphichnus*, Palaeophycus, Planolites, Rusophycus*, Selenichnites*, Taphrhelminthoides*

Three-dimensional burrows with a W-shaped backfill, attributed to myriapods, were reported by Retallack & Feakes (1987) from Late Ordovician palaeosols in Pennsylvania (USA) (Retallack 2001). Narrow trackways and trails, from the dried-out margins of lacustrine sediments within the Borrowdale Volcanic Group of Cumbria (UK), represent the activities of freshwater margin myriapods capable of surviving temporary subaerial excursions (Johnson *et al.* 1994).

The Tumblagooda Sandstone of Western Australia, originally regarded as Late Silurian in age, has recently been re-dated as Late Ordovician (Iasky *et al.* 1998; McNamara & Trewin 2002). A highly diverse ichnofauna from a fluvial, aeolian sandsheet and flooded interdune facies was described by Trewin & McNamara (1995).

Silurian

An Early Silurian ichnofauna from China contains three discrete ichnocoenoses, including an uppermost "*Sintanichnus*" association, representing a subtidal to deltaic setting (Yang 1984). This ichnofauna contains a number of unusual trace fossil taxa that may be referred to more well-known taxa if this material is subjected to a modern review (see Table 3 for provisional revisions, based on S. Haseman 2004: pers. comm.).

Late Silurian diverse intertidal and non-marine ichnofaunas are known from Arctic Canada (Narbonne 1984), Norway (Hanken & Størmer 1975; Whitaker 1979), and Newfoundland (Wright *et al.* 1995) (Table 3). The Clam Bank Formation of Newfoundland, in particular, has produced subaerial *Diplichnites* trackways, attributed to eoarthropleurid myriapods. In addition to the ichnofauna reported by Wright *et al.* (1995), *Beaconites* burrows, and scratch-arrays, similar to *Striatichnium*, representing substrate-foraging arthropods, have recently been discovered at this locality.

Discussion

Why did the arthropods invade the land? Undoubtedly, different groups had different reasons. The occurrence of the earliest non-marine trace fossils may be explained by perturbations in marginal environments, enabling normally marine animals to periodically inhabit non-marine settings via a salt wedge, i.e. a periodic influx of marine waters into a tidal channel (Maples & Archer 1989). Many extant marginal-marine animals display amphidromy (i.e. the ability to periodically migrate into non-marine settings), particularly for food or during spawning periods; indeed, a salt wedge does not always need to be present (Maples & Archer 1989).

Trewin & McNamara (1995) suggested that predatory pressure in aquatic environments (e.g. from early fish and large predatory arthropods) drove many groups onto land. Another potential explanation is linked to the exploitation of empty or under-utilised ecospace (Buatois

Table 3. Silurian non-marine ichnoassemblages, inferred environment (bold text denotes evidence for subaerial exposure), and ichnotaxa present (asterisks denote arthropod-produced trackways, trails and resting traces).

Age	Stratigraphy; locality; reference	Setting	Ichnotaxa present
Late	Cape Storm and Leopold Formations; *Polarichnus–Bergaueria* association; Canadian Arctic; Narbonne (1984)	High intertidal	*Arenicolites, Arthraria, Beaconites, Bergaueria, Chondrites, Cochlichnus, Cruziana*, Diplocraterion, Gordia, Helicodromites, Monomorphichnus*, Palaeophycus, Petalichnus*, Polarichnus, Rusophycus*, Skolithos, Syncoprulus*
	Ringerike Formation; Norway; Hanken & Størmer (1975)	Fluvial to marginal marine	*Beaconites, Diplocraterion, Merostomichnites*, Palmichnum*, Polarichnus, Stiaria*, Stienfordichnus*
	Clam Bank Formation; Newfoundland; Wright *et al.* (1995)	**Muddy brackish to freshwater margin**	***Beaconites, Diplichnites*, Striatichnium***
Early	Shamao Formation; *Sintanichnus* association; central China; Yang (1984)	Subtidal to deltaic	*Bipodomorpha** [= *Paleohelcura**], *Gordia, Lockeia, Oniscoidichnus** [= *Protovirgularia*], *Psammichnites, Pteridichnites, Shamaoichnus, Sintanichnus, Torrowangea, Xizangichnus, Ziguichnus*

& Mángano 1993; Buatois *et al.* 1998). The evolution of land plants in the Ordovician represents a major ecological shift. For the first time organic detritus accumulated in non-marine settings; these previously nutrient-poor settings would have represented an under-exploited resource. Any organism capable of enduring the physiological stresses associated with terrestrial life (e.g. air-breathing, temperature and salinity fluctuations; Selden 2001) would have had an advantage. This was probably the most important factor associated with the terrestrialisation of the herbivorous myriapods. It is unlikely, however, that the eurypterids would have periodically left the water to scavenge for food, as they lacked suitable adaptations for terrestrial feeding, e.g. the advanced mouthparts (pre-oral cavity) of the terrestrial arachnids.

Another potential explanation may be linked to the periodic tolerance (as part of their life cycle) of semi-aquatic arthropods to terrestrial environments. Many amphidromous organisms cluster together to maximise reproductive potential. Indeed, various lines of evidence support a "mass-moult-mate" hypothesis, that some arthropods (e.g. eurypterids) migrated *en-masse* into near-shore and marginal environments (e.g. lagoons), as part of their life cycle to moult and mate (Braddy 2001b):

1. The occurrence of abundant accumulations of eurypterid exuviae from, for example, the Late Silurian Bertie Waterlime assemblage of New York State, interpreted as a near-shore lagoon;
2. The functional morphology of eurypterid palaeobiology (reproduction and respiration); eurypterids possessed a dual respiratory system, and females were able to store spermatophores, delaying spawning until the environmental conditions would have ensured the survival of larvae.

3. The occurrence of juvenile eurypterids in near-shore settings such as tidal mudflats and lagoons.
4. The occurrence of abundant, subparallel eurypterid trackways in marginal environments.
5. Comparisons with the only extant aquatic chelicerates, the xiphosurans (e.g. *Limulus*), which undertake seasonal mass migrations to concentrate populations, ensuring increased reproductive success (Braddy 2001b).

Conclusions

The evidence from trace fossils supports the view that the arthropod invasion of land was staggered, although it occurred much earlier than is evident from the body fossil record. Ichnological evidence indicates that the colonisation of non-marine settings began in the Cambrian, with periodic migrations of marine arthropods into marginal aquatic habitats in response to environmental perturbations. The first tentative steps onto land were taken by the euthycarcinoids in the Late Cambrian. Narrow trackways, trails and *Scoyenia* burrows indicate that myriapods occupied the early bryophyte soils, via freshwater margin routes, in the Late Ordovician, probably feeding as detritivores on the early land plants. Eurypterid trackways from Late Ordovician intertidal settings indicate movement of this group onto land directly from the sea, although it is unlikely that these animals would have remained for long in these habitats (Braddy 2001b).

Trace fossil diversity and distribution indicate that the major colonisation of coastal marine and fluvial settings did not occur until the Early Devonian, approximately coincident with the earliest body fossils. In the Early Devonian, trackways occur in lakeshore settings within continental interiors (Pollard & Walker 1984). The wide-spread colonisation of land progressed throughout the Devonian, until all non-marine habitats were colonised by the Carboniferous (Buatois *et al.* 1998).

Acknowledgements

I thank John Pollard and Ricardo Melchor for their review of this paper, and the Leverhulme Trust (grant F/82/AZ: to D.E.G. Briggs) for financial support.

References

Baldwin, C.T. 1977: The stratigraphy and facies association of trace fossils in some Cambrian and Ordovician rocks of northwestern Spain. *Geological Journal Special Issue 9*, 9–40.

Braddy, S.J. 2001a: Trackways – arthropod locomotion. *In* Briggs, D.E.G. & Crowther, P.R. (eds): *Palaeobiology II*, 389–393. Blackwell Scientific Publications, Oxford.

Braddy, S.J. 2001b: Eurypterid palaeoecology: palaeobiological, ichnological and comparative evidence for a 'mass-moult-mate' hypothesis. *Palaeogeography, Palaeoclimatology, Palaeoecology 172*, 115–132.

Braddy, S.J. & Almond, J. 1999: Eurypterid trackways from the Table Mountain Group (Lower Ordovician) of South Africa. *Journal of African Earth Sciences 29*, 165–177.

Buatois, L.A. & Mángano, M.G. 1993: Ecospace utilization, paleoenvironmental trends, and the evolution of early nonmarine biotas. *Geology 21*, 595–598.

Buatois, L.A., Mángano, M.G., Genise, J.F. & Taylor, T.N. 1998: The ichnologic record of the continental invertebrate invasion: evolutionary trends in environmental expansion, ecospace utilization, and behavioral complexity. *Palaios 13*, 217–240.

Crimes, T.P., Legg, I., Marcos, A. & Arboleya, M. 1977: ?Late Precambrian–Lower Cambrian trace fossils from Spain. *Geological Journal Special Issue 9*, 91–138.

Draganits, E., Braddy, S.J. & Briggs, D.E.G. 2001: A Gondwanan coastal arthropod ichnofauna from the Muth Formation (Lower Devonian, northern India): paleoenvironment and tracemaker behavior. *Palaios 16*, 126–147.

Hanken, N.-M. & Størmer, L. 1975: The trail of a large Silurian eurypterid. *Fossils and Strata 4*, 255–270.

Iasky, R.P., Mory, A.J., Ghori, K.A.R. & Shevchenko, S.I. 1998: Structure and petroleum potential of the southern Merlinleigh Sub-basin, Carnarvon Basin, Western Australia. *Geological Survey of Western Australia, Report 61*, 1–63.

Johnson, E.W., Briggs, D.E.G., Suthren, R.J., Wright, J.L. & Tunnicliff, S.P. 1994: Non-marine arthropod traces from the subaerial Ordovician Borrowdale Volcanic Group, English Lake District. *Geological Magazine 131*, 395–406.

MacNaughton, R.B., Cole, J.M., Dalrymple, R.W., Braddy, S.J., Briggs, D.E.G. & Lukie, T.D. 2002: First steps on land: arthropod trackways in Cambrian–Ordovician eolian sandstone, southeastern Ontario, Canada. *Geology 30*, 391–394.

MacNaughton, R.B. & Narbonne, G.M. 1999: Evolution and ecology of Neoproterozoic–Lower Cambrian trace fossils, NW Canada. *Palaios 14*, 97–115.

Maples, C.G. & Archer, A. 1989: The potential of Paleozoic non-marine trace fossils for paleoecological interpretations. *Palaeogeography, Palaeoclimatology, Palaeoecology 73*, 185–195.

McNamara, K.J. & Trewin, N.H. 2002: Late Ordovician/Early Silurian arthropod colonisation of the land – evidence from the Tumblagooda Sandstone, Western Australia. First International Palaeontological Congress (IPC2002), Sydney. *Geological Society of Australia, Abstracts 68*, 112–113.

Mikuláš, R. 1995: Trace fossils from the Paseky Shale (Early Cambrian, Czech Republic). *Journal of the Czech Geological Society 40*, 47–54.

Morrissey, L.B. & Braddy, S.J. In press: Terrestrial trace fossils from the Lower Old Red Sandstone, south-west Wales. *Geological Journal Special Issue 39*.

Narbonne, G.M. 1984: Trace fossils in Upper Silurian tidal flat to basin slope carbonates of Arctic Canada. *Journal of Paleontology 58*, 398–415.

Pollard, J.E. 1985: Evidence from trace fossils. *Philosophical Transactions of the Royal Society London, Series B 309*, 241–242.

Pollard, J.E. & Walker, E. 1984: Reassessment of sediments and trace fossils from Old Red Sandstone (Lower Devonian) of Dunure, Scotland, described by John Smith (1909). *Geobios 17*, 567–576.

Retallack, G.J. 2001: *Scoyenia* burrows from Ordovician palaeosols of the Juniata Formation in Pennsylvania. *Palaeontology 44*, 209–235.

Retallack, G.J. & Feakes, C.R. 1987: Trace fossil evidence for Late Ordovician animals on land. *Science 235*, 61–63.

Selden, P.A. 2001: Terrestrialization of animals. *In* Briggs, D.E.G. & Crowther, P.R. (eds): *Palaeobiology II*, 71–74. Blackwell Scientific Publications, Oxford.

Selley, R.C. 1970: Ichnology of Palaeozoic sandstones in the southern desert of Jordan: a study of trace fossils in their sedimentological context. *Geological Journal Special Issue 3*, 477–488.

Sharpe, S.C.F. 1932: Eurypterid trails from the Ordovician. *American Journal of Science 24*, 355–361.

Smith, A., Braddy, S.J., Marriott, S.B. & Briggs, D.E.G. 2003: Arthropod trackways from the Early Devonian of South Wales: a functional analysis of producers and their behaviour. *Geological Magazine 140*, 63–72.

Trewin, N.H. & McNamara, K.J. 1995: Arthropods invade the land: trace fossils and palaeoenvironments of the Tumblagooda Sandstone (?late Silurian) of Kalbarri, Western Australia. *Transactions of the Royal Society of Edinburgh: Earth Sciences 85*, 177–210.

Weber, B. & Braddy, S.J. 2004: A marginal marine ichnofauna from the Blaiklock Glacier Group (?Lower Ordovician) of the Shackleton Range, Antarctica. *Transactions of the Royal Society of Edinburgh: Earth Sciences 94*, 1–20.

Whitaker, J.H. McD. 1979: A new trace fossil from the Ringerike Group, southern Norway. *Proceedings of the Geologists' Association 91*, 85–89.

Wilson, H.M. & Anderson, L.I. 2004: Morphology and taxonomy of Paleozoic millipedes (Diplopoda : Chilognatha : Archipolypoda) from Scotland. *Journal of Paleontology 78*, 169–184.

Wright, J.L., Quinn, L., Briggs, D.E.G. & Williams, S.H. 1995: A subaerial arthropod trackway from the Upper Silurian Clam Bank Formation of Newfoundland. *Canadian Journal of Earth Science 32*, 304–313.

Yang, S. 1984: Silurian trace fossils from the Yangzi Gorges and their significance to depositional environments. *Acta Palaeontogica Sinica 23*, 705–718.

The trace fossil record of burrowing decapod crustaceans: evaluating evolutionary radiations and behavioural convergence

NOELIA B. CARMONA, LUIS A. BUATOIS & M. GABRIELA MÁNGANO

Carmona, N.B., Buatois, L.A. & Mángano, M.G. 2004 10 25: The trace fossil record of burrowing decapod crustaceans: evaluating evolutionary radiations and behavioural convergence. *Fossils and Strata*, No. 51, pp. 141–153. Argentina. ISSN 0300-9491.

Trace fossils assigned to the activity of decapod crustaceans are well known from the stratigraphic record. Changes in abundance and ichnodiversity of these structures through the Phanerozoic have been analysed. A database summarising trace fossil occurrences was compiled. This information is interpreted with respect to the life history of burrowing decapods, as inferred from the body fossil record. The Palaeozoic records are sparse and difficult to interpret in terms of producers. The presence of burrow systems (i.e. *Thalassinoides*) in Early Palaeozoic rocks most probably records burrowing by groups other than decapod crustaceans, therefore reflecting behavioural convergence. The possibility that decapods may have produced burrow systems in the Late Palaeozoic cannot be ruled out completely, but it is more likely that other malacostracans were involved in the construction of these structures. Post-Palaeozoic records are more confidently attributed to decapods and the slow rise in the abundance of these structures through the Mesozoic shows a good correlation with the trends recorded in the body fossil record. During the Palaeogene, the number of decapod ichnofossils reported is considerably lower than that for the Cretaceous. This may reflect the effect of the end-Cretaceous mass extinction and/or monographic effects related to the larger amounts of ichnological work carried out on Cretaceous shallow-marine deposits. During the Neogene, the abundance of decapod trace fossils underwent a remarkable increase, and crustacean burrows became dominant elements of the shallow-marine ichnofaunas, commonly a part of communities displaying complex endobenthic tiering patterns. Finally, the data are related to the history of infaunalisation. In the colonisation of infaunal ecospace, decapods played an important role, becoming dominant components of the modern faunas.

Key words: Ichnology; decapods; burrows; infaunalisation; radiations; evolutionary paleoecology.

Noelia B. Carmona [ichnolog@infovia.com.ar], Conicet-Insugeo, Casilla de correo (Correo Central), 4000 San Miguel de Tucumán, Argentina

Luis A. Buatois [luis.buatois@usask.ca] & M. Gabriela Mángano [gabriela.mangano@usask.ca], Department of Geological Sciences, University of Saskatchewan, 114 Science Place, Saskatoon, SK S7N 5E2, Canada

Introduction

Ichnological studies provide valuable information about the animal–sediment relationships that cannot be inferred from the study of body fossils alone. Accordingly, trace fossils have been widely used in sedimentological analyses because they commonly refine palaeoenvironmental interpretations (e.g. Pemberton *et al.* 2001). However, there are relatively few studies that use ichnofossils as tools for palaeobiological inferences about evolutionary trends undergone by the trace makers. Representatives of the Order Decapoda (Superclass Crustacea, Class Malacostraca) construct distinctive structures, which are particularly abundant in the fossil record (e.g. Fürsich 1973; Schlirf 2000). Accordingly, burrow systems produced by decapod crustaceans may be of use in understanding evolutionary trends. However, similar structures, particularly those recorded in the Palaeozoic, are most likely the product of other trace makers, representing an example of behavioural convergence. The aim of this paper is to evaluate possible trace makers of burrow systems commonly attributed to decapod crustaceans and to compare the patterns with the information obtained from the body fossil record of decapods through the Phanerozoic. This analysis is especially focused on the record of radiation and extinction events of decapod crustaceans and on the possibility of detecting these events through the trace fossil record. Obviously, this analysis is problematic because a comparison between the ichnofossil and body fossil patterns is not always straight-forward. However, it is important to explore the question of whether or not a possible correlation between these two sources of information actually exists. Furthermore, the study may reveal if the trace fossil record can provide new evidence that is not elucidated from the analysis of the body fossil record alone. Integration of the palaeoecological information offered by trace fossils with the origin, radiation, and extinction patterns inferred from the body fossil record is important for the analysis of those organisms that do not have good preservation potential, especially the decapods with a weakly calcified exoskeleton (Förster 1985). The presence of these organisms is expressed indirectly by the record of their life activities.

The overall trend indicated by the record of biogenic structures currently attributed to decapods through the Phanerozoic is analysed through each geological period, and related to information in the body fossil record. A database was compiled of all the trace fossil occurrences related to the activity of decapods through the Phanerozoic. Potential biases in the analysis of the database were evaluated. Finally, an attempt was made to explore the role played by decapods in the acquisition of the infaunal habit through their evolutionary development.

Methods

The database includes 451 records of biogenic structures commonly attributed to the activity of burrowing decapods (mazes and boxworks), and was constructed based on a detailed literature survey of primary sources and other data from authors. The majority of the records belong to the most common ichnogenera *Thalassinoides* and *Ophiomorpha* and, to a lesser extent, *Psilonichnus*, *Spongeliomorpha*, *Pholeus*, *Gyrolithes*, *Macanopsis* and *Sinusichnus* (Fig. 1). Table 1 summarises the first appearances of these ichnotaxa. The database was constructed at the ichnogeneric level. Information considered important included burrow morphology, trophic type, burrowing depth, tiering position and environment of deposition. The analysis was restricted to soft-ground, shallow-marine ichnofaunas and, therefore, examples of decapod burrows in deep-marine turbidite systems or in omission surfaces were not considered. The data were used to develop a general abundance diagram (Fig. 2) showing the number of burrow systems similar to those constructed by modern decapods through the Phanerozoic.

The trends shown in the database have been analysed in terms of the temporal duration of the stratigraphic intervals (Fig. 3), the volume of shallow-marine rocks deposited during each period (Fig. 4), and the area covered by the sea in each interval of time (Fig. 5). Durations of stratigraphic intervals are based on the International Stratigraphic Chart of the International Union of Geological Sciences and UNESCO (Remane *et al.* 2000). An estimation of sea-covered area and rock volume was obtained from Ronov *et al.* (1980). For the volume of rocks, the carbonate, carbonate and clastic, and marine clastic subdivisions of Ronov *et al.* (1980) have been used. For sea-covered areas, the platforms and continents are taken to reflect both continental platforms and epeiric seas.

Ronov *et al.* (1980) postulated that changes in volumes of sediment were directly related to the changes in areas of marine sedimentation, and concluded that both measures were controlled by global processes. These two measures show the same trend, so it seems they reflect the same tendency. Also, with increasing age, the sedimentary rocks would have had more chance to be destroyed, so the younger rocks would probably have greater representation than the older ones. However, Ronov *et al.* (1980) plotted relative masses of sedimentary rocks for each period against time and concluded that there was no regular decrease in the relative mass of rocks, but a periodic fluctuation in the distribution of sediments through the Phanerozoic.

Crustacean burrows through geological time

The analysis of the decapod trace fossil records for Early, Middle and Late Palaeozoic, Triassic, Jurassic, Cretaceous, Palaeogene and Neogene is presented here. Additionally, the possible correlations with the body fossil record are explored. The scale of analysis was determined by the quality and number of records available.

Palaeozoic

Early Palaeozoic

There are few Cambrian and Ordovician burrow systems that resemble those currently attributed to younger decapod crustaceans. Almost all the examples correspond to the ichnogenus *Thalassinoides*, and they are preserved generally in carbonate deposits. Structures in Lower Cambrian carbonates referred to commonly as *Aulophycus* (Zhuravleva *et al.* 1982; Astashkin 1985) and compared with *Ophiomorpha* (Balsam 1984; Burzin *et al.* 2001) have been recorded in Russia and the eastern USA. However, their origins are still uncertain. *Ophiomorpha* has also been recorded in the Cambrian of the Czech

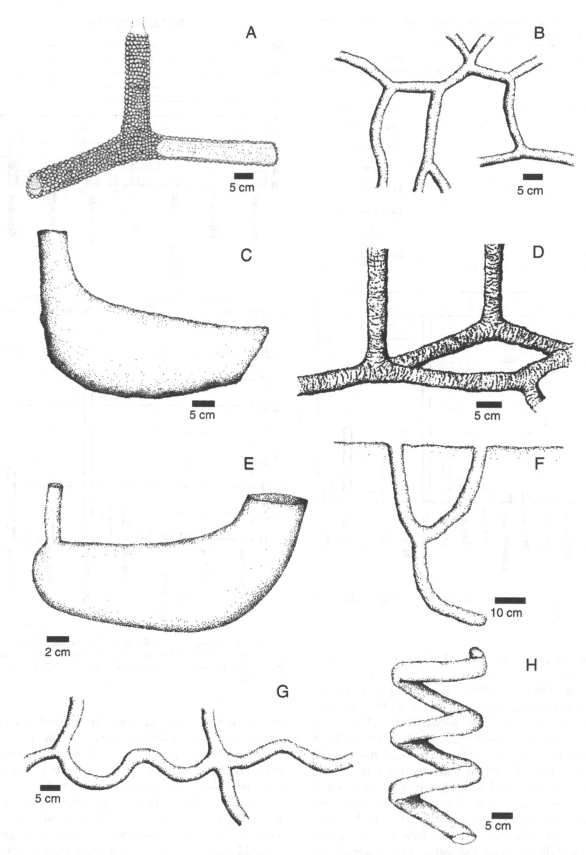

Fig. 1. Morphological diversity of trace fossils attributed to burrowing decapods. A: *Ophiomorpha*. B: *Thalassinoides*. C: *Macanopsis*. D: *Spongeliomorpha*. E: *Pholeus*. F: *Psilonichnus*. G: *Sinusichnus*. H: *Gyrolithes*. Drawings based on de Gibert (1996), Schlirf (2000), Muñiz & Mayoral (2001a, b), and Knaust (2002).

Table 1. First stratigraphic appearances of the ichnogenera discussed. *Ardelia* is only known from Late Permian.

Ichnogenera	First occurrence
Gyrolithes	Early Cambrian
Thalassinoides	Late Cambrian
Ophiomorpha	Late Carboniferous
Spongeliomorpha	Early Permian
Ardelia	Late Permian
Pholeus	Middle Triassic
Macanopsis	Early Cretaceous
Sinusichnus	Late Cretaceous
Psilonichnus	Early Eocene

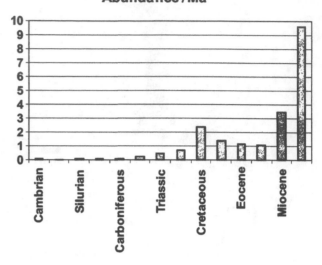

Abundance / Ma

Fig. 3. Abundance of ichnofossils similar to those produced by burrowing decapods, with reference to the duration of periods. Abundance/million years (Ma).

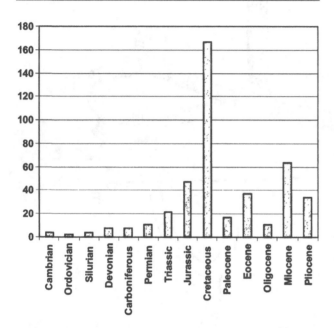

Fig. 2. Abundance of ichnofossils similar to those produced by burrowing decapods through the Phanerozoic.

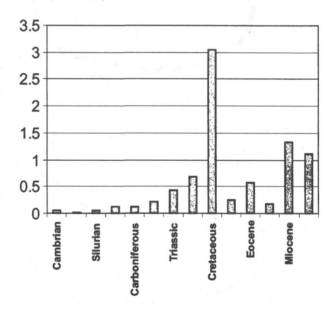

Fig. 4. Abundance of ichnofossils similar to those produced by burrowing decapods, with reference to the volume of sediment deposited in shallow-marine environments during the Phanerozoic [data from Ronov *et al.* (1980)].

Republic (Gaba & Pek 1980), but the interpretation is questionable.

Myrow (1995) described the ichnospecies *Thalassinoides horizontalis* from the Late Cambrian Peerless Formation and the Early Ordovician Manitou Formation in Colorado. This ichnospecies is characterised by a bedding-plane orientation of excavations, a smooth wall, considered to be diagenetic halos representing "a zone of mucus-impregnated sediment around the original burrow", and it tends to develop a polygonal architecture (Myrow 2003, pers. comm.). The ichnospecies has been regarded as a junior synonym of *Thalassinoides suevicus* by Schlirf (2000). Possible constructors of these ichnofossils are unknown, and the features that could relate them to a decapod architecture are absent (e.g. scratch

marks, turnarounds and three-dimensional branching systems). Myrow (1995) postulated other marine organisms as possible producers of these structures, for example, phyllocarids or enteropneusts, but there is no conclusive identification of the animal producer.

Recently, Ekdale & Bromley (2003) described a new and complex ichnospecies of *Thalassinoides* from the Lower Ordovician of Sweden. This new ichnospecies, named *Thalassinoides bacae*, is characterised by horizontal tunnels, mazes, and numerous vertical shaft openings. Probable trace makers are also unknown, and the authors

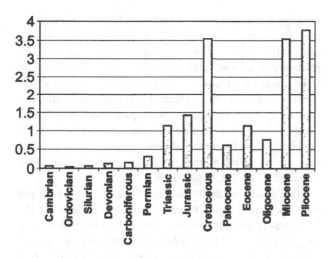

Fig. 5. Abundance of ichnofossils similar to those produced by burrowing decapods, with reference to sea-covered areas during the Phanerozoic [data from Ronov *et al.* (1980)].

discussed some candidates as possible constructors of these structures (Ekdale & Bromley 2003).

The ichnogenus *Thalassinoides* became well established in the Middle and Upper Ordovician (Droser & Bottjer 1988) in inner and middle shelf carbonate deposits, and exhibiting bioturbation of specimens to a depth of 30 cm. However, identification of the potential trace maker also remains uncertain. These structures have the typical boxwork architecture, but assignment to the activity of decapod crustaceans is not possible because unquestioned malacostracan scratch marks – what Seilacher (1970) called "fingerprints" – have not been identified. Sheehan & Schiefelbein (1984) reported *Thalassinoides* structures from the Upper Ordovician of the Great Basin. These burrows have a three-dimensional structure and display classical T- and Y-branching and turnaround points with larger diameters, that are morphologically similar to Recent *Thalassinoides*.

The Ordovician increase in bioturbation has been linked to an increase in the size of the discrete structures and to a change in the architecture of *Thalassinoides* burrows from networks to mazes (Droser & Bottjer 1989). Mángano & Droser (2004) noted that although *Thalassinoides* does occur in Cambrian and Lower Ordovician rocks, examples are typically less than 10 mm in burrow diameter, architecturally simple, and form two-dimensional networks. By contrast, Upper Ordovician *Thalassinoides* favours comparison with three-dimensional burrow systems produced by decapod crustaceans, recording intense sediment reworking that led to the formation of pervasive mottled textures. This has been regarded as a major behavioural and ecological innovation that is not apparently accompanied by an increase in body fossil diversity (Mángano & Droser 2004). However, Cañas (1995) reported *Thalassinoides*

with a clearly defined three-dimensional morphology from Late Cambrian–Tremadocian carbonates from the Precordillera of Argentina. This record may reflect an earlier origin of the boxwork architecture (Mángano & Buatois 2003). While most Lower Palaeozoic examples of *Thalassinoides* are recorded from carbonates, burrow mazes of this ichnotaxa also occur in the Arenig shallow-marine clastics of northwest Argentina (Aceñolaza & Fernández 1984; Mángano & Buatois 2003).

The ichnogenus *Gyrolithes* is also very common in Cambrian strata (e.g. Fritz 1980; Fedonkin 1981, 1983; Liñán 1984; Crimes & Anderson 1985; Hein *et al.* 1991; Jensen 1997; Jensen & Grant 1998; Stanley & Feldmann 1998). However, a crustacean origin is highly unlikely. Compared with the younger representatives, Cambrian *Gyrolithes* is characterised by its small size and the absence of a terminal expansion, scratch marks, and carapace impressions (Jensen 1997). Stanley & Feldmann (1998) described annulations in specimens of *Gyrolithes polonicus* as evidence of peristaltic movements, which suggests an annelid producer. Additionally, Cambrian *Gyrolithes* burrows usually consist only of a few whorls and are interpreted as occupying considerably shallower conditions than their Mesozoic–Cenozoic counterparts.

To summarise, *Thalassinoides* and *Gyrolithes* are known in the Lower Palaeozoic, but no undisputed examples of *Ophiomorpha* have been recorded. Cambrian specimens of *Gyrolithes* differ significantly from younger representatives of the ichnogenus, and were probably formed by annelid worms. Lower Palaeozoic occurrences of *Thalassinoides* are relatively common in the literature, but the identification of the producers remains problematic. The general morphology of some of the Cambrian–Ordovician examples of *Thalassinoides* resembles modern structures produced by decapods. However, the fine morphology shows some differences and no definite decapod scratch marks have been found. Further detailed studies are needed in order to determine the degree of similarity of the Cambrian–Ordovician burrow systems with their younger counterparts. In addition, these early records of *Thalassinoides* largely predate the first occurrence of decapod crustacean body fossils, which are first recorded from the Devonian (Schram *et al.* 1978). It seems unlikely, therefore, that these Cambrian–Ordovician networks and boxworks were made by primitive burrowing decapods. Producers may have included other malacostracans (e.g. phyllocarids) or unrelated clades (e.g. enteropneusts), and represent examples of behavioural convergence in their burrowing activity.

Middle Palaeozoic

The number of Middle Palaeozoic occurrences is slightly higher than for Early Palaeozoic times. Silurian

Thalassinoides from carbonate deposits of Wisconsin (Watkins & Coorough 1997) show some of the typical characteristics of modern excavations, such as boxwork architecture and swollen areas at burrow junctions. Possible trace makers are unknown, but a possible arthropod origin has been postulated (Watkins & Coorough 1997). Articulated trilobites and moulted exoskeletons were found within these burrows. In modern environments, diverse organisms, such as fish, worms and other crustaceans, live within decapod excavations in order to seek protection, food and irrigation (Bromley 1996). This Silurian example may be interpreted to suggest that the commensalism observed in modern burrows may have developed by this time (Watkins & Coorough 1997).

The abundance of trace fossils resembling those constructed by decapods underwent a slight rise during the Devonian. As in the earlier Palaeozoic, most of the Devonian records of burrow systems belong to the ichnogenus *Thalassinoides*. Also, a Late Devonian *Ophiomorpha* has been reported from China (Qi & Bin 2000). However this material has not been illustrated, so the morphological characteristics cannot be confirmed. In terms of behavioural complexity, *Ophiomorpha* records construction of a pelletoidal burrow wall with active manipulation of the sediment. This behavioural strategy, more complex than that involved in the construction of simple smooth lined walls, probably did not originate until the Late Palaeozoic. The oldest confirmed decapod body fossil record comes from the Upper Devonian of the US Midwest where three genera of shrimps have been described (Schram *et al.* 1978). However, although they are clearly shrimps, specimens are poorly preserved and their affinities are uncertain (Briggs & Clarkson 1990). Briggs & Clarkson (1990) suggested that the Devonian was a period of rapid cladogenesis. Consequently, a link between the slight rise in abundance of the Middle Palaeozoic burrow systems and the origin of decapods has yet to be demonstrated.

Late Palaeozoic

The number of specimens found in Carboniferous and Devonian strata is almost the same for both periods, but a rise in occurrences of burrow systems is detected in the Permian. For the Carboniferous, the majority of records are referred to *Thalassinoides*. However, *Ophiomorpha* burrows were found in Pennsylvanian deposits of Utah, Colorado (Driese & Dott 1984), and Kansas (Buatois et al. 2002). A diversification in burrowing morphologies is detected in the Permian with the addition of the ichnogenera *Spongeliomorpha* and *Ardelia*. The first record of *Spongeliomorpha* comes from Early Permian deposits of Australia (Carey 1979). The only record of the ichnogenus *Ardelia* comes from Late Permian strata of Utah (Chamberlain & Baer 1973). However, this ichnogenus is still poorly understood, and requires further study. Specimens of *Ophiomorpha nodosa* are associated with *Ardelia*. These *Ophiomorpha* specimens are relatively small and display poorly developed pelletoidal walls (except for records from Utah and Colorado which are strikingly similar to Recent *Ophiomorpha*). Although burrow systems similar to those produced by infaunal decapods are not usually dominant components of Carboniferous–Permian shallow-marine ichnofaunas, they are relatively abundant in carbonate deposits (e.g. Maerz et al. 1976; Chaplin 1996).

Body fossils of decapods are relatively rare in Upper Palaeozoic strata, but become more common after the Permian–Triassic (Glaessner 1969). In fact, it has been suggested that the low number of decapod representatives between their first appearance (Late Devonian) and their major radiation (Jurassic) points to a macro-evolutionary lag (Briggs & Clarkson 1990). In general, the low number of Late Palaeozoic burrows similar to those constructed by modern decapods is consistent with the poor representation of this taxonomic group in the body fossil record (Förster 1985; Sepkoski 2000). However, other orders within the Malacostraca have a moderately diverse body fossil record (Schram 1981; Briggs & Clarkson 1990). Although the possibility that decapods may have produced burrow systems in the Late Palaeozoic cannot be ruled out completely, it is more likely that other malacostracans were involved in the construction of these structures. For example, eocarids and stomatopods are common in Upper Palaeozoic marginal- to shallow-marine habitats (Briggs & Clarkson 1990). Burrowing behaviour of eocarids is poorly known, but the lack of chelae, specialised appendages, and calcified cuticles may have been disadvantageous to their active burrowing. Stomatopod crustaceans are probably the best candidates because they construct burrow systems (incipient *Thalassinoides*) on Recent shorelines that are very similar to those produced by decapods (Howard & Frey 1975).

Mesozoic

The scenario established in the Palaeozoic changed significantly during the Mesozoic, showing an increased acceleration in the number of decapod records throughout this era. This remarkable increase in abundance probably reflects the Mesozoic decapod radiation as also indicated by the body fossil record (Förster 1985; Feldmann 2003).

Triassic

The Triassic represents a transition between the limited records of the Palaeozoic and the markedly abundant occurrences through the rest of the Mesozoic era. The occurrences in Triassic strata are only slightly higher than those recorded in the Late Palaeozoic (Figs. 2, 3). However, when the volume of rocks (Fig. 4) and, particularly, sea surface area (Fig. 5) are considered, the increase seems to be more significant. Additionally, there was a significant difference in crustacean ichnodiversity from Late Palaeozoic times to the Triassic, with higher levels comparable with that of Late Mesozoic and Cenozoic times. *Thalassinoides* and *Ophiomorpha* are still the most commonly recorded ichnotaxa, but crustacean-produced *Gyrolithes* (resembling modern records) and *Pholeus* are added to the list (De *et al.* 1996; Knaust 2002). *Spongeliomorpha* is also well established in shallow-marine environments by the Early Mesozoic (Mayer 1981; Hary *et al.* 1981; Dahmer & Hilbrecht 1986; Zonneveld *et al.* 1997). This suggests an expansion of behavioural types through the Triassic period. The increase in ichnodiversity is most evident by Middle Triassic times.

Jurassic

The number of decapod burrows present in Jurassic deposits shows a slight increase in abundance with respect to previous periods across all the analyses performed (Figs. 2–5). This is clearly related to the appearance of the body fossils of callianassid decapods in the Late Jurassic (Glaessner 1969). These organisms construct complex burrow systems in modern marine environments, and the general morphology of these excavations does not differ from those recognised in the Jurassic. Accordingly, it is possible to assign these ichnofossils to the activity of callianassid decapods, or taxonomically related groups. The ethological programme recorded by the construction of these dwelling systems has not significantly changed since Jurassic times. The body fossil record reaches a peak of diversity for the decapod faunas in the Late Jurassic, but this record is probably biased owing to an abundance of documented decapod occurrences in the Solnhofen Lagerstätten (Sepkoski 2000).

Cretaceous

The abundance of decapod trace fossils underwent a major increase towards the end of the Mesozoic (Figs. 2–5). In the Cretaceous, crustacean dominance in the marine realm is indicated by the great diversity of body fossils, as well as by the abundance of crustacean burrowing activity in shallow-marine deposits. The biological affinities of these burrow systems with a decapod origin are supported by the presence of crustacean claws within some of these ichnofossils (e.g. Mángano & Buatois 1991; Swen et al. 2001) and by the presence of diagnostic features, such as pelleted walls, striations reflecting the use of hard appendages [the "fingerprints" of Seilacher (1970)] and enlargements at turnaround points. Decapod excavations are particularly abundant in middle and lower shoreface clastic environments, as well as in shallow carbonate settings. The increase in the number of decapod trace fossil occurrences accelerates by the end-Cretaceous. In addition, the oldest examples of the burrow system of *Sinusichnus* have been discovered in the Cretaceous of Antarctica (Buatois, unpublished data).

Cenozoic

Palaeogene

During the Palaeogene, the number of decapod ichnofossils recorded is considerably reduced compared with the Cretaceous (Figs. 2–5). Figure 2 shows that the Eocene has higher abundances of ichnofossils than in the Oligocene. This general tendency is maintained in comparisons based on the volume of shallow-marine clastics and carbonates and the sea surface (Figs. 4, 5). In contrast, when the length of periods is considered, the maximum abundances recorded in the Palaeocene are only slightly greater than for the rest of the Palaeogene.

Neogene

During the Neogene, the abundance of decapod trace fossils underwent another major important increase, probably reaching the highest peak for the whole Phanerozoic. When the length of stratigraphic intervals is considered, the Neogene abundances clearly exceed those of the Cretaceous (Fig. 3). Although still showing a maximum, this trend is somewhat attenuated when sea surface area is introduced into the analysis (Fig. 5). In contrast, the peak in Neogene abundances is secondary with respect to that of the Cretaceous when the volume of shallow-marine clastics and carbonates are considered (Fig. 4). In any case, a peak in abundance is apparent in all analyses and it is therefore likely to be considered as a real trend, rather than a sampling artefact. This pattern is consistent with the record of body fossils, which shows that decapods underwent important radiations after the Cretaceous, becoming extremely important components of the modern evolutionary fauna (Sepkoski 2000). In addition, crustacean burrows are among the dominant elements of Neogene shallow-marine ichnofaunas, commonly forming part of communities that display complex endobenthic tiering patterns (Buatois et al. 2003; Carmona & Buatois 2003).

Discussion

Although behavioural convergence may be invoked to explain the Palaeozoic record of burrowing activity by groups other than decapods, the increase in abundance of decapod trace fossils from the Mesozoic onwards may well reflect the evolutionary diversification of the decapod crustaceans themselves. The rises and falls in ichnofossil abundance detected in the database may be related to diversification and extinction patterns during the evolution of the decapod crustacean group. However, some possible sources of error may somewhat bias the analysed results (Johnson & McCormick 1999). These may include the effects of differing temporal duration of the stratigraphic intervals, and the changes in the volume of shallow-marine rocks available and the area covered by the sea through these same intervals of time.

Diversification and extinction patterns inferred from the database

The Early Palaeozoic ichnofossil record is difficult to correlate with the pattern of diversification shown by decapod crustacean body fossils, because the first record of this group apparently does not appear until the latest Devonian (Schram *et al.* 1978). As discussed above, the nature of the possible producers of the Cambrian–Ordovician structures are unknown, and further detailed studies are required. Undoubtedly, the presence of *Thalassinoides* burrow systems in Early Palaeozoic rocks records bioturbation by groups other than decapod crustaceans, therefore reflecting behavioural convergences. This fact complicates the analysis of the database because some of these groups may also have been continuing to construct similar structures in later times.

In general, the ichnofossil record of decapods is sparse during the rest of the Palaeozoic, whereas the group essentially constitutes one of the main components of the post-Palaeozoic marine fauna. From the Devonian to the Permian there is a gradual increase in the number of ichnofossils that resemble decapod burrow systems. This trend probably reflects the incipient expansion of different orders of malacostracans (some of which may have been rather efficient burrowers). The Permo-Triassic mass extinction seems to have had little effect on the decapods themselves, but it may have affected the structure of Late Palaeozoic shallow-marine communities due to the decline of many epifaunal components and, thereby, created the ecospace for the subsequent radiations of decapods through the Mesozoic.

Although other organisms continued to construct burrow systems after the Palaeozoic (e.g. stomatopods), through the Early Mesozoic, records are more confidently assigned to the activity of decapods and they seem to reflect in a more accurate way the evolutionary history of the group. During the Mesozoic, the increase in abundance of ichnogenera accelerated. From the Triassic to the Jurassic there is a continuous increase in trace fossils attributed to decapods, and then follows an important rise during the Cretaceous, as is evident from all the analyses presented (Figs. 2–5). This rise to a peak

undoubtedly reflects the Late Mesozoic radiation of decapods.

In the Palaeogene, the number of records appears to decline, a situation originally thought to be related to the Cretaceous/Palaeocene mass extinction because this drop is represented in all the analyses undertaken (Figs. 2–5). However, it is difficult to interpret the effects (or role) of the end-Cretaceous mass extinction on the infaunal decapods. Sepkoski (2000) showed that this mass extinction affected some groups within the Decapoda, such as brachyurans, palinurans and astacideans, but in general the impact of this event was not too severe on the decapods. In a recent study, Feldmann (2003) also concluded that at least 29 decapod families had their origins in the Cretaceous and continued through into the Cenozoic. Moreover, the possibility that the Cretaceous peak reflects, at least in part, the extensive amount of ichnological work carried out on shallow-marine rocks of this age (i.e. the monographic effects *sensu* Raup 1976) cannot be discounted.

The record of crustacean body fossils indicates that the Eocene was an interval of evolutionary radiation for decapods, at least for those groups with North Pacific distributions (Schweitzer 2001). Moreover, the abundance of trace fossils is apparently higher during this epoch than for the rest of the Palaeogene period. During the Oligocene, the number of records seems to have decreased again. Analysis of the fossil record of Oligocene decapod fauna from the North Pacific suggests that this epoch was a time when extinction rates were higher than origination rates for genera (Schweitzer 2001). Although this information is biogeographically and taxonomically restricted (almost all the studied records belong to brachyurans), it may reveal a more general pattern for the majority of groups within the decapods. The situation from late Eocene to Oligocene time may be related to a worldwide cooling that produced a biotic crisis with a high extinction rate of decapods (Schweitzer 2001).

During the Neogene, a remarkable increase in the abundance of decapod burrows is shown in all the graphs (Figs. 2–5). Information from the body fossil record indicates that the Miocene was a time of high origination rates within the decapod fauna, with the appearance of modern genera that dominate the present seas (Schweitzer 2001). Neogene data collected during this study are consistent with the body fossil information, showing that the primacy of trace fossils attributed to decapod crustaceans during this epoch is probably not a sampling artefact, but represents part of a real evolutionary event in shallow-marine environments.

Infaunalisation

The history of infaunalisation may be viewed as an evolutionary process in which new and under-utilised ecospace is exploited (Bambach 1983). There was a slow development of infaunalisation through the Palaeozoic, and then a marked increase during Mesozoic and Cenozoic times (Thayer 1979). In order to colonise the infaunal environment, the development of specific morphological traits and physiological mechanisms that allowed decapods to live within the sediment was necessary. For example, some morphological adaptations include hardened and flattened articulated appendages, used to manipulate the sediment; special cuticular structures that increase the friction with the substrate (Haj & Feldmann 2002); and tegumentary glands that reinforce the excavations (Dworschak 1998). In addition, to colonise the infaunal realm, organisms must have evolved a series of adaptations enabling them to survive in an environment of low oxygen levels and high carbon dioxide conditions (Bellwood 2002), and to maintain a constant flow of water through the excavation, in order to obtain food and oxygenated water. These adaptations may have been gradually developed in malacostracans during the Late Palaeozoic and it is probable that they had become well established by the beginning of the Mesozoic era. This is suggested by the important and rapid increase in the abundance and depth of infaunal structures in post-Palaeozoic deposits. Thayer (1979) postulated that the decline in immobile suspension feeders living on the surface of soft substrata during the Phanerozoic was due to the increased activity of bioturbation by deposit-feeder organisms since the Late Palaeozoic. Figure 6 shows the diversification in adaptive strategies reflected by the Malacostraca in the Phanerozoic. Although this information is referred to all the members of the malacostracan class, it is important to compare the data from the Middle and Upper Palaeozoic with that of post-Palaeozoic times, when the development of new adaptations – that allowed colonisation of the available ecospace – took place.

In this study, we have tried to detect whether a general pattern of intensification in ecospace exploitation is also supported by the ichnological record. Information obtained in this analysis shows a slow and gradual increase in the abundance of infaunal structures through the Palaeozoic. However, the biological affinity of Palaeozoic burrow systems is not fully understood, especially those records from the Early Palaeozoic. The depth and extent of bioturbation currently attributed to the activity of decapods increase during the Mesozoic (particularly towards the end of this era). This trend is consistent with the hypothesis that infaunal suspension feeders and deposit feeders diversified during Mesozoic times (Thayer 1979). In Mesozoic rocks, crustacean burrows became elite trace fossils (*sensu* Bromley 1996) in shallow-marine communities.

There are several possible explanations for the development of the infaunal habit. Crustaceans could

Middle & Upper Paleozoic Malacostraceans

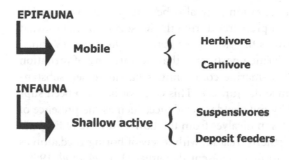

Mesozoic & Cenozoic Malacostraceans

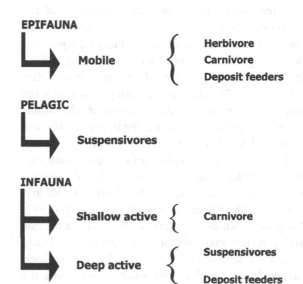

Fig. 6. Comparison between the adaptive strategies of Middle and Upper Palaeozoic malacostracans, and those of Mesozoic and Cenozoic Malacostraca [modified from Bambach (1983), and with information based on trace fossil data herein].

have developed this life habit to obtain protection, especially during moulting periods, when the exoskeleton is soft and the organisms are very vulnerable to attacks (Förster 1985). Bromley & Asgaard (1972) found fragments of *Glyphea* exuviae in *Thalassinoides* burrows from the Lower Jurassic of east Greenland. These decapods may have used their burrows for protection during moulting. This novelty in the life habit may well have expanded within the group, allowing organisms to live permanently within the sediment, producing a gradual reduction of the integument calcification (except for the chelae, which were used for excavation) (Förster 1985). Bellwood (2002) postulated that developing the capacity to bury themselves was widespread in basal brachyuran groups, and that this characteristic arose independently at least three times in the evolutionary history of the groups. Infaunaliol:sation also allowed new and under-exploited resources of food to be obtained, especially for deposit feeders (Bambach 1983).

Infaunalisation may also be a response to increasing predation pressures during the Mesozoic, owing to what is known as the "Mesozoic marine revolution" (Vermeij 1977). Vermeij postulated that the intensity of predation in shallow-marine communities has increased substantially since the Jurassic. This concept has also been supported by independent evidence, such as the presence of boreholes in bivalves from the Lower Jurassic of England and Northern Ireland, with levels of boring predation as high as in many present day areas (Harper *et al.* 1998). Organisms developed diverse strategies to resist predation, including modifications of shell structure, migration to restricted environments and the acquisition of the infaunal condition (Vermeij 1977). For example, bivalves underwent a dramatic increase in infaunal siphonate groups during the Mesozoic (Stanley 1977). A similar pattern was observed in echinoids and gastropods (Kier 1974; Stanley 1977), with various groups colonising the infaunal environment. The development of adaptations that allowed organisms to colonise the infaunal realm was not restricted to crustaceans; these selective pressures affected many other groups of organisms in the same way.

Cenozoic patterns support the same general tendency towards continuity and accelerated levels of infaunalisation in shallow-marine deposits, especially during the Neogene, when burrow systems of decapods became dominant in the shallow-marine environments. While the Mesozoic radiation of decapods produced an ecological change in the benthic communities, increasing the degree and depth of bioturbation (Carmona *et al.* 2002), in the lower Miocene deeper decapod structures were developed that are dominant in lower shoreface deposits of Patagonia (Carmona & Buatois 2003). These Miocene benthic communities also reveal the development of a complex partitioning of the infaunal ecospace, reaching levels of complexity similar to those observed in modern environments (Buatois *et al.* 2003). Thus, the evolutionary history of infaunal decapod crustaceans may be viewed as related not only to a search for shelter, but also to the development of adaptations and trophic types that enabled them to exploit previously under-utilised ecospace and to become dominant components of infaunal communities.

Conclusions

1. The abundance of trace fossils currently attributed to decapod activities shows an increase through the Phanerozoic. The Palaeozoic trend is not clearly understood due to the uncertainties about the identity of the organisms that constructed these structures. The presence of *Thalassinoides* in Early Palaeozoic rocks probably records burrowing by groups other than decapod crustaceans, therefore reflecting the early appearance of a burrowing behaviour convergently similar to patterns produced by later decapods. The Upper Palaeozoic ichnofossils are more similar to decapod burrow systems, and are more numerous than in the earlier record, but it is likely that these were also constructed by other malacostracans than by decapods.

2. Mesozoic data are more clearly correlated with the decapod body fossil record. During the Triassic and Jurassic, the trace fossils underwent a gradual increase, and this pattern can be correlated directly with the appearances of callianassid decapod body fossils, especially in the Jurassic. Triassic crustacean ichnodiversity reached a level similar to that of Late Mesozoic and Cenozoic times. In the Cretaceous, crustacean dominance in the marine realm is indicated by the great diversity of body fossils, as well as by the abundance of crustacean structures in the shallow-marine deposits.

3. During the Palaeogene, the number of decapod ichnofossils recorded is considerably less than that for the Cretaceous. This may reflect the effect of the end-Cretaceous mass extinction and/or the monographic effects related to the large amount of ichnological work performed in Cretaceous shallow-marine deposits. Most of the analysis undertaken suggests that the Eocene displays the highest abundance, and then the number of ichnofossils decreases again in the Oligocene, a general tendency that is consistent with data from the body fossil record. The drop in abundance of decapod trace fossils during the Oligocene is coincident with a worldwide cooling event. During the Neogene, the abundance of decapod trace fossils underwent a remarkable increase and crustacean burrows became the dominant elements of the shallow-marine ichnofaunas, commonly forming part of communities that display complex endobenthic tiering patterns.

4. The timing of acquisition of the infaunal habit by decapods remains controversial, because the producers of the Palaeozoic structures could not be identified. In the Mesozoic, however, the trace fossil records can be more confidently assigned to the decapods. The trends in ichnodiversity and abundance of decapod burrows reflect that the use of infaunal ecospace was well established by crustaceans in the Mesozoic. The evolutionary history of infaunal decapod crustaceans was a response to the need for shelter and the acquisition of adaptations and trophic types. The development of the infaunal habit was a key innovation that allowed

decapods to become dominant components of modern marine communities.

Acknowledgements

We thank Beatriz Aguirre-Urreta, Jordi de Gibert, Radek Mikulás, Paul Myrow and Carrie Schweitzer for providing literature and for valuable comments on different aspects. Euan Clarkson is thanked for critically reading an early draft of the manuscript and Ron Feldmann and Murray Gingras for their detailed reviews. This project was supported by the Antorchas Foundation.

References

Aceñolaza, F.G. & Fernández, R. 1984: Nuevas trazas fósiles en el Paleozoico inferior del noroeste argentino. *3er Congreso Argentino de Paleontología y Bioestratigrafía, Actas 3*, 13–28.

Astashkin, V.A. 1985: Problematichnye organizmy – porodoobrazovateli v nizhnem kembrii Sibiroskoy platformy [Problematic rock-forming organisms in the Lower Cambrian of the Siberian Platform]. *Trudy, Institut geologii i geofiziki, Sibirskoe otdelenie, Akademiya nauk SSSR 632*, 144–149 [in Russian, with English abstract].

Bambach, R.K. 1983: Ecospace utilization and guilds in marine communities through the Phanerozoic. *In* Tevesz, M.J.S. & McCall, P.L. (eds): *Biotic Interactions in Recent and Fossil Benthic Communities*, 719–746. Plenus Press, New York.

Balsam, W.L. 1984: Reinterpretation of an Archaeocyathid reef: Shady Formation, Southwestern Virginia. *Southeastern Geology 16*, 121–129.

Bellwood, O. 2002: The occurrence, mechanics and significance of burying behaviour in crabs (Crustacea: Brachyura). *Journal of Natural History 36*, 1223–1238.

Bromley, R.G. 1996: *Trace Fossils. Biology, Taphonomy and Applications*. Chapman & Hall, London.

Bromley, R.G. & Asgaard, U. 1972: The burrows and microcopolites of *Glyphea rosenkrantzi*, a Lower Jurassic palinuran crustacean from Jameson Land, East Greenland. *Grønlands Geologiske Undersøgelse Rapport 49*, 15–21.

Buatois, L.A., Mángano, M.G., Alissa, A. & Carr, T.R. 2002: Sequence stratigraphy and sedimentologic significance of biogenic structures from a late Paleozoic marginal- to open-marine reservoir, Morrow Sandstone, subsurface of southwest Kansas, USA. *Sedimentary Geology 152*, 99–132.

Buatois, L.A., Mángano, M.G., Bromley, R.G., Bellosi, E. & Carmona, N.B. 2003: Ichnology of shallow marine deposits in the Miocene Chenque Formation of Patagonia: complex ecologic structure and niche partitioning in the Neogene ecosystems. *Publicación Especial de la Asociación Paleontológica Argentina 9*, 85–95.

Burzin, M.B., Debrenne, F. & Zhuravlev, A.Y. 2001: Evolution of shallow-water level-bottom communities. *In* Zhuravlev, A.Y. & Riding, R. (eds): *The Ecology of the Cambrian Radiation*, 200–216. Columbia University Press, New York.

Cañas, F. 1995: Early Ordovician carbonate platform facies of the Argentine Precordillera: restricted shelf to open platform evolution. In: Cooper, J.D., Droser, M.L. & Finney, S.C. (eds.): *Ordovician Odyssey: Short papers for the Seventh International Symposium on the Ordovician System*. The Pacific Section Society for Sedimentary Geology (SEPM) Book 77, 221–224.

Carmona, N.B. & Buatois, L.A. 2003: Estructuras de crustáceos en el Mioceno de la Cuenca del Golfo San Jorge: Implicancias Paleobiológicas y Evolutivas. In: Buatois, L.A. & Mángano, M.A. (eds.): *Icnología: hacia una convergencia entre Geología y Biología. Publicación Especial de la Asociación Paleontológica Argentina 9*, 97–108.

Carmona, N.B., Buatois, L.A. & Mángano, M.G. 2002: The trace fossil record of the decapod crustacean radiations. *First International Palaeontological Congress (IPC 2002), Geological Society of Australia, Abstracts, 68*, 29–30.

Carey, J. 1979: Sedimentary environments and trace fossils of the Permian Snapper Point Formation, southern Sydney Basin. *Journal of the Geological Society of Australia 25*, 433–458.

Chamberlain, C.K. & Baer, J.L. 1973: *Ophiomorpha* and a new Thalassinid burrow from the Permian of Utah. *Brigham Young University Research Studies, Geology Series 20*, 79–94.

Chaplin, J.R. 1996: Ichnology of transgressive–regressive surfaces in mixed carbonate–siliciclastic sequences, Early Permian Chase Group, Oklahoma. *Geological Society of America Special Paper 306*, 399–418.

Crimes, T.P. & Anderson, M.M. 1985: Trace fossils from the Late Precambrian–Early Cambrian strata of southeastern Newfoundland (Canada): temporal and environmental implications. *Journal of Paleontology 59*, 310–343.

Dahmer, D.D. & Hilbrecht, H. 1986: Facies dynamics of the lower Muschelkak (Middle Triassic) near Bad Hersfeld (northern Hesse) with comments on the origin of the micrites. *Neues Jahrbuch für Geologie und Paläontologie, Monatshefte 9*, 513–528.

De, C.B., Pei, X.G. & Mai, P.Q. 1996: Trace fossils from Triassic Qinglong Formation of Lower Yangzi Valley. *Acta Paleontologica Sinica 35*, 714–729.

de Gibert, J.M. 1996: A new decapod burrow system from the NW Mediterranean Pliocene. *Revista Española de Paleontología 11*, 251–254.

Driese, S.G. & Dott, R.H. 1984: Model for sandstone–carbonate "cyclothems" based on upper member of Morgan Formation (Middle Pennsylvanian) of northern Utah and Colorado. *American Association of Petroleum Geologists Bulletin 68*, 574–597.

Droser, M.L. & Bottjer, D.J. 1988: Trends in extent and depth of Early Paleozoic bioturbation in the Great Basin (California, Nevada, and Utah). *Geological Society of America, Cordilleran Section, Field Trip Guidebook*, 123–135.

Droser, M.L. & Bottjer, D.J. 1989: Ordovician increase in extent and depth of bioturbation. Implications for understanding early Paleozoic ecospace utilization. *Geology 17*, 850–852.

Dworschak, P.C. 1998: The role of tegumental glands in burrow construction by two Mediterranean callianassid shrimp. *Senckenbergiana Maritima 28*, 143–149.

Ekdale, A.A. & Bromley, R.G. 2003: Paleoethologic interpretation of complex *Thalassinoides* in shallow-marine limestones, Lower Ordovician, southern Sweden. *Palaeoclimatology, Palaeoecology, Palaeoecology 192*, 221–227.

Fedonkin, M.A. 1981: Belomorskaja biota venda. [The Vendian White Sea biota]. *Trudy Akademii Nauk SSSR 342*, 1–100.

Fedonkin, M.A. 1983: Besskeletnaya fauna podol'skogo pridnestrov'ya. [Non-skeletal fauna of Podolia (Diniestr River Valley)]. *In* Velikanov, V.A., Aseeva, M.A. & Fedonkin, M.A. (eds): *Vend Ukrainy: Naukova Dunka*, 128–139. Naukova Dumka Kiev.

Feldmann, R.M. 2003: The Decapoda: new initiatives and novel approaches. *Journal of Paleontology 77*, 1021–1039.

Förster, R. 1985: Evolutionary trends and ecology of Mesozoic decapod crustaceans. *Transactions of the Royal Society of Edinburgh 76*, 299–304.

Fritz, W.H. 1980: International Precambrian–Cambrian Boundary Working Group's 1979 field study to Mackenzie Mountains, Northwest Territories, Canada. *Geological Survey of Canada Paper 10-1A*, 41–45.

Fürsich, F.T. 1973: A revision of the trace fossil *Spongeliomorpha, Ophiomorpha* and *Thalassinoides*. *Neues Jahrbuch für Geologie und Paläontologie, Monatshefte 1973*, 719–735.

Gaba, Z. & Pek, I. 1980: Lebensspuren aus den Geschieben des tschechischen Schlesiens. [Life traces in the boulder detritus of the Czech Silesia]. *Der Geschiebe-Sammler 14*, 13–30.

Glaessner, M.F. 1969: Decapoda. *In* Moore, R.C. (ed.): *Treatise on Invertebrate Paleontology, Part R, Arthropoda 4, Volume 2*, 400–533. Geological Society of America and University of Kansas Press Lawrence, Kansas.

Haj, A.E. & Feldmann, R.M. 2002: Functional morphology and taxonomic significance of a novel cuticular structure in Cretaceous Raninid crabs (Decapoda: Brachyura: Raninidae). *Journal of Paleontology 76*, 472–485.

Harper, E.M., Forsythe, G.T.W. & Palmer, T. 1998: Taphonomy and the Mesozoic marine revolution: preservation state masks the importance of boring predators. *Palaios 13*, 352–360.

Hary, A., Hendriks, F. & Muller, A. 1981: Lithofaciès, stratofaciès et ichnofaciès du Rhétien gréso-pélitique de la carrière d'Aboncourt (NE de la Lorraine). *Publications du Service Geologique du Luxembourg 9*, 1–35.

Hein, F.J., Robb, G.A., Wolberg, A.C. & Longstaffe, F.J. 1991: Facies descriptions and associations in ancient reworked (?transgressive) shelf sandstones: Cambrian and Cretaceous examples. *Sedimentology 38*, 405–431.

Howard, J.D. & Frey, R.W. 1975: Regional animal–sediment characteristics of Georgia estuaries. *Senckenbergiana Maritima 7*, 33–103.

Jensen, S. 1997: Trace fossils from the Lower Cambrian Mickwitzia sandstone, south-central Sweden. *Fossils and Strata 42*, 1–111.

Jensen, S. & Grant, S.W.F. 1998: Trace fossils from the Divdalen Group, northern Sweden: implications for Early Cambrian biostratigraphy of Baltica. *Norsk Geologisk Tidsskrift 78*, 305–317.

Johnson, K.G. & McCormick, T. 1999: The quantitative description of the biotic change using palaeontological databases. *In* Harper, D.A.T. (ed.): *Numerical Palaeobiology*, 227–247. John Wiley and Sons, London.

Kier, P.M. 1974: Evolutionary trends and their functional significance in the post-Paleozoic echinoids. *Paleontological Society Memoir 5*, 48, 1–95.

Knaust, D. 2002: Ichnogenus *Pholeus* Fiege, 1944, revisited. *Journal of Paleontology 76*, 882–891.

Liñán, E. 1984: Los icnofósiles de la Formación Torrearboles (Precámbrico?–Cámbrico inferior) en los alrededores de fuente de Cantos, Badajoz. *Cuadernos do Laboratoria Xeologico de Laxe 8*, 47–74.

Maerz, R.H. Jr, Kaesler, R.L. & Hakes, W.G. 1976: Trace fossils from the Rock Bluff Limestone (Pennsylvanian, Kansas). *The University of Kansas, Paleontological Contributions 80*, 1–6.

Mángano, M.G. & Buatois, L.A. 1991: Discontinuity surfaces in the Lower Cretaceous of the High Andes (Mendoza, Argentina): trace fossils and environmental implications. *Journal of South American Earth Sciences 4*, 215–229.

Mángano, M.G. & Buatois, L.A. 2003: Trace fossils. *In* Benedetto, J.L. (ed.): *Ordovician Fossils from Argentina*, 507–553. Secretaría de Ciencia y Tecnología, Universidad Nacional de Córdoba.

Mángano, M.G. & Droser, M.L. 2004: The ichnologic record of the Ordovician radiation. *In* Webby, B.D., Droser, M.L., Paris, F. &

Percival, I.G. (eds.): *The Great Ordovician Biodiversification Event*, 369–379. Columbia University Press, New York.

Mayer, G. 1981: Spongeliomorphe Gebilde aus dem Unteren Muschelkalk von Bad Kissingen und Bad Driburg. [Spongeliomorph material from the lower Muschelkalk of Bad Kissingen and Bad Driburg]. *Der Aufschluss 32*, 505–508.

Muñiz, F. & Mayoral, E. 2001a: *Macanopsis plataniformis* nov. ichnosp. from the Lower Cretaceous and Upper Miocene of the Iberian Peninsula. *Geobios 34*, 91–98.

Muñiz, F. & Mayoral, E. 2001b: El icnogénero *Spongeliomorpha* en el Neógeno superior de la Cuenca de Guadalquivir (área de Lepe-Ayamonte, Huelva, España). *Revista Paleontológica de Paleontología 16*, 115–130.

Myrow, P.M. 1995: *Thalassinoides* and the enigma of Early Paleozoic open-framework burrow systems. *Palaios 10*, 58–74.

Pemberton, S.G., Spila, M., Pulham, A.J., Saunders, T., MacEachern, J.A., Robbins, D. & Sinclair, I.K. 2001: Ichnology and sedimentology of shallow to marginal marine systems: Ben Nevis & Avalon Reservoirs, Jeanne d'Arc Basin. *Geological Association of Canada Short Course Notes 15*, 1–343.

Qi, Y.H. & Bin, G. Z. 2000: The influence of bioturbation structures containing *Ophiomorpha* on petrophysical properties of Donghe sandstone reservoir in central Tarim Basin China. *Abstracts 31st International Geological Congress*. [CD-Rom]

Raup, D.M. 1976: Species diversity in the Phanerozoic: an interpretation. *Paleobiology 2*, 289–297.

Remane, J. *et al.* 2000: *International Stratigraphic Chart; with Explanatory Notes, 2000 Edition*, 1–16. International Union of Geological Sciences (International Commission on Stratigraphy)/UNESCO, Paris.

Ronov, A.B., Khain, V.E., Balukhovsky, A.N. & Seslavinsky, K.B. 1980: Quantitative analysis of Phanerozoic sedimentation. *Sedimentary Geology 25*, 311–325.

Schlirf, M. 2000: Upper Jurassic trace fossils from the Boulonnais (northern France). *Geologica et Palaeontologica 34*, 145–213.

Schram, F.R. 1981: Late Paleozoic crustacean community. *Journal of Paleontology 55*, 126–137.

Schram, F.R., Feldmann, R.M. & Copeland, M.J. 1978: The Late Devonian Palaeopalaemonidae and the earliest decapod crustaceans. *Journal of Paleontology 52*, 1375–1378.

Schweitzer, C.E. 2001: Paleobiogeography of Cretaceous and Tertiary decapod crustaceans of the North Pacific Ocean. *Journal of Paleontology 75*, 808–826.

Seilacher, A. 1970: *Cruziana* stratigraphy of "non-fossiliferous" Palaeozoic sandstones. *Geological Journal Special Issue 3*, 447–476.

Sepkoski, J.J., Jr. 2000: Crustacean biodiversity through the marine fossil record. *Contributions to Zoology 69*, 213–222.

Sheehan, P.M. & Schiefelbein, D.R.J. 1984: The trace fossil *Thalassinoides* from the Upper Ordovician on the Eastern Great Basin: deep burrowing in the Early Paleozoic. *Journal of Paleontology 58*, 440–447.

Stanley, S.M. 1977: Trends, rates and patterns of evolution in the Bivalvia. *In* Hallam, A. (ed.): *Patterns of Evolution*, 209–250. Elsevier Science, Amsterdam.

Stanley, T.M. & Feldmann, R.M. 1998: Significance of nearshore trace-fossil assemblages of the Cambro-Ordovician Deadwood Formation and Aladdin Sandstone, South Dakota. *Annals of the Carnegie Museum 67*, 1–51.

Swen, K., Fraaije, R.H.B. & van der Zwaan, G.J. 2001: Polymorphy and extinction of the Late Cretaceous burrowing shrimp *Protocallianassa faujasi* and the first record of the genera *Corallianassa* and *Calliax* (Crustacea, Decapoda, Thalassinoidea) from the Cretaceous. *Contributions to Zoology 70*, 85–98.

Thayer, C.W. 1979: Biological bulldozers and the evolution of marine benthic communities. *Science 203*, 458–461.

Vermeij, G.J. 1977: The Mesozoic marine revolution: evidence from snails, predators and grazers. *Paleobiology 3*, 245–258.

Watkins, R. & Coorough, P.J. 1997: Silurian *Thalassinoides* in an offshore carbonate community, Wisconsin, USA. *Palaeogeography, Palaeoecology, Palaeoecology 129*, 109–117.

Zhuravleva, I.T., Meshlova, N.P., Luchinina, V.A. & Kashina, L.N. 1982: Biofacies of the Anabar Sea in the Late Precambrian and Early Cambrian. *Trudy Instituta Geologii i Geofiziki 510*, 74–103.

Zonneveld, J.P., Moslow, T.F. & Henderson, C.M. 1997: Lithofacies associations and depositional environments in a mixed siliciclastic-carbonate coastal depositional system, upper Liard Formation, Triassic, northeastern British Columbia. *Bulletin of Canadian Petroleum Geology 45*, 553–575.